HYDRAULIC DESIGN OF HIGHWAY CULVERTS

1. Report No. FHWA-HIF-12-026 HDS 5	2. Government Accession No.	3. Recipient's Catalog No.

4. Title and Subtitle	5. Report Date
HYDRAULIC DESIGN OF HIGHWAY CULVERTS Third Edition	April 2012
	6. Performing Organization Code

7. Author(s)	8. Performing Organization Report No.
James D. Schall, Philip L. Thompson, Steve M. Zerges, Roger T. Kilgore, and Johnny L. Morris	

9. Performing Organization Name and Address	10. Work Unit No. (TRAIS)
Ayres Associates 3665 JFK Parkway Building 2, Suite 200 Fort Collins, Colorado 80525	
	11. Contract or Grant No. DTFH61-06-D-00010

12. Sponsoring Agency Name and Address	13. Type of Report and Period Covered
Office of Bridge Technology National Highway Institute Federal Highway Administration 1310 North Courthouse Rd., Suite 300 1200 New Jersey Ave, SE Arlington, Virginia 22201 Washington, D.C. 20590	
	14. Sponsoring Agency Code

15. Supplementary Notes

Technical Project Managers: Eric R. Brown (FHWA Resource Center), Scott A. Hogan (FHWA Federal Lands Highway Division) and Brian L. Beucler (FHWA Office of Bridge Technology)

Technical Assistance: Dusty W. Robinson (Ayres Associates) and Bart S. Bergendahl (FHWA Federal Lands Highway Division)

16.

Hydraulic Design Series Number 5 (HDS 5) originally merged culvert design information contained in Hydraulic Engineering Circulars (HEC) 5, 10, and 13 with other related hydrologic, storage routing and special culvert design information. This third edition is the first major rewrite of HDS 5 since 1985, updating all previous information and adding new information on software solutions, aquatic organism passage, culvert assessment, and culvert repair and rehabilitation. The result is a comprehensive culvert design publication. The appendices of the publication contain the equations and methodology used in developing the design charts (nomographs) and software programs, information on hydraulic resistance of culverts, the commonly used design charts, and Design Guidelines (DG) illustrating various culvert design calculation procedures. The number of design charts provided has been reduced recognizing the increased use of software solutions; however, the full set of culvert design charts will continue to be available in the archived second edition of HDS 5.

17. Key Words	18. Distribution Statement
Culverts, inlet control, outlet control, tapered inlets, broken-back culverts, culvert repair and rehabilitation, culvert lining, aquatic organism passage, hydrology, hydraulics, storage routing	This document is available to the public on the FHWA website and through the National Technical Information Service, Springfield, VA 22161

19. Security Classif. (of this report)	20. Security Classif. (of this page)	21. No. of Pages	22. Price
Unclassified	Unclassified	326	

TABLE OF CONTENTS

(page intentionally left blank)

LIST OF FIGURES

(page intentionally left blank)

LIST OF TABLES

(page intentionally left blank)

LIST OF SYMBOLS

a	Cross-sectional area of orifice, ft^2 (m^2)
a'	Constant in rainfall intensity formula
A	Full cross-sectional area of culvert barrel or channel, ft^2 (m^2)
A_b	Area of bend section of slope-tapered inlet, ft^2 (m^2)
A_f	Area of inlet face section of tapered inlet, ft^2 (m^2)
A_P	Area of flow prism, ft^2 (m^2)
A_t	Area of tapered inlet throat section, ft^2 (m^2)
A_w	Watershed area, acres (hectares)
b	Face dimension of side bevel, in (mm)
b'	Constant in rainfall intensity formula
B	Span of culvert barrel, ft (m)
B_b	Width of bend section of a slope-tapered inlet, ft (m)
B_f	Width of face section of a tapered inlet, ft (m)
c	Coefficient for submerged inlet control equation
C	Runoff coefficient for use in the Rational equation
C_b	Discharge coefficient for bend section control
C_d	Coefficient of discharge for flow over an embankment
C_f	Discharge coefficient for face section control
C_r	Free flow coefficient of discharge for flow over an embankment
C_t	Discharge coefficient for throat section control
CMP	Corrugated metal pipe
d	Face dimension of top bevel, in (mm)
d_c	Critical depth, ft (m)
d_n	Normal depth, ft (m)
D	Interior height of culvert barrel, ft (m)
D_{50}	Size of streambed material which exceeds 50% of the material by weight; i.e., the median size, in or ft (mm or m)
E	Height of face of tapered inlet, excluding bevel, ft (m)
EL_C	Elevation of weir crest, ft (m)
EL_f	Invert elevation at face, ft (m)
El_{ha}	Allowable headwater elevation, ft (m)
EL_{hc}	Headwater elevation required for flow to pass crest in crest control, ft (m)

EL_{hf}	Headwater elevation required for flow to pass face section in face control, ft (m)
EL_{hi}	Headwater elevation required for culvert to pass flow in inlet control, ft (m)
EL_{ho}	Headwater elevation required for culvert to pass flow in outlet control, ft (m)
EL_{ht}	Headwater elevation required for flow to pass throat section in throat control, ft (m)
EL_o	Invert elevation at outlet, ft (m)
EL_{sf}	Stream bed elevation at face of culvert, ft (m)
EL_{so}	Stream bed elevation at outlet of culvert, ft (m)
EL_t	Invert elevation at throat, ft (m)
EL_{tw}	Tailwater elevation, ft (m)
f	Darcy resistance factor
F_r	Froude number
G	The number of different materials (roughnesses) in the perimeter of a conduit with composite roughness
g	Acceleration due to gravity, 32.2 ft/s/s (9.81 m/s/s)
HGL	Hydraulic grade line
h	Height of hydraulic grade line above centerline of orifice, ft (m)
h_f	Friction head loss, ft (m)
h_o	Height of hydraulic grade line above outlet invert, ft (m)
h_t	Height of tailwater above crown of submerged road, ft (m)
H	Sum of inlet loss, friction loss, and velocity head in a culvert, ft (m)
H_L	Total energy required to pass a given discharge through a culvert, ft (m)
H_b	Head loss at bend, ft (m)
H_C	Specific head at critical depth ($d_C + V_C^2/2g$), ft (m)
H_e	Entrance head loss, ft (m)
H_f	Friction head loss in culvert barrel, ft (m)
H_g	Head loss at bar grate, ft (m)
H_j	Head loss at junction, ft (m)
H_l	Friction head loss in tapered inlet, ft (m)
H_o	Exit head loss, ft (m)

H_v	Velocity head $= V^2/2g$, ft (m)
HW	Depth from inlet invert to upstream total energy grade line, ft (m)
HW_a	Allowable headwater depth, ft (m)
HW_b	Headwater depth above the bend section invert, ft (m)
HW_c	Headwater depth above the weir crest, ft (m)
HW_d	Design headwater depth, ft (m)
HW_i	Headwater depth above inlet control section invert, ft (m)
HW_f	Headwater depth above the culvert inlet face invert, ft (m)
HW_o	Headwater depth above the culvert outlet invert, ft (m)
HW_r	Total head of flow over embankment, ft (m) (Measured from roadway crest to up stream surface level.)
HW_t	Depth from throat invert to upstream total energy grade line, ft (m)
I	Rate of inflow into a storage basin, ft^3/s (m^3/s) Rainfall intensity, in/hr
k	Flow constant for an orifice, $Q = kah^{0.5}$, ft$^{0.5}$/s (m$^{0.5}$/s)
k_e	Entrance loss coefficient
k_t	Correction factor for downstream submergence during roadway overtopping
K	Coefficient for unsubmerged inlet control equation
K_b	Dimensionless effective pressure term for bend section control
K_f	Dimensionless effective pressure term for inlet face section control
K_9	Dimensionless bar shape factor for calculating grate head losses
K_t	Dimensionless effective pressure term for inlet throat control
L	Actual culvert length, ft (m)
L_a	Approximate length of culvert, including tapered inlet, but excluding wing walls, ft (m)
L_r	Width of roadway prism crest, ft (m)
$L_1, L_2,$ L_3, L_4	Dimensions relating to tapered inlets, ft (m)
M	Exponent in unsubmerged inlet control equation
n	Manning's roughness coefficient
n_c	Composite or weighted Manning's n value
N	Number of barrels
O	Rate of outflow from a storage basin, ft^3/s (m^3/s)

LIST OF SYMBOLS (Cont.)

p	Wetted perimeter, ft (m)
p_f	Wetted perimeter of tapered inlet face, ft (m)
p_t	Wetted perimeter of tapered inlet throat, ft (m)
P	Length from crest of depression to face of culvert, ft (m)
q_0	Discharge over segment of embankment, ft^3/s (m^3/s)
Q	Discharge, ft^3/s (m^3/s)
Q_b	Flow through culvert barrel(s) as opposed to flow over embankment, ft^3/s (m^3/s)
Q_c	Discharge at critical depth, ft^3/s (m^3/s)
Q_d	Design discharge, ft^3/s (m^3/s)
Q_o	Overtopping discharge over total length of embankment, ft^3/s (m^3/s)
Q_P	Peak flow rate, ft^3/s (m^3/s)
Q_r	Routed (reduced) peak flow, ft^3/s (m^3/s)
Q_t	Total of $Q_b + Q_o$, ft^3/s (m^3/s)
Q_{50}	Discharge for 50-year return period (similar for other return periods), ft^3/s (m^3/s)
R	Hydraulic radius = cross-sectional area of flow through culvert or channel divided by wetter perimeter, ft (m)
RCB	Reinforced concrete box
RCP	Reinforced concrete pipe
s	Storage in a storage basin, ft^3 (m^3)
S	Slope of culvert barrel, ft/ft (m/m)
S_e	Slope of embankment or face of excavation, expressed as S_e:1, horizontal:vertical, ft/ft (m/m)
S_D	Slope of throat depression at culvert inlet, expressed as S_D:1, horizontal:vertical, ft/ft, (m/m)
S_f	Friction slope of full flow HGL, ft/ft (m/m)
S_o	Slope of channel bed, ft/ft (m/m)
t	Time, min or sec
t_i	Time of concentration for Rational equation, min
t_p	Time to peak of a runoff hydrograph, min or sec

T	Depression of inlet control section below the stream bed, measured from stream bed to face invert for culvert, to throat invert for culvert with tapered inlet, ft (m)
	Rainfall duration, min
T_c	Critical storm duration, min
T_P	Top width of flow prism, m (ft)
TAPER	Cotangent of angle of sidewalls in tapered inlet with respect to an extension of the culvert sidewalls, ft/ft (m/m)
TW	Tailwater depth measured from culvert outlet invert, ft (m)
V	Mean velocity of flow, ft/s (m/s)
V_c	Velocity at critical depth, ft/s (m/s)
V_d	Channel velocity downstream of culvert, ft/s (m/s)
V_g	Velocity of flow between bars in a grate, ft/s (m/s)
V_o	Velocity at outlet of culvert, ft/s (m/s)
V_U	Approach velocity upstream of culvert, ft/s (m/s)
w	Maximum cross-sectional width of the bars facing the flow, ft (m)
W	Length of weir crest for slope tapered inlet with mitered face, ft (m)
W_P	Length of weir crest of fall, excluding sides of depression, ft (m)
WW	Wingwall of culvert entrance
x	Minimum clear spacing between bars, ft (m)
X_1, X_2, X_3	Lengths of overflow sections along embankment, ft (m)
y	Depth of flow, ft (m)
y'	Change in hydraulic grade line through a junction, ft (m)
Y	Additive term in submerged inlet control equation
Y_h	Hydraulic depth = A_P/T_P, ft (m)
Z	The difference in elevation between the crest and face section of a slope tapered inlet with a mitered face, ft (m)
Θ_g	Angle of bar grate with respect to the horizontal, degrees
Θ_s	Flare angles of side walls of tapered inlet with respect to extension of culvert side wall, degrees
Θ_t	Angle of departure of the top slab from a plane parallel to the bottom slab, degrees
Θ_W	Flare angle of wingwalls with respect to extension of culvert side wall, degrees
Θ_j	Angle between outfall and lateral at a junction, degrees

(page intentionally left blank)

ACKNOWLEDGMENTS

First Edition

The first edition of this Hydraulic Design Series was published in 1985 (FHWA-IP-85-15). The authors were J.M. Normann, R.J. Houghtalen and W.J. Johnston with Jerome M. Normann and Associates, Norfolk, VA. The FHWA Project Manager was John M. Kurdziel and the Technical Assistants were Phil Thompson, Dennis Richards and Sterling Jones.

Second Edition

The second edition was published in 2001 (FHWA-NHI-01-020). This edition corrected minor errors, added several new design charts and provided dual units for all equation and charts. There were no changes in technical content and the original authorship was retained. The editor of this edition was Johnny Morris with Ayres Associates Inc., Fort Collins, CO. The FHWA Project Manager was Phil Thompson. In 2005 a revised version of the second edition was printed to correct minor errors related to some of the nomographs.

DISCLAIMER

(page intentionally left blank)

GLOSSARY

A listing of terms related to culvert hydraulics, highways and the river environment is provided below:

abrasion:	Removal of material due to entrained sediment, ice, or debris rubbing against the boundary.
aggradation:	General and progressive buildup of the longitudinal profile of a channel bed due to sediment deposition.
allowable headwater:	Maximum possible headwater, or ponding, at the upstream side of a culvert.
alluvial stream:	Stream which has formed its channel in cohesive or noncohesive materials that have been and can be transported by the stream.
annual flood:	Maximum flow in 1 year (may be daily or instantaneous).
AOP:	Aquatic organism passage refers to designing hydraulic structures so as not to impede movement or survival of aquatic species
apron:	Protective material placed on a streambed to resist scour.
armor (armoring):	Surfacing of channel bed, banks, or embankment slope to resist erosion and scour. (A) Natural process whereby an erosion-resistant layer of relatively large particles is formed on a streambed due to the removal of finer particles by streamflow; (B) placement of a covering to resist erosion.
articulated concrete mattress:	Rigid concrete slabs which can move without separating as scour occurs; usually hinged together with corrosion-resistant cable fasteners; primarily placed for lower bank protection.
average velocity:	Velocity at a given cross section determined by dividing discharge by cross sectional area.
backfill:	Material used to refill a ditch or other excavation, or the process of doing so.
backwater:	Increase in water surface elevation relative to elevation occurring under natural channel and floodplain conditions. It is induced by a culvert, bridge or other structure that obstructs or constricts the free flow of water in a channel.
backwater area:	Low-lying lands adjacent to a stream that may become flooded due to backwater.

GLOSSARY (cont.)

bank:	Sides of a channel between which the flow is normally confined.
bank, left (right):	Sides of a channel as viewed in a downstream direction.
bankfull discharge:	Discharge that, on average, fills a channel to the point of overflowing.
bank protection:	Engineering works for the purpose of protecting streambanks from erosion.
bank revetment:	Erosion-resistant materials placed directly on a streambank to protect the bank from erosion.
bar:	Elongated deposit of alluvium within a channel, not permanently vegetated.
base floodplain:	Floodplain associated with the flood with a 100-year recurrence interval.
bed:	Bottom of a channel bounded by banks.
bed form:	Recognizable relief feature on the bed of a channel, such as a ripple, dune, plane bed, antidune, or bar. Bed forms are a consequence of the interaction between hydraulic forces (boundary shear stress) and the bed sediment.
bed load:	Sediment that is transported in a stream by rolling, sliding, or skipping along the bed or very close to it; considered to be within the bed layer (contact load).
bed load discharge (or bed load):	Quantity of bed load passing a cross section of a stream in a unit of time.
bed material:	Material found in and on the bed of a stream (May be transported as bed load or in suspension).
bedrock:	Solid rock exposed at the surface of the earth or overlain by soils and unconsolidated material.
bed shear stationary (tractive force):	Force per unit area exerted by a fluid flowing past a boundary.
bed slope:	Inclination of the channel bottom.
blanket:	Material covering all or a portion of a streambank to prevent erosion.
boulder:	Rock fragment whose diameter is greater than 10 in. (250 mm).

GLOSSARY (cont.)

braided stream:	Stream whose flow is divided at normal stage by small mid-channel bars or small islands; the individual width of bars and islands is less than about three times water width; a braided stream has the aspect of a single large channel within which are subordinate channels.
bridge owner:	Any Federal, State, Local agency, or other entity responsible for a structure defined as a highway bridge by the National Bridge Inspection Standards (NBIS).
broken-back culvert:	An alternative to a steeply sloped culvert by breaking the culvert slope into steeper and flatter sections.
bulkhead:	Vertical, or near vertical, wall that supports a bank or an embankment; also may serve to protect against erosion.
catchment:	See drainage basin.
causeway:	Rock or earth embankment carrying a roadway across water.
caving:	Collapse of a bank caused by undermining due to the action of flowing water.
cellular-block mattress:	Interconnected concrete blocks with regular cavities placed directly on a streambank or filter to resist erosion. The cavities can permit bank drainage and the growth of vegetation where synthetic filter fabric is not used between the bank and mattress.
channel:	Bed and banks that confine surface flow of a stream.
channelization:	Straightening or deepening of a natural channel by artificial cutoffs, grading, flow-control measures, or diversion of flow into an engineered channel.
channel diversion:	Removal of flows by natural or artificial means from a natural length of channel.
channel pattern:	Aspect of a stream channel in plan view, with particular reference to the degree of sinuosity, braiding, and anabranching.
check dam:	Low dam or weir across a channel used to control stage or degradation.
choking (of flow):	Excessive constriction of flow which may cause severe backwater effect.
clay (mineral):	Particle whose diameter is in the range of 0.00024 to 0.004 mm.
cobble:	Fragment of rock whose diameter is in the range of 2.5 to 10 in. (64 to 250 mm).

GLOSSARY (cont.)

concrete revetment:	Unreinforced or reinforced concrete slabs placed on the channel bed or banks to protect it from erosion.
confluence:	Junction of two or more streams.
constriction:	Natural or artificial control section, such as a bridge crossing, channel reach or dam, with limited flow capacity in which the upstream water surface elevation is related to discharge.
corrosion:	Chemical or electro-chemical reaction between the soil and/or water and the culvert or bridge material
countermeasure:	Measure intended to prevent, delay or reduce the severity of hydraulic problems.
critical shear stress:	Minimum amount of shear stress required to initiate soil particle motion.
cross section:	Section normal to the trend of a channel or flow.
current:	Water flowing through a channel.
current meter:	Instrument used to measure flow velocity.
cutoff wall:	Wall, usually of sheet piling or concrete, that extends down to scour-resistant material or below the expected scour depth.
daily discharge:	Discharge averaged over 1 day (24 hours).
debris:	Floating or submerged material, such as logs, vegetation, or trash, transported by a stream.
degradation (bed):	General and progressive (long-term) lowering of the channel bed due to erosion, over a relatively long channel length.
depression:	Lowering the inlet or throat of a culvert to increase the effective head on the flow control section.
depth of scour:	Vertical distance a streambed is lowered by scour below a reference elevation.
design flow (design flood):	Discharge that is selected as the basis for the design or evaluation of a hydraulic structure.
discharge:	Volume of water passing through a channel during a given time.
drainage basin:	Area confined by drainage divides, often having only one outlet for discharge (catchment, watershed).
drift:	Alternate term for vegetative "debris."

eddy current:
Vortex-type motion of a fluid flowing contrary to the main current, such as the circular water movement that occurs when the main flow becomes separated from the bank.

embedded culvert:
A culvert that depressed into the streambed to create a natural bottom often in relation to AOP design.

entrenched stream:
Stream cut into bedrock or consolidated deposits.

ephemeral stream:
Stream or reach of stream that does not flow for parts of the year. As used here, the term includes intermittent streams with flow less than perennial.

erosion:
Displacement of soil particles due to water or wind action.

erosion control matting:
Fibrous matting (e.g., jute, paper, etc.) placed or sprayed on a streambank for the purpose of resisting erosion or providing temporary stabilization until vegetation is established.

fabric mattress:
Grout-filled mattress used for streambank protection.

fill slope:
Side or end slope of an earth-fill embankment. Where a fill-slope forms the streamward face of a spill-through abutment, it is regarded as part of the abutment.

filter:
Layer of fabric (geotextile) or granular material (sand, gravel, or graded rock) placed between bank revetment (or bed protection) and soil for the following purposes: (1) to prevent the soil from moving through the revetment by piping, extrusion, or erosion; (2) to prevent the revetment from sinking into the soil; and (3) to permit natural seepage from the streambank, thus preventing the buildup of excessive hydrostatic pressure.

filter blanket:
Layer of graded sand and gravel laid between fine-grained material and riprap to serve as a filter.

filter fabric (cloth):
Geosynthetic fabric that serves the same purpose as a granular filter blanket.

fine sediment load:
That part of the total sediment load that is composed of particle sizes finer than those represented in the bed (wash load). Normally, the fine-sediment load is finer than 0.062 mm for sand-bed channels. Silts, clays and sand could be considered wash load in coarse gravel and cobble-bed channels.

flashy stream:
Stream characterized by rapidly rising and falling stages, as indicated by a sharply peaked hydrograph. Typically associated with mountain streams or highly disturbed urbanized catchments. Most flashy streams are ephemeral, but some are perennial.

GLOSSARY (cont.)

flood-frequency curve:	Graph indicating the probability that the annual flood discharge will exceed a given magnitude, or the recurrence interval corresponding to a given magnitude.
floodplain:	Nearly flat, alluvial lowland bordering a stream, that is subject to frequent inundation by floods.
fluvial geomorphology:	Science dealing with morphology (form) and dynamics of streams and rivers.
freeboard:	Vertical distance above a design stage that is allowed for waves, surges, drift, and other contingencies.
Froude number:	Dimensionless number that represents the ratio of inertial to gravitational forces in open channel flow.
gabion:	Basket or compartmented rectangular container made of wire mesh. When filled with cobbles or other rock of suitable size, the gabion becomes a flexible and permeable unit with which flow- and erosion-control structures can be built.
general scour:	General scour is a lowering of the streambed across the stream or waterway at the culvert or bridge. This lowering may be uniform across the bed or non-uniform. That is, the depth of scour may be deeper in some parts of the cross section. General scour may result from contraction of the flow or other general scour conditions such as flow around a bend.
geomorphology/ morphology:	That science that deals with the form of the Earth, the general configuration of its surface, and the changes that take place due to erosion and deposition.
grade-control structure (sill, check dam):	Structure placed bank to bank across a stream channel (usually with its central axis perpendicular to flow) for the purpose of controlling bed slope and preventing scour or headcutting.
gravel:	Rock fragment whose diameter ranges from 2 to 64 mm.
grout:	Fluid mixture of cement and water or of cement, sand, and water used to fill joints and voids.
headcutting:	Channel degradation associated with abrupt changes in the bed elevation (headcut) that generally migrates in an upstream direction.
headwater:	The water surface elevation on the upstream side of a culvert providing the energy to force water through the culvert.
hydraulics:	Applied science concerned with behavior and flow of liquids, especially in pipes, channels, structures, and the ground.

GLOSSARY (cont.)

hydraulic model:
Small-scale physical or mathematical representation of a flow situation.

hydraulic radius:
Cross-sectional area of a stream divided by its wetted perimeter.

hydraulic structures:
Facilities used to impound, accommodate, convey or control the flow of water, such as dams, weirs, intakes, culverts, channels, and bridges.

hydrograph:
The graph of stage or discharge against time.

hydrology:
Science concerned with the occurrence, distribution, and circulation of water on the earth.

incised stream:
Stream which has deepened its channel through the bed of the valley floor, so that the floodplain is a terrace.

inlet control:
One of two basic types of flow control in culvert hydraulics where the culvert barrel is capable of conveying more flow than the inlet will accept.

invert:
Lowest point in the channel cross section or at flow control devices such as weirs, culverts, or dams.

lateral erosion:
Erosion in which the removal of material is extended horizontally as contrasted with degradation and scour in a vertical direction.

levee:
Embankment, generally landward of top bank, that confines flow during high-water periods, thus preventing overflow into lowlands.

load (or sediment load):
Amount of sediment being moved by a stream.

local scour:
Removal of material from around piers, abutments, spurs, and embankments caused by an acceleration of flow and resulting vortices induced by obstructions to the flow.

long-span culvert:
A culvert that exceeds typical maximum sizes for a given shape, with openings that are 20 to 40 ft (7 to 14 m) wide or larger.

longitudinal profile:
Profile of a stream or channel drawn along the length of its centerline. In drawing the profile, elevations of the water surface or the thalweg are plotted against distance as measured from the mouth or from an arbitrary initial point.

low-water crossing:
A crossing designed to provide safe-passage during low flows, often with small culverts, but overtops at high flows and is therefore closed to traffic.

GLOSSARY (cont.)

lower bank: That portion of a streambank having an elevation less than the mean water level of the stream.

mathematical model: Numerical representation of a flow situation using mathematical equations (also computer model).

mattress: Blanket or revetment of materials interwoven or otherwise lashed together and placed to cover an area subject to scour.

meandering stream: Stream having a sinuosity greater than some arbitrary value. The term also implies a moderate degree of pattern symmetry, imparted by regularity of size and repetition of meander loops. The channel generally exhibits a characteristic process of bank erosion and point bar deposition associated with systematically shifting meanders.

median diameter: Particle diameter of the 50th percentile point on a size distribution curve such that half of the particles (by weight, number, or volume) are larger and half are smaller (D_{50}).

migration: Change in position of a channel by lateral erosion of one bank and simultaneous accretion of the opposite bank.

mud: A soft, saturated mixture mainly of silt and clay.

nonalluvial channel: Channel whose boundary is in bedrock or non-erodible material.

normal stage: Water stage prevailing during the greater part of the year.

outlet control: One of two types of flow control in culvert hydraulics where the barrel is not capable of conveying as much flow as the inlet opening will accept.

overbank flow: Water movement that overtops the bank either due to stream stage or to overland surface water runoff.

performance curve: A plot of the headwater depth or elevation verses flow rate for a given culvert.

perennial stream: Stream or reach of a stream that flows continuously for all or most of the year.

piping: Removal of soil material through subsurface flow of seepage water that develops channels or "pipes" within the soil bank.

probable maximum flood: Very rare flood discharge value computed by hydro-meteorological methods, usually in connection with major hydraulic structures.

quarry-run stone:	Stone as received from a quarry without regard to gradation requirements.
rapid drawdown:	Lowering the water against a bank more quickly than the bank can drain without becoming unstable.
reach:	Segment of stream length that is arbitrarily bounded for purposes of study.
recurrence interval:	Reciprocal of the annual probability of exceedance of a hydrologic event (also return period, exceedance interval).
relief bridge:	An opening in an embankment on a floodplain to permit passage of overbank flow.
revetment:	Rigid or flexible armor placed to inhibit scour and lateral erosion. (See bank revetment).
riffle:	Natural, shallow flow area extending across a streambed in which the surface of flowing water is broken by waves or ripples. Typically, riffles alternate with pools along the length of a stream channel.
riparian:	Pertaining to anything connected with or adjacent to the banks of a stream (corridor, vegetation, zone, etc.).
riprap:	Layer or facing of rock or broken concrete dumped or placed to protect a structure or embankment from erosion; also the rock or broken concrete suitable for such use. Riprap has also been applied to almost all kinds of armor, including wire-enclosed riprap, grouted riprap, partially grouted riprap, sacked concrete, and concrete slabs.
roughness coefficient:	Numerical measure of the frictional resistance to flow in a channel, as in the Manning's or Chezy's formulas.
rubble:	Rough, irregular fragments of materials of random size used to retard erosion. The fragments may consist of broken concrete slabs, masonry, or other suitable refuse.
runoff:	That part of precipitation which appears in surface streams of either perennial or intermittent form.
sand:	Rock fragment whose diameter is in the range of 0.062 to 2.0 mm.
scour:	Erosion of streambed or bank material due to flowing water; often considered as being localized (see local scour, contraction scour, total scour).

GLOSSARY (cont.)

sediment discharge: Quantity of sediment that is carried past any cross section of a stream in a unit of time. Discharge may be limited to certain sizes of sediment or to a specific part of the cross section.

sediment load: Amount of sediment being moved by a stream.

seepage: Slow movement of water through small cracks and pores of the bank material.

shear stress: See unit shear force.

silt: Particle whose diameter is in the range of 0.004 to 0.062 mm.

sinuosity: Ratio between the thalweg length and the valley length of a stream.

slope (of channel or stream): Fall per unit length along the channel centerline or thalweg.

slope protection: Any measure such as riprap, paving, vegetation, revetment, brush or other material intended to protect a slope from erosion, slipping or caving, or to withstand external hydraulic pressure.

sloughing: Sliding or collapse of overlying material; same ultimate effect as caving, but usually occurs when a bank or an underlying stratum is saturated.

slope-area method: Method of estimating unmeasured flood discharges in a uniform channel reach using observed high-water levels.

slump: Sudden slip or collapse of a bank, generally in the vertical direction and confined to a short distance, probably due to the substratum being washed out or having become unable to bear the weight above it.

soil-cement: A designed mixture of soil and Portland cement compacted at a proper water content to form a blanket or structure that can resist erosion.

spread footing: Pier or abutment footing that transfers load directly to the earth.

stable channel: Condition that exists when a stream has a bed slope and cross section which allows its channel to transport the water and sediment delivered from the upstream watershed without aggradation, degradation, or bank erosion (a graded stream).

stage: Water-surface elevation of a stream with respect to a reference elevation.

GLOSSARY (cont.)

stream simulation design:
An AOP design method that simulates the conditions of the natural stream within the culvert.

stone riprap:
Natural cobbles, boulders, or rock dumped or placed as protection against erosion.

stream:
Body of water that may range in size from a large river to a small rill flowing in a channel. By extension, the term is sometimes applied to a natural channel or drainage course formed by flowing water whether it is occupied by water or not.

streambank erosion:
Removal of soil particles or a mass of particles from a bank surface due primarily to water action. Other factors such as weathering, ice and debris abrasion, chemical reactions, and land use changes may also directly or indirectly lead to bank erosion.

streambank failure:
Sudden collapse of a bank due to an unstable condition such as removal of material at the toe of the bank by scour.

streambank protection:
Any technique used to prevent erosion or failure of a streambank.

suspended sediment discharge:
Quantity of sediment passing through a stream cross section above the bed layer in a unit of time suspended by the turbulence of flow (suspended load).

subcritical, supercritical flow:
Open channel flow conditions with Froude Number less than and greater than unity, respectively.

tailwater:
The depth of water on the downstream side of a culvert measured from the outlet invert, and an important factor in outlet control culvert hydraulics.

tapered inlet:
a more gradual transition into a culvert provided by two standard designs, either a side-tapered or slope-tapered inlet

thalweg:
Line extending down a channel that follows the lowest elevation of the bed.

toe of bank:
That portion of a stream cross section where the lower bank terminates and the channel bottom or the opposite lower bank begins.

total sediment load:
Sum of suspended load and bed load or the sum of bed material load and wash load of a stream (total load).

tractive force:
Drag or shear on a streambed or bank caused by passing water which tends to move soil particles along with the streamflow.

turbulence:
Motion of fluids in which local velocities and pressures fluctuate irregularly in a random manner as opposed to laminar flow where all particles of the fluid move in distinct and separate lines.

GLOSSARY (cont.)

uniform flow:	Flow of constant cross section and velocity through a reach of channel at a given time. Both the energy slope and the water slope are equal to the bed slope under conditions of uniform flow.
unit discharge:	Discharge per unit width (may be average over a cross section, or local at a point).
unit shear force (shear stress):	Force or drag developed at the channel bed by flowing water. For uniform flow, this force is equal to a component of the gravity force acting in a direction parallel to the channel bed on a unit wetted area. Usually in units of stress, lb/ft^2 or (N/m^2) .
unsteady flow:	Flow of variable discharge and velocity through a cross section with respect to time.
upper bank:	Portion of a streambank having an elevation greater than the average water level of the stream.
velocity:	Time rate of flow usually expressed in m/s (ft/sec). Average velocity is the velocity at a given cross section determined by dividing discharge by cross-sectional area.
vortex:	Turbulent eddy in the flow generally caused by an obstruction such as a bridge pier or abutment (e.g., horseshoe vortex).
wash load:	Suspended material of very small size (generally clays and colloids) originating primarily from erosion on the land slopes of the drainage area and present to a negligible degree in the bed itself.
watershed:	See drainage basin.
weep hole:	A hole in an impermeable wall or revetment to relieve the neutral stress or pore pressure in the soil.

CHAPTER 1

INTRODUCTION

1.1 GENERAL

The purpose of this publication is to provide information for the planning and hydraulic design of culverts (Figures 1.1 and 1.2). Chapter 2 provides a summary of design considerations including hydrology, site data and site assessments. Chapter 3 provides detailed information on the hydraulic design of the barrel (size, shape, material) and the inlet configuration (pipe end section, headwalls, wingwalls, bevels, and tapers). Chapter 4 provides an overview of aquatic organism passage (AOP) design concepts. A wide range of assorted design topics including bends, junctions, erosion, sedimentation, site modifications, structural considerations, broken back culverts, storage routing, and failure modes is summarized in Chapter 5. Finally, Chapter 6 discusses culvert repair and rehabilitation.

Figure 1.1. Typical small corrugated metal pipe culvert (from WI DOT).

Figure 1.2. Typical concrete box culvert.

The design methodology in this publication provides a simple, consistent approach to culvert design. The experienced designer is assumed to understand the variety of flow conditions that are possible in hydraulic structures and make appropriate adjustments as needed. The inexperienced designer and those unfamiliar with hydraulic phenomena should use this publication with caution. All readers would benefit by being familiar with the basic hydraulic fundamentals found in Hydraulic Design Series Number 4 (HDS 4), Introduction to Highway Hydraulics (FHWA 2008a).

Check lists, design charts and tables, and calculation forms in this publication will provide the designer with necessary tools to perform a wide range of culvert hydraulic analysis and design. However, it is important to recognize that comprehensive design of a culvert crossing involves many diverse considerations beyond hydraulic design. Other factors include proper location and alignment, debris loading, channel stability and sediment movement, minimization of long-term maintenance requirements, outlet channel protection, safety, structural, economic and life-cycle costs. Using the hydraulic design information provided in this document, with information available from other Federal Highway Administration (FHWA) documents, a comprehensive culvert design can be completed. These other documents include Hydraulic Engineering Circular Number 9 (HEC-9), Debris Control Structures Evaluation and Countermeasures (FHWA 2005a); HEC-14, Hydraulic Design of Energy Dissipators for Culverts and Channels (FHWA 2006a); HEC-20, Stream Stability at Highway Structures (FHWA 2012a); and HEC-26, Culvert Design for Aquatic Organism Passage (FHWA 2010a). Figure 1.3 is an overall flowchart identifying major design choices often considered in culvert design. Other considerations include hydrologic routing, broken-back culverts, energy dissipaters, etc.

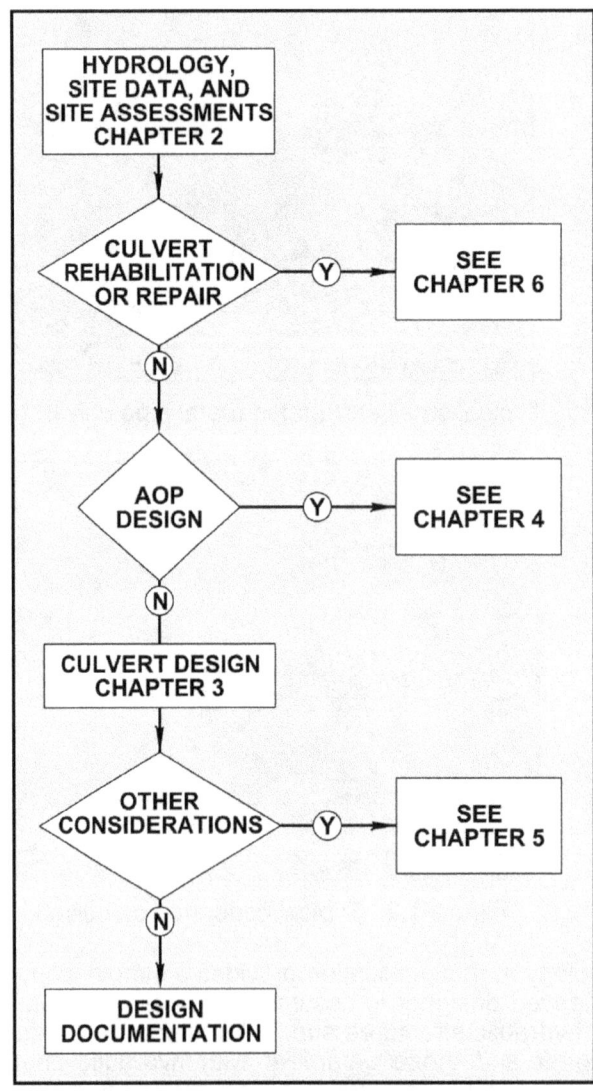

Figure 1.3. Culvert design flowchart.

1.2 COMPARISONS BETWEEN CULVERTS, BRIDGES, AND STORM DRAINS

Culverts, bridges and storm drains all provide for management and conveyance of storm water runoff throughout a roadway system. The designer must decide which analysis approach to use at a crossing or to design alternatives. For most highway crossings, the choice is between a culvert and a bridge. Storm drain analysis is sometimes required and discussed at the end of this section.

Comparing culverts to bridges the designer must determine which type of structure is best for a particular location, and then decide how to analyze the crossing. For example, in many respects a large box culvert begins to resemble a small single-span bridge with vertical wall abutments, and so:

- Which one is preferred hydraulically, aesthetically and economically?

- How should the structure be analyzed (as a bridge with free surface flow using gradually varied, open channel flow concepts or as a culvert with headwater based on traditional culvert analysis)?

Culverts are used:

- Where bridges are not hydraulically required
- Where debris and ice potential are tolerable
- Where more economical than a bridge (including guardrail and safety concerns)

Bridges are used:

- Where culverts are impractical
- Where more economical than a culvert
- To satisfy land-use and access requirements
- To mitigate environmental concerns not satisfied by a culvert
- To avoid floodway encroachments
- To accommodate ice and large debris

Traditionally, economic considerations were of primary importance in deciding between the use of a bridge or a culvert at stream crossings where either will satisfy hydraulic and structural requirements. The initial cost for a culvert is usually less than a bridge since the use of increased headwater at a culvert installation normally permits the use of a smaller opening (Figure 1.4), compared to a bridge which is normally designed with freeboard at the design discharge. However, this advantage must be balanced against possible flood damages associated with increased headwater, especially at higher discharges, and more recently against AOP issues that often dictate a larger opening and a natural invert.

Maintenance costs for culverts may result from channel erosion at the inlet and outlet, erosion and deterioration of the culvert invert, sedimentation, ice and debris buildup, and embankment repair in case of overtopping. Bridge maintenance is typically more costly, however, including such aspects as maintenance of the bridge deck and superstructure, erosion around piers and abutments, and possible sediment and debris accumulation.

Safety and aesthetic considerations are also involved in the choice of a bridge or culvert. Safety considerations for culverts include the use of guardrails or safety grates. It is important to recognize that culverts exceeding a 20 ft (6.1 m) span width (either as a single barrel or the total width of a multiple barrel crossing) are considered bridges in the National Bridge Inspection Standards (NBIS) and therefore subject to routine inspection according to NBIS requirements. Bridge decks often constrict shoulder and median widths and are subject to icing which can present traffic safety problems. A bridge may be considered more aesthetically pleasing in traversing a scenic valley or canyon.

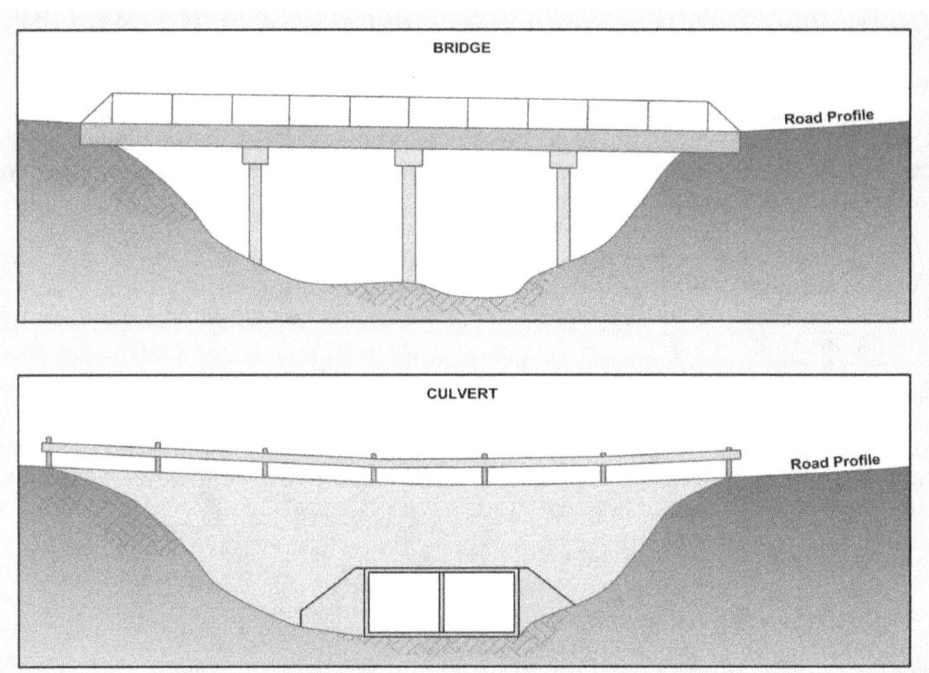

Figure 1.4. Bridge versus culvert at same location.

There are differences in the hydraulic assumptions and analyses used for culverts and bridges. The analysis of open channel flow in any structure can be based on relatively simple normal depth assumptions or on more complicated gradually varied flow calculations. The tailwater condition in the channel downstream of a culvert is typically based on normal depth analysis. Additionally, culvert analysis often assumes no velocity approaching the culvert or in the channel immediately downstream of the culvert, which overestimates entrance and exit energy losses. In contrast, bridge hydraulic analysis is typically based on gradually varied flow calculations providing a more accurate water surface profile throughout the crossing, and which accounts for the velocity approaching and leaving the structure.

Ultimately, large culverts with free surface flow through the structure (i.e., no headwater) are typically better analyzed based on the gradually varied open channel flow concepts used in bridge analysis than the calculation procedures detailed in this publication. Smaller structures are more easily designed using culvert based hydraulic analysis and procedures. Based on NBIS regulations, as well as hydraulic issues, a reasonable guideline is to use bridge based modeling for a single culvert with a span of 20 ft (6.1 m) or more, given that such structures will typically operate with free surface flow. For small stream crossings that might use large barrel sizes, such as an AOP culvert, bridge based hydraulic modeling is again probably more appropriate given the expected free surface flow conditions.

At the other extreme, when does a long culvert begin to resemble a "short" storm drain system? If there are multiple inflow points along the culvert (such as roadway or median inlets), multiple geometric changes (bends, pipe size changes, slope changes, etc), or potential changes in flow regime within the barrel, the typically more robust storm drain analyses and computer programs might be preferred over the use of culvert analysis procedures as described in this publication.

Information presented throughout the rest of this document assumes that the decision has been made to design a culvert, rather than a bridge, and that the culvert to be designed or analyzed is not so large that it should be considered a bridge, nor so long or complicated that it should be considered a storm drain. For detailed information on bridge hydraulics and analysis refer to HDS 7, "Hydraulic Design of Safe Bridges" (FHWA 2012b). For detailed information on storm drain design see HEC-22, Urban Drainage Design (FHWA 2009a).

1.3 OVERVIEW OF CULVERTS

A culvert is a conduit which conveys stream flow through a roadway embankment or past some other type of flow obstruction. Culverts are constructed from a variety of materials and are available in many different shapes and configurations. Culvert selection factors include roadway profiles, channel characteristics, flood damage evaluations, construction and maintenance costs, and estimates of service life.

1.3.1 Shapes

Numerous cross-sectional shapes are available for both closed conduit and open-bottom culverts. The most common closed conduit shapes are circular, box (rectangular), elliptical, and pipe-arch (Figure 1.5a). These typical manufactured culvert shapes have the same material on the entire perimeter. Shape selection is based on the cost of construction, the limitation on upstream water surface elevation, roadway embankment height, and hydraulic performance. These shapes can be constructed with embedment which is a depression below the streambed of both the inlet and outlet inverts. Design aids are only provided in the appendix for circular and box shapes. Typical open bottom culvert shapes are various box and arch configurations shown in Figure 1.5b. The cross section shapes shown in Figures 1.5a and b along with a user-defined shape comprise the standard shapes available in the FHWA culvert design computer program HY-8.

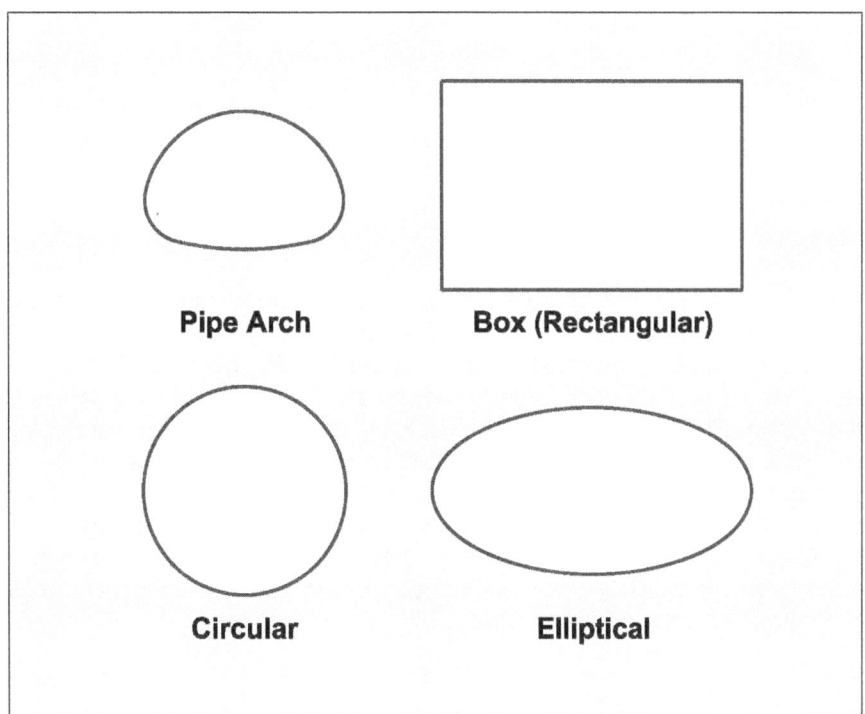

Pipe Arch   **Box (Rectangular)**

Circular   **Elliptical**

Figure 1.5a. Commonly used culvert shapes.

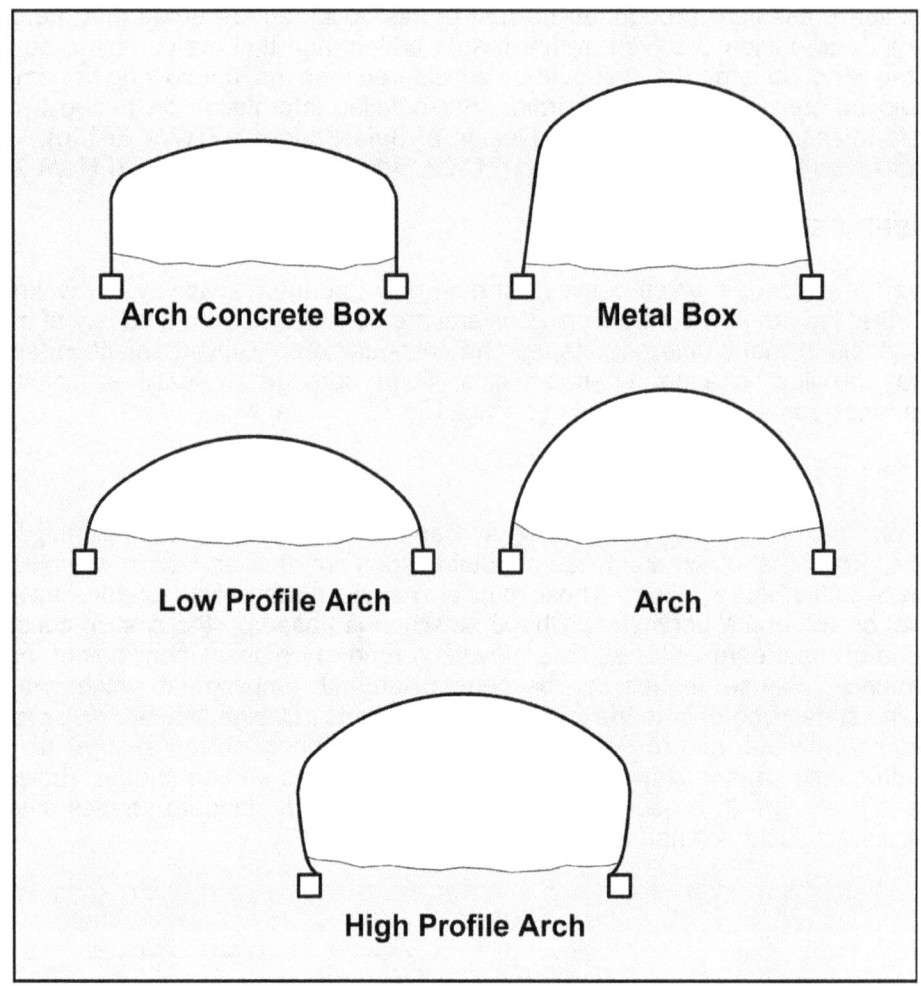

Figure 1.5b. Commonly used open-bottom culvert shapes.

1.3.2 Materials

The selection of a culvert material may depend upon structural strength, hydraulic roughness, durability (corrosion and abrasion resistance), and constructability. The most commonly used culvert materials are concrete (both reinforced and non-reinforced), corrugated metal (aluminum or steel) and plastic (high-density polyethylene (HDPE) or polyvinyl chloride (PVC)). A concrete box culvert and a corrugated metal pipe are depicted in Figures 1.6 and 1.7, respectively. Less commonly used materials include clay, stone and wood, as might be found in historic culvert structures. Materials for culverts continue to be developed and in the future could include various types of plastics, fiberglass, and composite materials. Culverts may also be lined with other materials to inhibit corrosion and abrasion, or to reduce hydraulic resistance. For example, corrugated metal culverts may be lined with asphaltic concrete or a polymer material.

Figure 1.6. Concrete box culvert.

Figure 1.7. Corrugated metal pipe.

1.3.3 Inlets

A multitude of different inlet configurations are utilized on culvert barrels. These include both prefabricated and constructed-in-place installations. Commonly used inlet configurations include projecting culvert barrels, cast-in-place concrete headwalls, precast or prefabricated end sections, and culvert ends mitered to conform to the fill slope (Figure 1.8). Hydraulic performance, structural stability, aesthetics, erosion control, and fill retention are considerations in the selection of various inlet configurations.

Figure 1.8. Four standard inlet types.

The hydraulic capacity of a culvert may be improved by appropriate inlet selection. The channel is often wider than the culvert barrel, causing a contraction at the culvert inlet which may be the primary flow control. The provision of a more gradual flow transition will lessen the energy loss and thus create a more hydraulically efficient inlet condition (Figure 1.9). Beveled edges are therefore more efficient than square edges.

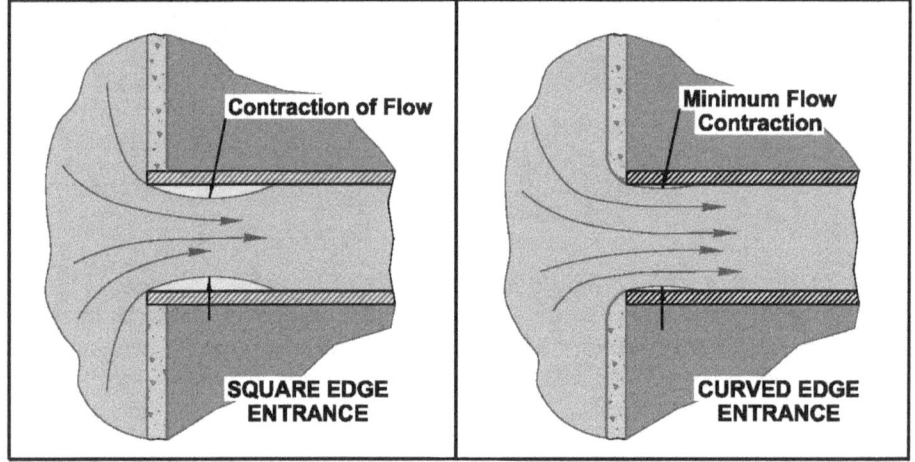

Figure 1.9. Entrance contraction.

Side-tapered and slope-tapered inlets, commonly referred to as tapered inlets, further reduce the flow contraction. Figure 1.10 illustrates both side-tapered and a slope-tapered inlets. Depression can increase the effective head on the flow control section, thereby further increasing the culvert efficiency. The depression can be either inlet depression or throat depression.

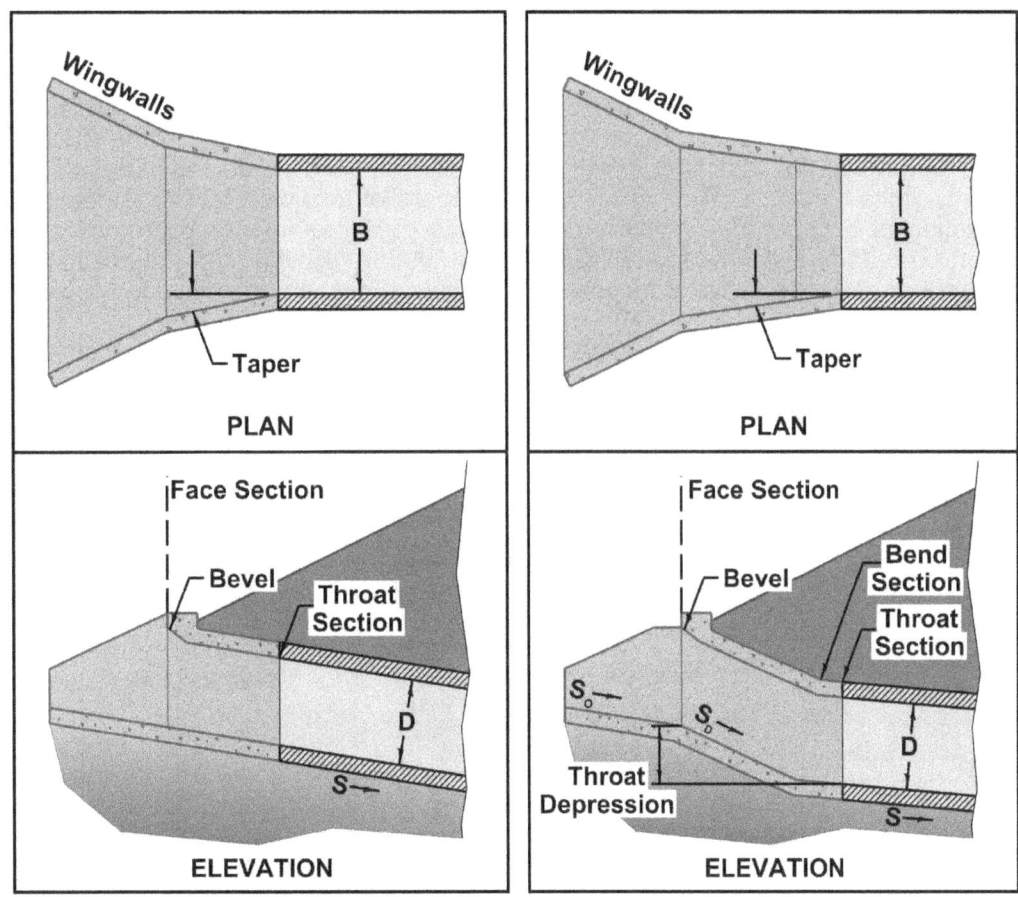

a. Side-tapered inlet. b. Slope-tapered inlet.

Figure 1.10. Schematic of side-tapered and slope-tapered inlets.

When considering the use of a tapered inlet the designer should evaluate the increased construction cost of the taper section against the saving in the barrel cost. Additionally, if AOP is important, the smaller barrel size possible with a tapered inlet is usually not an acceptable design alternative. However, when tapered inlets are feasible, the improvement in hydraulic performance can be significant in some cases. A typical example might be a relatively long culvert on a steep slope without AOP issues, as might occur in an urban area. Another application might be a roadway widening project where the existing barrel is in good condition, but undersized as a result of urbanization that has increased the design discharge. Rather than replacing the existing culvert with the associated cost and disruption to traffic, a tapered inlet could be designed as part of the culvert extension to improve hydraulic efficiency. Rehabilitation of a culvert under a high fill is another example where a tapered inlet at the entrance might help compensate for a liner that reduces the inlet area and barrel size.

1.3.4 Embedded and Open-Bottom Culverts

Embedded and open-bottom culverts provide a natural invert, which may be desirable for a variety of reasons. For example, embedded and open-bottom culverts are commonly used in crossings designed to address AOP issues. A natural invert may also be preferable in channels with high sediment transport, particularly with course materials (gravels and cobbles) where abrasion caused by moving sediment can quickly destroy the culvert invert. Other reasons for a natural invert can include aesthetics.

An open-bottom culvert is often a box or arch shape built on a vertical wall foundation (Figure 1.11). The foundation should be designed considering potential scour of the streambed that might undermine the structure, which is a common failure mode for this type of culvert. There is limited information on scour from flow parallel to a wall, as might result along the foundation of an open-bottom culvert. A design guideline for riprap protection for open-bottom culverts is available in HEC-23 (FHWA 2009b). However, if the scour potential is great, a less risky alternative to an open-bottom culvert may be an embedded culvert.

An embedded culvert can be any shape, but is most often a circular, box or pipe arch that has been buried into the ground typically 20-40% of its height (Figure 1.12). One advantage of an embedded culvert is the culvert invert can provide grade control and protection against extreme scour compared to an open-bottom culvert.

Figure 1.11. Open-bottom culvert.

Figure 1.12. Embedded culvert (from WI DOT).

The increased interest in open-bottom and embedded culverts, particularly for AOP applications, has resulted in additional research on their hydraulic characteristics. That research includes a project conducted by the National Cooperative Highway Research Program (NCHRP Project 15-24) to refine existing hydraulic loss coefficients and to develop new hydraulic loss coefficients for conventional and nontraditional, environmentally sensitive culvert installations (NCHRP 2011). The research included evaluating entrance and exit loss coefficients for embedded culverts and culverts that have been re-lined (Appendix A). It also included evaluating composite hydraulic loss methods for open-bottom and embedded culverts that have different roughness between the invert and culvert material. Chapter 3 provides the traditional method for determining composite roughness that was confirmed by the NCHRP research.

1.3.5 Low Water Crossings

A low-water crossing is designed to provide safe passage during low flows, but will overtop at high flows and therefore be closed to traffic. Low-water stream crossings can provide safe, cost-efficient alternatives to bridge and culvert crossings for certain low-volume roads, provided the streamflow and road-use conditions are suitable. Standards and criteria are provided in the Project Development and Design Manual (FHWA 2008b) and in the U.S. Forest Service (USFS) publication Low-water Crossings: Geomorphic, Biological, and Engineering Design Considerations (USFS 2006).

Those standards define two classes of low-water crossings, vented crossings and unvented crossing. A vented crossing has a hydraulic opening beneath the road for low flows, while an unvented crossing has no opening. An unvented low-water crossing is essentially a ford that might be as simple as a dip in the roadway to match the streambed elevation, or can be slightly elevated above the streambed. They are often used in arid climates where stream channels are dry most of the year (Figure 1.13). A vented low-water crossing is essentially a culvert sized to allow low flow passage but overtops the roadway at high flows (Figure 1.14). As designed, the crossing will operate as a culvert for flows up to the low flow design discharge, and as a broad crested weir combined with culvert flow for greater discharges. During significant runoff events the current in either an unvented or vented low-water crossing can be strong enough to sweep vehicles downstream. Warning signs approaching the crossing are important and should ideally restrict entry when water exceeds a specified depth, as might be identified by a post in, or near, the crossing.

Protecting the roadway and embankment from erosion and scour during overtopping is a critical design element for a low-water crossing. The potential drop in water surface elevation and acceleration of flow as water overtops a roadway can create erosion and scour on the downstream side of the roadway requiring some type of protection, typically rock riprap along the embankment or constructing the embankment with paving. Another issue that must be considered is potential impacts on AOP.

1.3.6 Aquatic Organism Passage (AOP)

Traditionally, culverts were designed based on hydraulic conveyance and flood capacity. The primary objective was to size the culvert to maintain an acceptable level of ponding upstream of the roadway as defined by the allowable headwater elevation. This typically resulted in a crossing structure that was significantly smaller and narrower than the stream channel, contracting the flow and increasing velocity and scour potential.

Figure 1.13 Low-water crossing.

Figure 1.14. Vented low-water crossing.

This contraction of flow altered the natural flow regime creating barriers to fish movement at or adjacent to the culvert. Concerns about the effect of these barriers, particularly with migrating fish, initiated new culvert design concepts based on the capabilities and behavior of fish at various life-stages. Although much of the original focus was on fish passage, it was soon recognized that many other organisms in the aquatic environment (e.g., frogs, salamanders, turtles) were impacted and the issue became better known as aquatic organism passage (AOP).

Desirable criteria for an AOP culvert initially focused on minimizing the amount of contraction at the entrance, maintaining reasonable velocities and flow depths within the culvert, avoiding abrupt changes in grade or elevation, and maintaining a natural channel bottom. As knowledge grew, the focus shifted to simulating the natural channel through the culvert to the degree possible. The FWHA AOP design procedure to better create these types of conditions is detailed in HEC-26 (FHWA 2010a). A summary of this information is provided in Chapter 4.

When AOP issues are a concern, the culvert is typically larger than necessary to meet hydraulic conveyance and flood capacity design standards. While the culvert will cost more initially, it has the potential for reducing maintenance costs over the life of the culvert installation. The FHWA procedure emphasizes the use of oversized, embedded culverts that provide a natural invert, but also allows some measure of grade control by the culvert invert. Many other structure types and configurations may also be used.

1.3.7 Long Span Culverts

Long span culverts are better defined on the basis of structural design aspects than on the basis of hydraulic considerations. According to the Specifications for Highway Bridges, long span structural plate structures (AASHTO 2002):

1. Exceed certain defined maximum sizes for pipes, pipe-arches, and arches, or

2. May be special shapes of any size that involve a long radius of curvature in the crown or side plates.

Special shapes include vertical and horizontal ellipses, underpasses, and low and high profile arches. Generally, the spans of long span culverts range from 20 to 40 ft (7m to 14m). Several typical long span culvert shapes were illustrated in Figure 1.5b. A typical long span installation is shown in Figure 1.15.

Figure 1.15. Long span culvert.

Long span culverts depend on interaction with the earth embankment for structural stability. Therefore, proper bedding and selection and compaction of backfill are of utmost importance. For multiple barrel structures, care must be taken to avoid unbalanced loads during backfilling. Some manufacturers of long span culverts will not sell their products to a client unless the design and installation is supervised by their engineers. If this is not required, the project should be coordinated with the manufacturer's engineering staff.

Various manufacturers utilize different techniques to achieve the desired long span configuration. In some instances, reinforcing ribs are used to strengthen the structure. In other cases, specially designed longitudinal structural stiffeners are installed on the top arch. Ribs and stiffeners which project into the barrel may increase the hydraulic resistance, particularly if the elements are perpendicular to the flow.

Anchorage of the ends of long span culverts is required to prevent flotation or damage due to high velocities at the inlet. This is especially true for mitered inlets. Severe miters and skews are not recommended.

The same hydraulic principles apply to the design of long span culverts as to other culverts. However, given their size they seldom have flow exceeding the top of the barrel and often flow partly full at the design discharge. Consequently, normal open channel flow backwater calculations are often the best analytical approach to evaluate hydraulic performance. For any of the open-bottom shapes, design and protection for scour along the foundation must be accounted for.

1.3.8 Culvert Function

Culverts perform a wide range of hydraulic and non-hydraulic functions. The most common hydraulic function is providing cross drainage for a stream channel, as presented earlier in this chapter. Other hydraulic functions include floodplain relief, where a culvert might be placed in the overbank of a wide floodplain to provide drainage of the overbank area during large flood events. Such culverts often have no defined channel upstream or downstream of the barrel and may be dry for years at a time. Smaller culvert structures often function to provide ditch relief for drainage ditches along a roadway, diverting some of the discharge from the ditch. Culverts are also often incorporated into the outlet control structures for detention ponds and various water management structures. Non-hydraulic functions include crossing structures for human or animal traffic, such as a pedestrian or trail crossing, cattle crossings, farm equipment access and crossings designed to facilitate wildlife movement.

1.4 CULVERT HYDRAULICS

A complete theoretical analysis of culvert hydraulics based on fundamental equations can be difficult. Flow conditions vary over time for any given culvert. The barrel of the culvert may flow full or partly full depending upon upstream and downstream conditions, barrel characteristics, and inlet geometry. This realization and the lack of any consistent culvert design methodology led to research by the Federal Highway Administration (FHWA), formerly the Bureau of Public Roads (BPR), at the National Bureau of Standards (NBS) laboratories starting in the 1950s. This research resulted in a series of seven research reports and the equations presented in Appendix A. The equations were used to develop nomographs for evaluating culvert hydraulics (Appendix C). These nomographs are the basis of the FHWA culvert design procedures that were first published in HEC-5 (FHWA 1965), HEC-13 (FHWA 1972a), and now in this publication.

The basic approach presented in HEC-5 was to analyze a culvert for various types of flow control and then design for the control which produces the minimum performance. Designing for minimum performance ignores transient conditions which might result in periods of better performance. The benefits of designing for minimum performance are ease of design and assurance of adequate performance under the least favorable hydraulic conditions.

1.4.1 Flow Conditions

A culvert barrel may flow full over all of its length or partly full. Full flow in a culvert barrel is rare. Generally, at least part of the barrel flows partly full. A water surface profile calculation is the only way to accurately determine how much of the barrel flows full.

a. Full Flow. The hydraulic condition in a culvert flowing full is called pressure flow. If the cross-sectional area of the culvert in pressure flow were increased, the flow area would expand. One condition which can create pressure flow in a culvert is the back pressure caused by a high downstream water surface elevation. A high upstream water surface elevation may also produce full flow. Regardless of the cause, the capacity of a culvert operating under pressure flow is affected by upstream and downstream conditions and by the hydraulic characteristics of the culvert.

b. Partly Full (Free Surface) Flow. Free surface flow or open channel flow may be categorized as subcritical, critical, or supercritical. A determination of the appropriate flow regime is accomplished by evaluating the dimensionless number, F_r, called the Froude number:

$$Fr = \frac{V}{\sqrt{gy}} \qquad (1.1)$$

In this equation, V is the average velocity of flow, g is the gravitational acceleration, and y is a representative depth, typically the equivalent depth or the hydraulic depth. The equivalent depth is often used to define the representative depth in a circular section and is defined as the square root of one-half of the cross-sectional flow area $(A/2)^{0.5}$. The hydraulic depth is used for other shapes and is calculated by dividing the cross-sectional flow area by the width of the free water surface (A/T). When Fr > 1.0, the flow is supercritical and is characterized as rapid. When Fr < 1.0, the flow is subcritical and is characterized as tranquil. If Fr = 1.0, the flow is defined as critical.

The three flow regimes are illustrated in the depiction of a small dam in Figure 1.16. Subcritical flow occurs upstream of the dam crest where the water is deep and the velocity is low. Supercritical flow occurs downstream of the dam crest where the water is shallow and the velocity is high. Critical flow occurs at the dam crest and represents the dividing point between the subcritical and supercritical flow regimes. To analyze free surface flow conditions, a point of known depth and flow (control section) must first be identified. A definable relationship exists between critical depth and critical flow at the dam crest, making it a convenient control section.

Identification of subcritical or supercritical flow is required to continue the analysis of free surface flow conditions. The example using the dam of Figure 1.16 depicts both flow regimes. Subcritical flow characteristics, such as depth and velocity, can be affected by downstream disturbances or restrictions. For example, if an obstruction is placed on the dam crest (control section); the water level upstream will rise. In the supercritical flow regime, flow characteristics are not affected by downstream disturbances. For example, an obstruction placed at the toe of the dam does not affect upstream water levels.

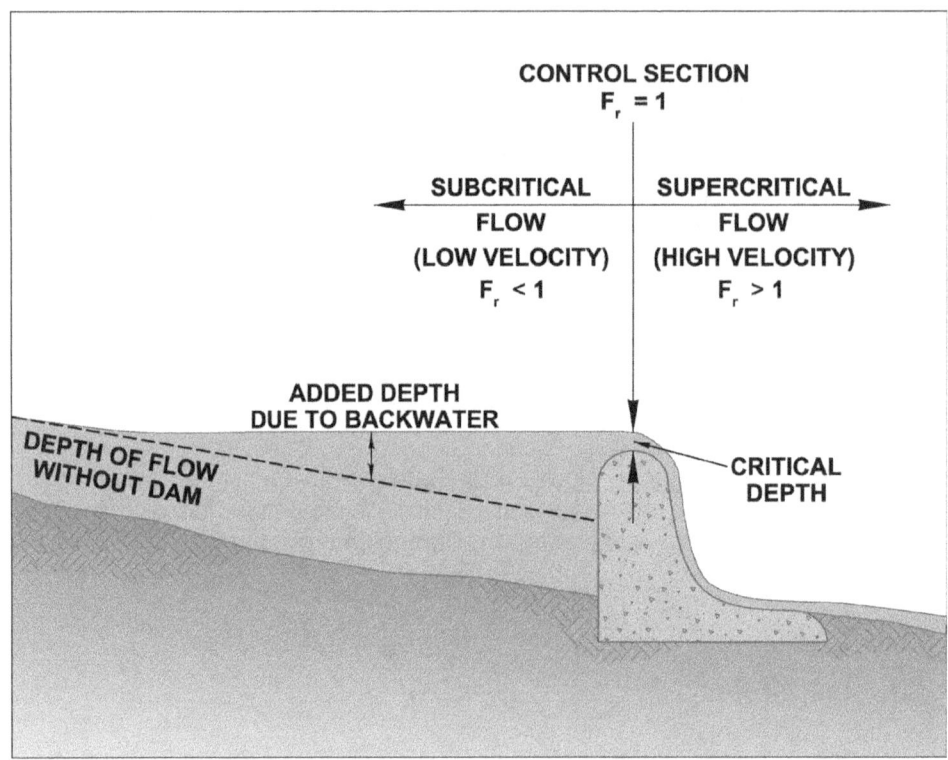

Figure 1.16. Flow conditions over a small dam.

The same type of flow illustrated by the small dam may occur in a steep culvert flowing partly full (Figure 1.17). In this situation, critical depth occurs at the culvert inlet, subcritical flow exists in the upstream channel, and supercritical flow exists in the culvert barrel.

A special type of free surface flow is called "just-full flow." This is a special condition where a pipe flows full with no pressure. The water surface just touches the crown of the pipe. The analysis of this type of flow is the same as for free surface flow.

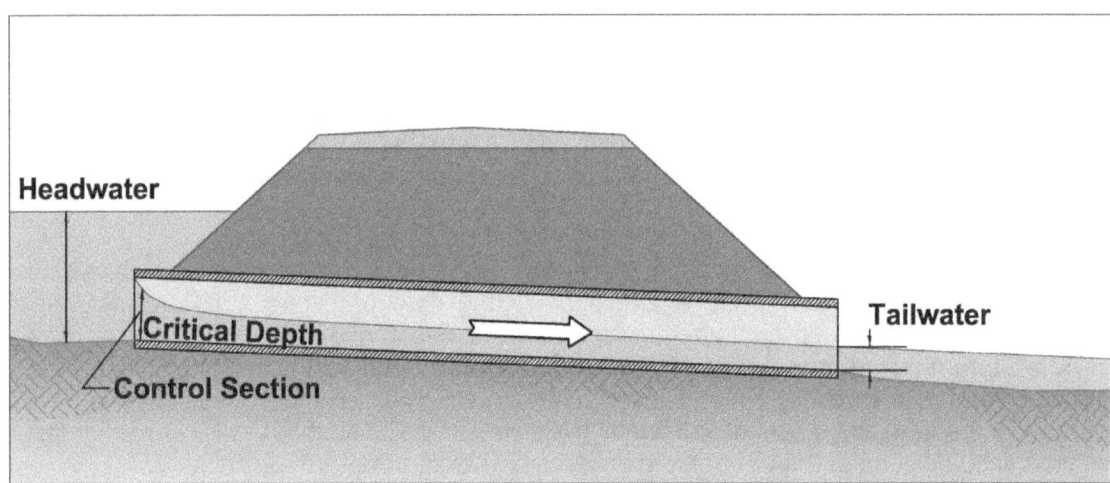

Figure 1.17. Typical inlet control flow section.

1.4.2 Types of Flow Control

Inlet and outlet control are the two basic types of flow control defined in the research conducted by the NBS and the FHWA (formerly BPR). The basis for the classification system was the location of the control section. The characterization of pressure, subcritical, and supercritical flow regimes played an important role in determining the location of the control section and thus the type of control. The hydraulic capacity of a culvert depends upon a different combination of factors for each type of control.

a. Inlet Control. Inlet control occurs when the culvert barrel is capable of conveying more flow than the inlet will accept. The control section of a culvert operating under inlet control is located just inside the entrance. Critical depth occurs at or near this location, and the flow regime immediately downstream is supercritical. Figure 1.17 shows one typical inlet control flow condition. Hydraulic characteristics downstream of the inlet control section do not affect the culvert capacity. The upstream water surface elevation and the inlet geometry represent the major flow controls. The inlet geometry includes the inlet shape, inlet cross-sectional area, and the inlet configuration (Table 1.1).

b. Outlet Control. Outlet control flow occurs when the culvert barrel is not capable of conveying as much flow as the inlet opening will accept. The control section for outlet control flow in a culvert is located at the barrel exit or further downstream. Either subcritical or pressure flow exists in the culvert barrel under these conditions. Figure 1.18 shows two typical outlet control flow conditions. All of the geometric and hydraulic characteristics of the culvert play a role in determining its capacity. These characteristics include all of the factors governing inlet control, the water surface elevation at the outlet, and the barrel characteristics (Table 1.1).

The factors in Table 1.1 distinguish between the geometric properties of the inlet versus the barrel to account for the effect of tapered inlets used on some culverts. For a culvert without a taper the inlet area and shape would be equal to the barrel area and shape. The slope of the culvert is called barrel slope to distinguish it from other slope parameters that may exist at the entrance, such as when a depressed inlet is used. Barrel slope is the primary factor influencing whether or not a culvert will be in inlet or outlet control. In the case of a mitered culvert, the length of the barrel is based on where the crown intersects the fill slope.

1.4.3 Headwater

Energy is required to force flow through a culvert. This energy takes the form of an increased water surface elevation on the upstream side of the culvert. The depth of the upstream water surface measured from the invert at the culvert entrance is generally referred to as headwater depth (Figures 1.17 and 1.18).

A considerable volume of water may be ponded upstream of a culvert under high fills or in areas with flat ground slopes. The pond which is created may attenuate flood peaks under such conditions, similar to the attenuation caused by a reservoir or lake. Analysis of flood peak attenuation is based on storage routing, as described in Chapter 5. Although this decrease in peak discharge may justify a reduction in the required culvert size, this is not a widely used practice in culvert design.

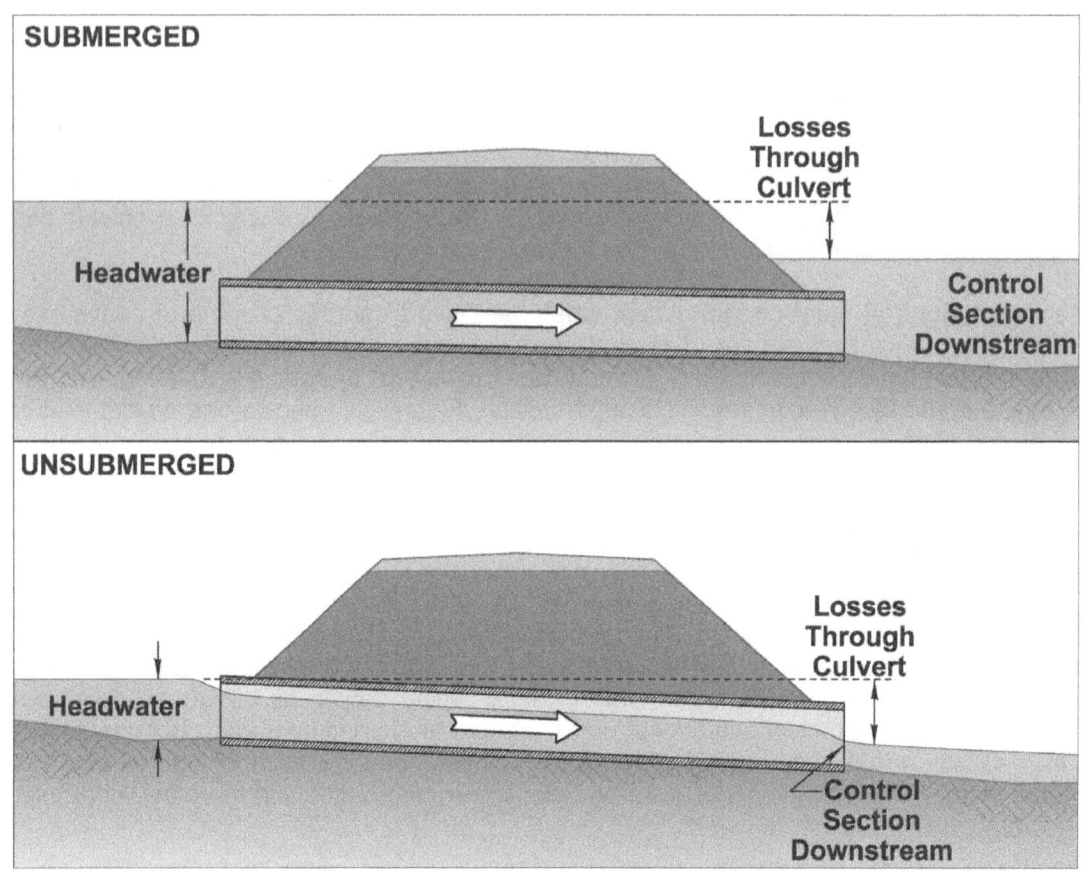

Figure 1.18. Typical outlet control flow conditions.

Table 1.1. Factors Influencing Culvert Design.		
Factor	Inlet Control	Outlet Control
Headwater	X	X
Area	X	X
Shape	X	X
Inlet Configuration	X	X
Barrel Roughness	-	X
Barrel Length	-	X
Barrel Slope	X	X
Tailwater	-	X
Note: For inlet control the area and shape factors relate to the inlet area and shape. For outlet control they relate to the barrel area and shape.		

1.4.4 Tailwater

Tailwater is defined as the depth of water downstream of the culvert measured from the outlet invert. It is an important factor in determining culvert capacity under outlet control conditions. The amount of tailwater is based on the characteristics of the downstream channel at the given design discharge and is evaluated based on traditional open channel flow calculations, often using normal depth approximations. Increased tailwater may be caused by an obstruction in the downstream channel, such as another highway crossing with a bridge or culvert, the confluence with another channel, the existence of a reservoir or beaver dam, etc. In such cases, backwater calculations from the downstream control point are required to precisely define tailwater. High tailwater alone is capable of making a culvert operate under outlet control, when it would otherwise be under inlet control.

1.4.5 Outlet Velocity

Since a culvert often constricts the available channel area, flow velocities in the culvert may be higher than in the channel. These increased velocities can cause streambed scour and bank erosion in the vicinity of the culvert outlet. Minor problems can occasionally be avoided by increasing the barrel roughness. Energy dissipaters and outlet protection devices are sometimes required to avoid excessive scour at the culvert outlet. When a culvert is operating under inlet control and the culvert barrel is not operating at capacity, it is often beneficial to flatten the barrel slope or add a roughened section to reduce outlet velocities (see broken-back culvert, Section 5.7). Methods of calculating outlet velocity for both inlet and outlet control are described in Section 3.1.6.

1.4.6 Performance Curves

A performance curve is a plot of headwater depth or elevation versus flow rate. The resulting graphical depiction of culvert operation is useful in evaluating the hydraulic capacity of a culvert for various headwaters. Among its uses, the performance curve displays the consequences of higher flow rates at the site.

In developing a culvert performance curve, both inlet and outlet control curves must be plotted. This is necessary because the dominant control at a given headwater is hard to predict. Also, control may shift from the inlet to the outlet or vice-versa over a range of flow rates. Figure 1.19 illustrates a typical culvert performance curve. Based on the concept of minimum performance (see Section 3.1.1), at the allowable headwater (see Section 2.2.5) the culvert operates under inlet control. With a better inlet configuration, the culvert performance can be increased to take better advantage of the culvert barrel capacity.

1.5 CULVERT REPAIR AND REHABILITATION

Culvert repair and rehabilitation is of interest as many structures nationwide are at or beyond their original design service life. Assessment of existing culvert conditions is an important first step in the design process for repair and rehabilitation efforts. Many existing culvert condition factors and indicators must be assessed and considered in selection of appropriate repair and rehabilitation methods and technologies. Many rehabilitative techniques intended to address culvert deterioration will also influence hydraulic performance of the culvert.

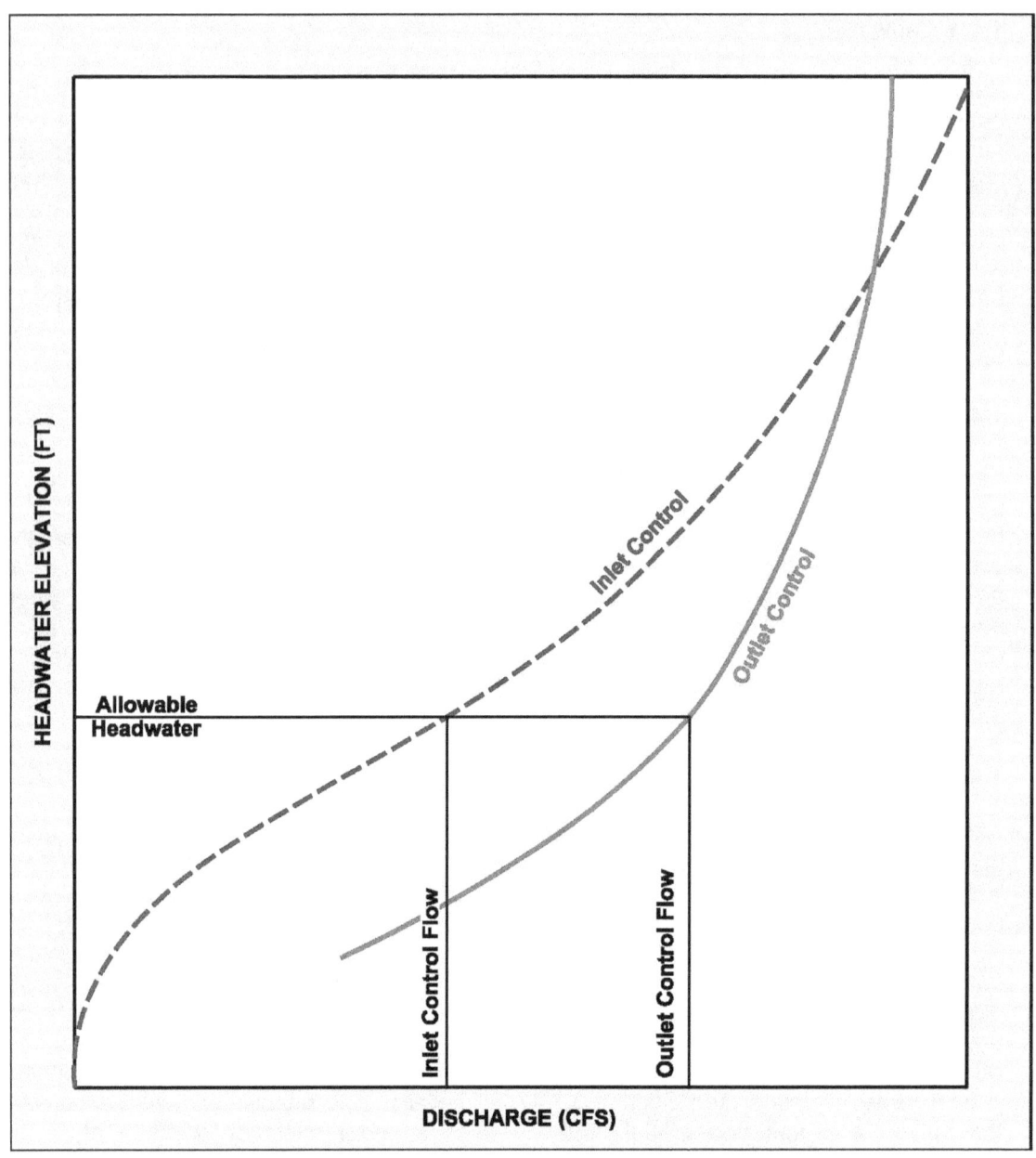

Figure 1.19. Culvert performance curve.

Given an accurate assessment of culvert condition, the decision between repair and replacement can be made. The typical outcomes of this decision making process are to repair with a lining, repair based on man-entry, or to simply replace. The Culvert Pipe Liner Guide and Specifications (FHWA 2005b) provides information on common lining techniques and methods. The Culvert Assessment and Decision-making Procedures Manual (FHWA 2010b) provides information on assessment procedures and repair options. An overview of this information is provided in Chapter 6, focusing on the potential changes repair and replacement can have on hydraulic performance of a culvert.

1.6 ECONOMICS

The design of a culvert installation should always include an economic evaluation. A wide spectrum of flood flows with associated probabilities will occur at the culvert site during its service life. The benefits of constructing a large capacity culvert to accommodate all of these events with no detrimental flooding effects are normally outweighed by the initial construction costs. Thus, an economic analysis of the tradeoffs is performed with varying degrees of effort and thoroughness.

The ideal culvert selection process minimizes the total annual cost of the installation over the life of the roadway. The annual cost includes capital expenditures, maintenance costs, and risks associated with flooding. The need to compare the cost of available shapes and sizes is well understood when designing a culvert. Perhaps less understood is the need to also consider durability, maintenance, and replacement costs along with the initial in-place construction cost. Selecting a culvert material that better withstands corrosion or abrasion may cost more initially, but the longer service life will lower total annual cost. Anticipating future maintenance requirements can also save money in the long run. For example, under a high fill it might be better to oversize the barrel initially to better accommodate potential re-lining in the future, in order to avoid the cost of replacement at that time.

1.6.1 Benefits and Costs

The purpose of a highway culvert is to convey water through a roadway embankment. The major benefits of the culvert are decreased traffic interruption time due to roadway flooding and increased driving safety. The major costs are associated with the construction of the roadway embankment and the culvert itself. Maintenance of the facility and flood damage potential must also be factored into the cost analysis.

1.6.2 Comparisons Between Materials and Shapes

Cost comparisons between various materials and shapes vary with region and with time. It is recommended that costs for culverts of equal hydraulic capacity be compared periodically to help guide material selection. Requesting alternative bids for several acceptable materials is economically beneficial on most projects.

Detailed economic analysis of culvert material selection requires site-specific considerations. Structural strength is a concern under high fills. Steep channel slopes produce high exit velocities which are further accelerated by using smooth pipes. Acidic drainage will promote corrosion of some materials. Certain materials cannot withstand the attack of abrasive bed loads. Water tightness at joints may be an important consideration. All of these factors have an impact on the annual cost of the culvert based upon the selected material.

Culvert shapes are as important in cost evaluations as culvert materials (Figure 1.5). Many shapes can be produced from a variety of materials; other shapes require certain materials. Circular culverts are the most common shape. They are generally reasonably priced, can support high structural loads, and are hydraulically efficient. However, limited fill height may necessitate the use of a pipe-arch or ellipse. Pipe-arches and ellipses are more expensive than circular pipes. Arches require special attention to their foundations, and failure due to scour is a concern. However, arches do provide a natural stream bed which is an advantage for AOP. Structural plate conduits can be constructed in a variety of shapes, quickly, with low transport and handling costs (Figure 1.20). Box culverts also possess flexibility in rise to span ratios by using multiple cells (Figure 1.21). Precast box sections overcome the disadvantage of longer construction times which are associated with cast-in-place installations; however handling costs are increased.

1.21

Figure 1.20. Structural plate culvert.

Figure 1.21. Multi-cell box culvert.

1.6.3 Service Life

The desired service life of the culvert should be considered in the selection process. If the culvert is in a location where replacement or relining would be impractical, the service life of the culvert should equal the service life of the highway. If rehabilitation is feasible, or if it is determined that the highway will be rebuilt in a relatively short time, a culvert with a shorter service life should be selected. The service life of the culvert should match the installation. There is no need to pay for an "eternal" culvert where a short lived one would suffice, and vice-versa.

1.6.4 Risk Analysis

Traditional economic evaluations for minor stream crossings have been somewhat simplistic. Culvert design flows are based on the importance of the roadway being served with little attention given to other economic and site factors. A more rigorous investigation, termed a risk analysis, is sometimes performed for large culvert installations or for locations with high potential flood damages. The objective of the risk analysis is to find the optimum culvert capacity based on a comparison of benefits and costs (Figure 1.22). The designer should be aware of the risk analysis process and consider using it to analyze alternatives where flood damage is large or culvert cost is significant.

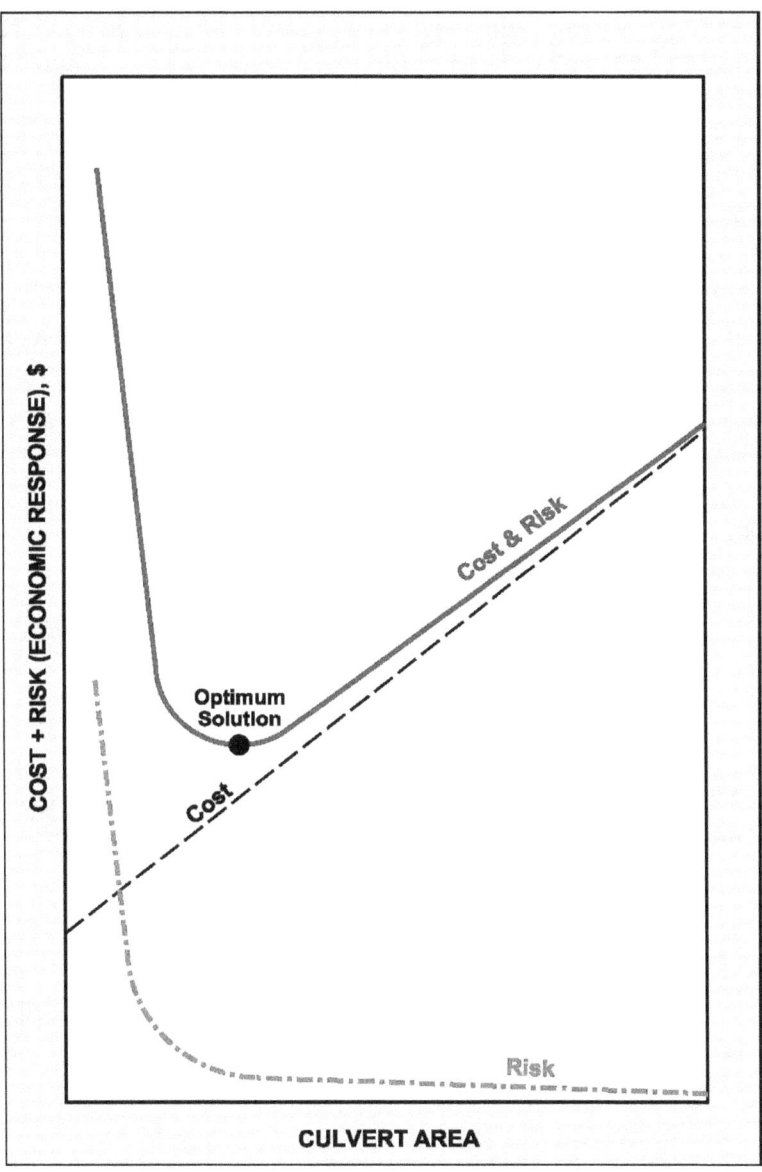

Figure 1.22. Risk analysis benefit versus cost curve.

The construction of a culvert represents a flood plain encroachment with the associated flood risks and initial construction costs. Each design strategy can be evaluated for an annual capital cost and an annual economic risk (cost), the sum of which is called the total expected cost (TEC). Optimization of the economic and engineering analyses will produce the least total expected cost (LTEC) design alternative (see HEC-17, FHWA 1981).

The influence of risk (cost) in the decision-making process represents the major distinction between traditional and LTEC design. In traditional design, the level of risk is an integral part of the establishment of design standards such as a specified design frequency flood or limitations on backwater. The influence of risk in the design of a specific culvert based on these design standards will vary with site conditions. In LTEC design, there is no arbitrary design frequency. The design process determines response of each alternate design to discrete points on the entire flood frequency curve. Flood frequency at which road overtopping occurs is more meaningful than design flood frequency.

A necessary part of the risk analysis process is the establishment of acceptable design alternatives. Engineering, legislative, and policy constraints may limit the range of alternatives. Examples of such constraints include:

- Prescribed minimum design flood criteria as in the case of interstate highways

- Limitations imposed by roadway geometrics such as maximum or minimum grade lines, site distance, and vertical curvature

- Flood plain ordinances or other legislative mandates limiting backwater or encroachment on the floodplain

- Channel stability considerations which would limit culvert velocity or the amount of constriction

CHAPTER 2
DESIGN CONSIDERATIONS

2.1 HYDROLOGY

2.1.1 General

Hydrologic analysis involves the estimation of a design flow rate based on climatological and watershed characteristics. This analysis is the first step in any culvert design. Information in this chapter provides a brief overview of hydrologic analysis. For more detailed information see HDS 2 (FHWA 2002).

A statistical concept often associated with hydrologic analysis is the return period of the discharge. Statistically, the return period is the reciprocal of frequency. For example, the flood which has a 2% chance of being equaled or exceeded (frequency) in any given year has a return period of 50 years; i.e., 1 / 0.02 = 50 years. Note that this does not mean that this flood will occur on a regular basis every 50 years. Two 50-year floods could occur in successive years or they may occur 500 years apart. The return period is only the long-term average number of years between occurrences.

Large and expensive culvert installations may warrant extensive hydrologic analysis. This increased level of effort may be necessary in order to perform risk analysis (see Section 1.6.4) where the size of the structure and/or the need for more accurate hydrologic analysis can be justified. More complex hydrologic methods are also required to define the entire flood event when completing storage routing calculations (see Section 5.8). However, these situations do not occur often and most highway culverts are designed using simpler hydrologic methods predicting peak discharge only.

While culverts have historically been designed to safely pass the peak flow of a major flood event, understanding low flow hydrology is also necessary when AOP must be considered. For example, minimum flow depths in a culvert during typical summer low flow conditions can impact fish movement and migration. Understanding low flow hydrology is also important in low-water crossings that might be designed to safely pass the average annual flow, but will overtop during the peak flow of a major flood event. HDS 2 describes flow duration curves and other tools to better understand and describe low flow hydrologic conditions.

2.1.2 Peak Design Flow

As a flood wave passes a point along a stream, the flow increases to a maximum and then recedes. The plot of this change in discharge over time is called a hydrograph, and the maximum flow rate is referred to as the peak flow. The peak flow has been, and continues to be, a major factor in culvert design.

In traditional culvert design, a structure is sized to pass a peak flow from one side of the roadway embankment to the other with an acceptable headwater elevation. The magnitude of the peak flow is dependent upon the selection of a return period. The assignment of a return period is generally based on the importance of the roadway and flood damage potential.

For gaged sites, statistical analyses can be performed on the recorded stream flow to provide an estimated peak design flow for a given return period. The accuracy of the estimate improves as the length of the record increases. HDS 2 provides complete guidance for this type of analysis. However, since measured streamflow data are typically only available for larger channels and rivers; this type of analysis is seldom employed in culvert design.

Ungaged sites are a more common culvert design problem. Determining the discharge at an ungaged site is based on available regression equations or one of several empirical hydrologic methods. Regional regression equations to predict peak discharge have been developed throughout the country using available stream gage data. These equations generally require basic watershed parameters such as drainage area and average stream slope. Given that the data available to develop regression equations represents larger watershed areas, these methods are typically best for larger drainage areas and culvert design projects. Deterministic methods are also available which model the rainfall-runoff process based on various empirical equations and coefficients. The key input parameter in these methods is rainfall which must be related to the desired return period. The amount of watershed data required is dependent upon the sophistication of the model. One of the simpler methods used for many smaller culvert designs is the Rational Method. Based on the assumptions in the Rational Method it is best applied to drainage areas smaller than 200 ac (80 hectares). Table 2.1 lists some of the commonly employed methods of peak flow generation for gaged and ungaged sites.

Table 2.1. Peak Determination Methods.	
Gaged Sites	Ungaged Sites
Log-Pearson Type III Distribution	U.S. Geological Survey (USGS) Regression Equations
	Natural Resource Conservation Service (NRCS) Peak Discharge Method
	Rational Method

2.1.3 Check Flows

Culvert operation should be evaluated for flows other than the peak design flow because: (1) It is good design practice to check culvert performance through a range of discharges to determine acceptable operating conditions; (2) regulations may require analysis at a larger discharge than that used to design the culvert, such as the 100-year discharge commonly used to define the regulatory floodplain; (3) in performing flood risk analyses, estimates of the damages caused by headwater levels due to floods of various frequencies are required.

Check flows are determined in the same manner as the peak design flow. The hydrologic procedures used should be consistent unless unusual circumstances dictate otherwise. For example, a stream gage record may be long enough to estimate a 10-year peak design flow but too short to accurately generate a 100-year check flow. Under these circumstances the check flow should be evaluated by another method.

2.1.4 Hydrographs

A flood hydrograph is a plot of discharge versus time. Figure 2.1 depicts a typical flood hydrograph showing the rise and fall of stream flow over time as the flood passes. One reason that the hydrograph might be required is when upstream storage will be considered in the culvert design. Actual flood hydrographs can be obtained using stream gage records. These measured storm events can then be used to develop design flood hydrographs. In the absence of stream gage data, empirical or mathematical methods, such as the Snyder and NRCS synthetic hydrograph methods, are used to generate a design flood hydrograph.

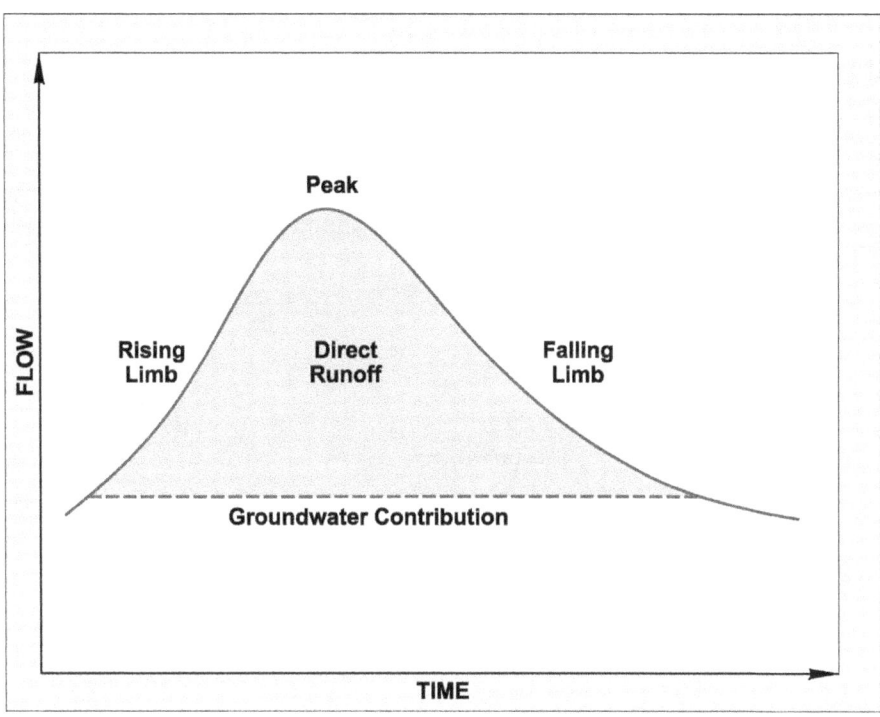

Figure 2.1. Flood hydrograph.

The unit hydrograph technique is a popular procedure for determining the response of a watershed to a specified design rainfall. A unit hydrograph represents the runoff response of a watershed to a uniform 1-inch (1-mm) rainfall of a given duration. A unit hydrograph may be generated from data for a gaged watershed or synthesized from rainfall and watershed parameters for an ungaged watershed. Both methods are briefly described below.

a. Unit Hydrograph Formulation - Gaged Watershed. To develop a unit hydrograph for a gaged watershed, the designer should have streamflow and rainfall records for a number of storm events. The rainfall data must be representative of the rainfall over the watershed during each storm event. In addition, the rainfall events should have relatively constant intensities over the duration of the storm. Procedures for unit hydrograph development are found in HDS 2.

b. Synthetic Unit Hydrograph. A synthetic unit hydrograph may be developed in the absence of stream gage data. The methods used to develop synthetic unit hydrographs are generally empirical and depend upon various watershed parameters, such as watershed size, slope, land use, and soil type. Two synthetic procedures which have been widely used are the Snyder Method and the NRCS Method. The Snyder Method uses empirically defined terms and physiographic characteristics of the drainage basin as input for empirical equations which characterize the timing and shape of the unit hydrograph. The NRCS method utilizes dimensionless hydrograph parameters based on the analysis of a large number of watersheds to develop a unit hydrograph. The only parameters required by the method are the peak discharge and the time to peak. A variation of the NRCS synthetic unit hydrograph is the NRCS synthetic triangular hydrograph. Procedures for synthetic unit hydrograph development are found in HDS 2.

2.1.5 Basics of Storage Routing

Measurement of a flood hydrograph at a stream location is analogous to recording the passage of a high amplitude, low frequency wave. As this wave moves downstream, its shape broadens and flattens provided there is no additional inflow along the reach of the stream. This change in shape is due to the channel storage between the upstream and downstream locations. If the wave encounters a significant amount of storage at a given location in the stream, such as a reservoir, the attenuation of the flood wave is increased. Figure 2.2 depicts the effects graphically.

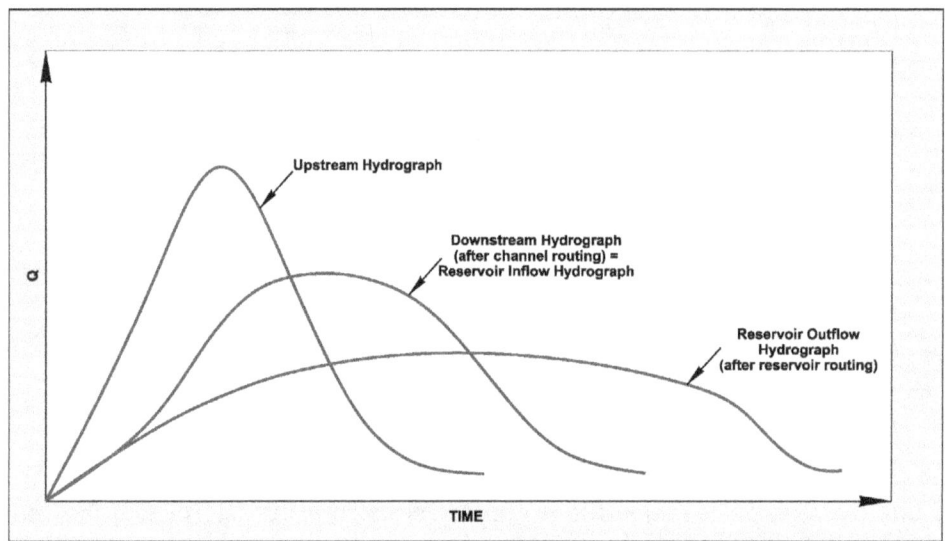

Figure 2.2. Flood hydrograph shape modification.

Storage routing is the numerical translocation of a flood wave (hydrograph). This process is applicable to reservoirs, channels, and watersheds. The effects of the routing are threefold: volume conservation, peak reduction, and time lag. Reservoir routing is dependent only upon storage in modifying a flood wave. Channel routing is dependent upon inflow and outflow, as well as storage in a stream reach. Watershed routing incorporates the runoff attenuating effects of the watershed and is of importance in some hydrograph generation methods.

A culvert with significant upstream ponding will act like a small reservoir and attenuate the hydrograph as it passes through the structure. However, in any storage routing the volume of storage needed to create significant reduction in peak discharge is typically quite large, and consequently, this effect is not often considered in practice when designing a culvert. Furthermore, ignoring this effect will result in a slightly more conservative barrel size and provide additional safety factor in the design.

One situation where a storage routing analysis might be particularly useful is evaluating how downstream conditions may change if a culvert is replaced by a bridge. A common misconception is that taking out a culvert, with the associated ponding and potential attenuation of discharge, will cause increased downstream flooding. A storage routing analysis of conditions with and without the culvert will quantify this impact. As mentioned above, unless there is a significant amount of storage occurring upstream of the culvert, the impacts on discharge will often be smaller than expected. Chapter 5 provides an overview of storage routing methods.

2.1.6 Computer Models

A variety of computer models are available for hydrologic analysis. Some models merely solve empirical hand methods more quickly. Other models are theoretical and solve the runoff cycle in its entirety using continuous simulation over short time increments. Results are obtained using mathematical equations to represent each phase of the runoff cycle such as interception, surface retention, infiltration, and overland flow.

In most simulation models, the drainage area is divided into subareas with similar hydrologic characteristics. A design rainfall is synthesized for each subarea, and abstractions, such as interception and infiltration, are removed. An overland flow routine simulates the movement of the remaining surface water. Adjacent channels receive this overland flow from the subareas. The channels of the watershed are linked together and the channel flow is routed through them to complete the basin's response to the design rainfall. All simulation models require calibration of modeling parameters using measured historical events to increase their accuracy. Most simulation models require a significant amount of input data and user experience to assure reliable results.

The use of geographical information systems (GIS) has greatly simplified the development of input data necessary for most hydrologic computer models. For example, StreamStats is a web-based GIS developed by the U.S. Geological Survey that allows users to easily obtain streamflow statistics, drainage-basin characteristics, and other information for user-selected sites on streams. It also provides an assortment of analytical tools for water-resources planning, management and design. Commercially available software is also available that uses GIS data to efficiently develop input data for many public domain hydrologic computer models (e.g., HEC-1, HEC-HMS, TR-20).

2.2 SITE DATA

2.2.1 General

The hydraulic design of a culvert installation requires the evaluation of a large amount of data including culvert location, waterway data, roadway data, and the allowable headwater. AOP design also requires data on sediment and bed material gradation, as well as discharge information in addition to the typical peak design flow for culvert design. Supplemental data and information may also be necessary for culverts designed to meet multiple-use purposes. For example, if a culvert will also be used for a bike path or pedestrian crossing, additional geometric data and information beyond that for hydraulic analysis may be necessary to design the culvert. Each of these items and its importance is discussed in the following paragraphs.

2.2.2 Culvert Location

A culvert should ideally be located in the existing channel bed to minimize costs associated with structural excavation and channel work. However, this is not always possible. Some streambeds are sinuous and cannot accommodate a straight culvert. In other situations, a stream channel may have to be relocated to avoid the installation of an inordinately long culvert. When relocating a stream channel, it is best to avoid abrupt stream transitions at either end of the culvert. Figure 2.3 displays two examples of culvert location procedures (Durow 1982). In one case, the culvert follows the natural channel alignment. In the second case, the channel has been relocated to reduce the culvert length. The U.S. Geological Survey (USGS) concluded that minor channel relocations for culvert alignments have been successful unless the natural channel was already unstable (USGS 1981). When considering channel relocation alternatives, the environmental impacts and permitting requirements, must be accounted for and may be a major factor in the decision making process.

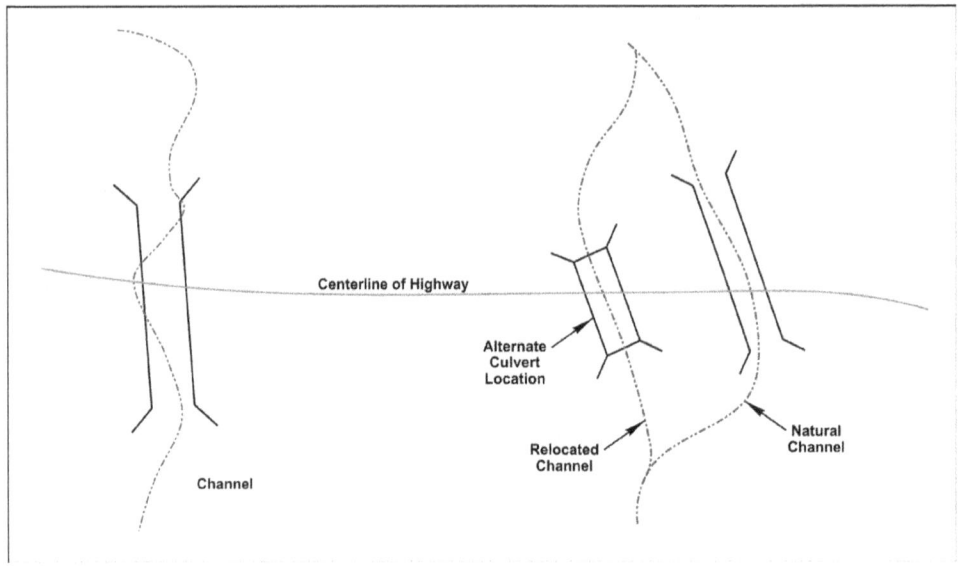

Figure 2.3. Culvert location methods.

2.2.3 Waterway Data

The installation of a culvert to convey surface water through a highway embankment often changes channel and waterway conditions upstream and downstream of the culvert. To predict the consequences of this alteration, accurate preconstruction waterway data must be collected. These data include cross-sectional information, stream slope, the hydraulic resistance of the stream channel and floodplain, channel stability and sediment transport conditions, any condition affecting the downstream water surface elevation, and the storage capacity upstream of the culvert. Photographs of site conditions are often beneficial.

a. <u>Cross Sections</u>. Stream cross sectional data acquired from a field survey at the site are highly desirable to supplement available topographic mapping. Ideally, a minimum of two cross sections should be taken, one upstream and one downstream. These cross sections should be representative of the reach. If significant ponding is likely and a routing analysis will be completed, additional sections may be necessary upstream of the culvert. Likewise, additional downstream sections may be necessary to establish downstream water level (tailwater) conditions, particularly when other downstream structures or channel conditions limit normal depth hydraulic analysis and a more detailed water surface profile analysis must be completed (see Tailwater discussion below).

If only one cross section of the natural channel is available, it will be used as the typical cross section. This assumption should be checked using topographic maps and aerial photos. Additional information on stream slope and upstream storage volume should also be obtained from the topographic maps.

b. <u>Stream Slope</u>. The longitudinal slope of the existing channel in the vicinity of the proposed culvert should be defined in order to properly position the culvert in vertical profile and to define flow characteristics in the natural stream. Often, the proposed culvert is positioned at the same longitudinal slope as the streambed. Surveyed channel cross sections and topographic mapping can be used to define the stream slope. With limited cross section or topographic data, a channel centerline survey data may be necessary to accurately define channel slope through the crossing.

c. Resistance. The hydraulic resistance coefficient of the natural channel must be evaluated in order to calculate channel flow conditions. This resistance coefficient is usually taken to be the Manning's n value. Various methods are available to evaluate resistance coefficients for natural streams, including comparisons with photographs of streams with known resistance values or tabular methods based on stream characteristics (USGS 1967, 1984). Table C.1 in Appendix C, provides Manning's n values for selected natural channels.

d. Channel Stability and Sediment Transport. Evidence of existing channel instability, both laterally and vertically, should be documented and considered in the design of a culvert crossing. This can include observations of eroding banklines, channel shifting, scour holes, and sediment depositional areas. For larger culvert crossings, acquisition and review of historical aerial photography can provide data and insight on long-term channel stability issues. Information on the sediment in the bed and banks of the channel can be very valuable in channel stability analysis, as well as AOP related evaluations. This typically involves field observations, field measurement and/or sediment grab samples that are sent to a laboratory for particle size analysis. HEC-20, Stream Stability at Highway Structures (FHWA 2012a) provides detailed information on evaluating channel stability and the type of data necessary to complete this analysis.

e. Tailwater. Culvert performance is likely to be affected by the downstream water surface elevation or tailwater. Defining the tailwater condition is an open channel flow calculation procedure, and is an important initial step in designing any culvert. HDS 4 (FHWA 2008a) provides an overview of open channel flow, and the normal depth procedure often used to define tailwater in culvert design. The assumption of normal depth begins to break down when conditions exist which might promote high tailwater elevations during flood events. Downstream impoundments, obstructions, channel constrictions, tidal effects, and junctions with other watercourses should be investigated, based on field observations and maps, in order to evaluate their impact on the resultant tailwater elevation. Lacking these conditions, tailwater elevations can be based on water surface elevations in the natural channel as defined by the normal depth. Otherwise, more detailed water surface profile calculations are required, typically using the standard step method applied through one of several widely used computer programs.

f. Upstream Storage. The storage capacity available upstream from a culvert may have an impact upon its design. Upstream storage capacity can be obtained from large scale contour maps of the upstream area, but a 2 ft (0.5 m) contour interval map is desirable. If such maps are not available, a number of cross sections should be obtained upstream of the proposed culvert. These sections must be referenced horizontally as well as vertically. The length of upstream channel to be cross-sectioned will depend on the headwater expected and the stream slope. The cross sections can be used to develop contour maps or the cross sectional areas can be used to compute storage. The topographic information should extend from the channel bed upward to an elevation equal to at least the allowable headwater elevation in the area upstream of the culvert.

2.2.4 Roadway Data

The proposed or existing roadway affects the culvert cost, hydraulic capacity, and alignment. Roadway profile and the roadway cross section information can be obtained from preliminary roadway drawings or from standard details on roadway sections. When the culvert must be sized prior to the development of preliminary plans, a best estimate of the roadway section can be used, but the culvert design must be checked after the roadway plans are completed.

a. Cross Section. The roadway cross section normal to the centerline is typically available from highway plans. However, the cross section needed by the culvert designer is the section at the stream crossing. This section may be skewed with reference to the roadway centerline. For a proposed culvert, the roadway plan, profile, and cross-sectional data should be combined as necessary to obtain this desired section. A schematic roadway plan and section with important elevations is shown in Figure 2.4.

b. Culvert Length. Important dimensions and features of the culvert will become evident when the desired roadway cross section is measured or established. The dimensions are obtained by superimposing the estimated culvert barrel on the roadway cross section and the streambed profile (Figure 2.4). This superposition establishes the inlet and outlet invert elevations. These elevations and the resulting culvert length are approximate since the final culvert barrel size must still be determined.

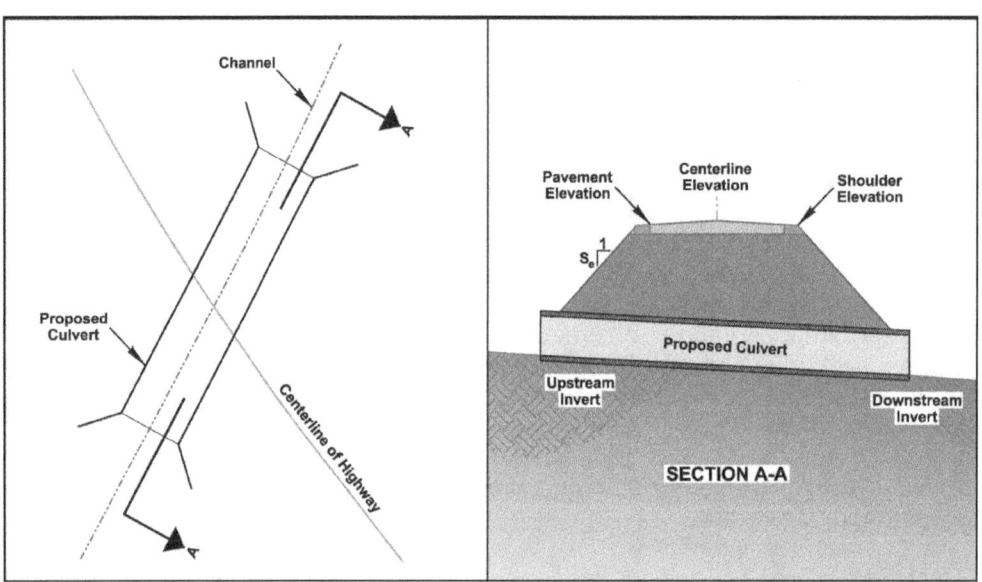

Figure 2.4. Roadway cross section and culvert length.

c. Longitudinal Roadway Profile. For cross drainage culverts the roadway profile represents an obstruction encountered by the flowing stream. The embankment containing the culvert acts much like a dam. The culvert is similar to the normal release structure, and the roadway crest acts as an emergency spillway in the event that the upstream pool (headwater) attains a sufficient elevation. The location of initial overtopping is dependent upon the roadway geometry (Figure 2.5).

The profile contained in highway plans generally represents the roadway centerline profile. Due to superelevation, these elevations may not represent the high point in the highway cross section. The culvert designer should extract the profile which establishes roadway flooding and roadway overflow elevations from the highway plans available. The low point of the profile is of critical importance, since this is the point at which roadway overtopping will first occur. Note that this low point can be a substantial distance away from the culvert location.

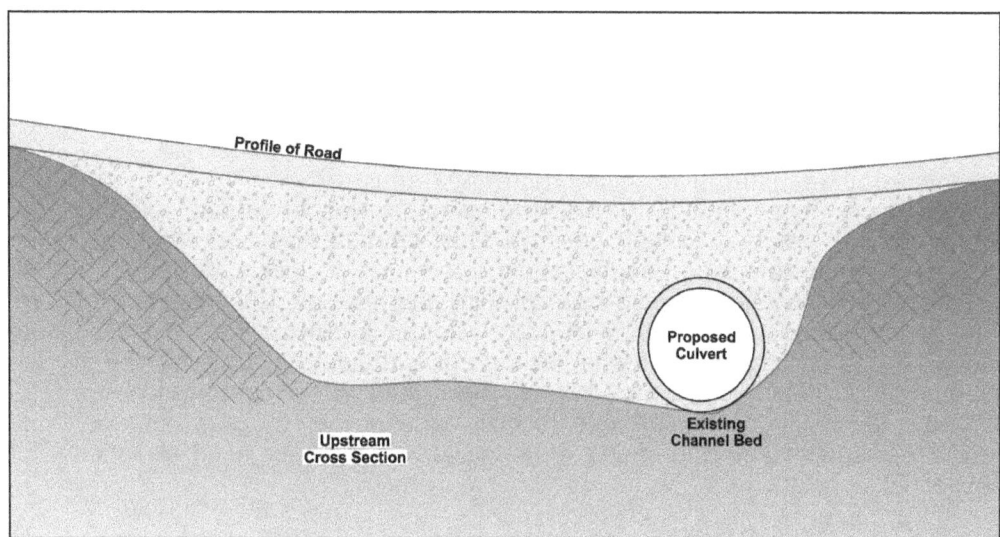

Figure 2.5. Road profile – valley section.

Other types of culverts, beside cross drainage culverts, are used in roadway drainage and may have different overtopping conditions and concerns. For example, ditch relief culverts are structures used to divert some of the runoff from the toe-of-slope ditch on a long downgrade to prevent the ditch from overtopping and/or to control flow velocity and erosion in the ditch. For this type of culvert there is no low-point or overtopping condition in the same context as a cross drainage culvert. Water not diverted simply continues down the ditch to the ditch relief culvert.

2.2.5 Allowable Headwater

The allowable headwater is the maximum possible headwater, or ponding depth, at the upstream side of the culvert. Note that this is different from the design headwater. The design headwater is actual headwater that will occur for the selected culvert as designed. The design headwater should not exceed the allowable headwater. If it does, additional culvert configurations must be selected and evaluated until a configuration is found that produces a design headwater equal to or less than the allowable headwater.

When barrel cost is the only consideration, the most economical culvert is one which would use all of the available headwater to pass the design discharge, since discharge capacity increases with increasing head. However, many other factors enter into establishing the allowable headwater. The definition of allowable headwater elevation generally hinges on one of four factors: economic considerations, regulatory constraints, AOP, or agency constraints.

Note that an increase in available headwater can be obtained at some sites by depressing the culvert inlet. This procedure is advantageous for steep culverts which operate under inlet control. Additional information on this procedure is contained in Chapter 3.

a. Economic Considerations. Although the use of ponding can reduce the barrel size required, detrimental economic consequences can occur from increased headwater elevations. For example, high headwater can lead to embankment piping around the culvert exterior causing damage and possible failure. Increased headwater can also cause higher outlet velocities and severe outlet scour that might require an energy dissipater. Areas with significant debris loading potential that might clog a culvert may use a lower allowable headwater to minimize potential damage from overtopping. Site specific constraints often define some designated elevation that is not to be exceeded within a specified return period to minimize risk and economic loss. This elevation may correspond to some critical point on the roadway such as the roadway shoulder or the roadway overtopping elevation. Another criteria might be the flood damage elevation of an upstream building or another upstream road crossing.

b. Regulatory Constraints. The requirements of the National Flood Insurance Program are a major consideration in culvert design. Most communities are now participating in this program. The limitation on floodplain construction as it affects the base (100-year) flood elevation is of primary importance in this program as administered by the Federal Emergency Management Agency (FEMA). Depending upon the culvert location, existing floodplain encroachments, and whether there is a regulatory floodway, the allowable water surface elevation increase varies from 0 to 1 ft (0.3 m). Regardless of the return period utilized in the culvert design for the particular roadway, the 100-year return period flood must be checked to evaluate the effects of the culvert on the base flood elevation (23 CFR 650 Subpart A).

c. AOP Considerations. A ponded condition creates rapidly changing flow conditions that are often turbulent with increased velocities as water transitions into and out of the culvert. These flow conditions are generally not favorable for fish and aquatic organism movement. In contrast, the most acceptable AOP design is often a crossing with flow characteristics (velocity, depth and width) that are very similar to the natural upstream and downstream channel. This usually implies larger culvert structures that create minimal headwater, except perhaps during very large floods (e.g., 50- or 100-year flood).

d. Agency Constraints. Some state or local highway agencies place limits on the headwater produced by a culvert. For example, the headwater depth may not be allowed to exceed the barrel height or some multiple of the barrel height, expressed as HW/D. The allowable HW/D ratio varies throughout the country, but commonly ranges from 1.0 to 1.5. Although very low HW/D constraints will severely limit the flexibility inherent in culvert design, they must be followed unless a design exemption is granted.

2.2.6 AOP Data

Evaluation or design of an AOP culvert may require additional site data. The primary issues relate to eliminating barriers to AOP and maintaining ecological connectivity between the upstream and downstream reaches. Information on channel velocities, depths, pools and riffles, potential fish resting areas, existing channel drops and channel substrate will be valuable in understanding existing stream crossing conditions and designing a culvert that meets AOP criteria. For information on the required data to complete AOP analyses see Chapter 4.

2.2.7 Cultural Data or Restrictions

In some locations, cultural or historic sites may impact culvert design. This could be the result of a cultural or historic site in the vicinity of a culvert, or it could be that the culvert appurtenances, typically headwalls and wingwalls, are classified as historic. If a cultural or historic site is upstream of a culvert, limitations on headwater may be necessary to avoid inundating the site. Areas downstream may justify additional erosion control measures to protect from potential high velocity and channel shifting that might occur downstream of the culvert. The State Historical Preservation Officer (SHPO) and state historical societies can provide data on sites and artifacts of concern in the area. In some cases, an archeological investigation may be warranted. Special permitting issues may be required that will require significant data.

2.2.8 Summary of Data Needs

Table 2.2 summarizes the various data needed for culvert design.

2.3 SITE ASSESSMENTS

An important first step in the design of a culvert is a comprehensive understanding of the site and conditions where the culvert will be located. The dynamic nature of watershed and river systems must be acknowledged and accounted for when designing a rigid, constructed structure such as a culvert. This includes watershed issues such as the amount and type of debris loading that may occur (Section 2.3.1), the changes in channel conditions (size, shape and location) that will occur from the natural processes of erosion and sedimentation (Section 2.3.2), and the impact of geology and soils on water quality and factors such as potential corrosion and abrasion of the selected culvert material (Section 2.3.4). Other concerns include how a culvert might impact roadside safety when a vehicle leaves the roadway, or create hazardous conditions for children in urban areas (Section 2.3.3).

2.3.1 Debris Assessment

Debris accumulation is a major problem at many culvert locations. Flood flows often carry both floating and submerged debris that can obstruct the culvert entrance and/or accumulate in the barrel. As a minimum, debris accumulation will increase maintenance costs and at the extreme can lead to increased upstream flooding, potential overtopping and roadway embankment failure. Consideration of debris accumulation and the need for debris control structures should be an integral part of any culvert design.

Both non-structural and structural methods have been used to prevent or reduce debris accumulation at culverts. Non-structural measures are primarily related to maintenance activities, both annual and on an emergency basis, to remove any debris that has collected at the entrance or in the barrel of the culvert. Structural measures include features that intercept debris upstream of the culvert, deflect debris near the culvert entrance, or orient debris to facilitate passage through the culvert. Regardless of the solution method employed, it may be desirable to provide a relief opening either in the form of a vertical riser or a relief culvert placed higher in the embankment. Typical culvert debris structures are shown in Figures 2.6 to 2.8. Selection of the appropriate countermeasure depends on the size, quantity, and type of debris, and the costs and potential hazard to life and property involved. HEC-9 (FHWA 2005a) provides more detailed information.

2.3.2 Stream Stability Assessment

A stream stability assessment is critical to the long-term performance of a culvert. Natural channel systems are very dynamic and always changing. Depending on the extent and nature of that change culverts may encounter unexpected problems in the future. A stream stability assessment can provide insight and understanding of potential problems. For example, a channel that is actively degrading may lead to a culvert that is perched with significant scour issues at the outlet or could undermine an open-bottom culvert unless the foundations are set deep enough. A channel reach that is depositional may quickly fill the barrel with sediment and limit hydraulic capacity and create on-going maintenance costs. Lateral instability can change the alignment into the barrel over time and create embankment erosion and instability. Channels with high sediment transport, particularly gravels and cobbles, may influence the type of culvert material selected in order to minimize abrasion damage along the invert.

DATA	SOURCE
HYDROLOGY	**HYDROLOGY**
Peak Flow	Stream gage analysis or calculated using Rational formula, SCS Method, regression equations, etc.
Check Flows	Same as for peak flow
Hydrographs (if storage routing is utilized	From stream gage information or synthetic development methods such as SCS, Method, Snyder Method, or computer models
SITE DATA	**SITE DATA**
Culvert Location	Based on site characteristics including natural stream section, slope, and alignment
Waterway Data	**Waterway Data**
Cross Sections	Field survey or topographic maps
Longitudinal Slope	Field survey or topographic maps
Resistance	Observation, photographs, or calculation Methods
Channel Stability	Observation, survey, maps
Tailwater (during field survey)	Field survey, maps
Upstream storage	Field survey, maps
Roadway Data	**Roadway Data**
Cross Section	Roadway plans
Profile	Roadway plans
Culvert Length	Roadway plans
AOP Data	**AOP Data**
Channel velocities and depths	Measurements or calculation
Pools and riffles	Observation, survey, photographs
Potential fish resting areas	Observation
Existing channel drops	Observation, survey, photographs
Channel substrate	Sampling, observation
Cultural Data	**Cultural Data**
Historic sites	State Historical Preservation Officer (SHPO) or
Artifacts	State historical society
Allowable Headwater	**Roadway Plans**
Critical points on roadway Surrounding buildings or structures	Aerial photographs, surveys, or topographic maps
Regulatory Constraints	Floodplain regulations for stream reach of interest
AOP	State or local fish and wildlife agencies
Agency Constraints	State or local regulations for culvert installations

Table 2.2. Data Requirements for Culvert Design.

Figure 2.6. Steel rail debris deflector (looking downstream).

Figure 2.7. Steel debris rack in urban area.

Figure 2.8. Concrete debris fins with sloping leading edge as extension of culvert walls.

HEC-20 (FHWA 2012a) provides detailed information on evaluating channel stability. Stream channel classification, stream reconnaissance techniques, and rapid assessment methods for channel stability are summarized. Quantitative techniques for channel stability analysis, including degradation analysis, are provided, and channel restoration concepts are introduced. If the stream is not found to be stable during the life of the culvert, countermeasures to improve channel stability might be necessary as described in HEC-23 (FHWA 2009b). At a minimum before any culvert is designed, the designer should visually confirm that the stream reach is both vertically and horizontally stable.

2.3.3 Safety Assessment

The primary safety considerations in the design and construction of a culvert are its structural and hydraulic adequacy. Assuming that these major considerations are appropriately addressed, attention should be directed toward supplementary safety considerations. These considerations include traffic safety and child safety. The safety of errant vehicles should be provided for by the appropriate location and design of culvert inlets and outlets. Safety barriers and grates may substitute or add to this protection. Safety grates also provide a degree of protection against inquisitive youngsters by inhibiting access to a culvert. For detailed information on roadside safety, refer to the Roadside Design Guide (AASHTO 2011) which recognizes the following:

- Small culverts (30 inch (750 mm) in diameter or less) can use an end section or slope paving.

- Culverts greater than 30 inch (750 mm) in diameter should receive one of the following:

  - Be extended to the appropriate "clear zone" distance.

  - Safety treated with a grate if the consequences of clogging and causing a potential flooding hazard are less than the hazard of vehicles impacting an unprotected end. If a grate is used, the net area of the grate (excluding the bars) should be 1.5 to 3.0 times the culvert entrance area.

– Shielded with a traffic barrier if the culvert is very large, cannot be extended, has a channel that cannot be safely traversed by a vehicle, or has a significant flooding hazard with a grate.

a. Inlet and Outlet Location and Design. The exposed end of a culvert or culvert headwall represents an unyielding barrier to vehicles leaving the roadway. Safety provisions must be made to protect occupants of such vehicles against injury or death. One technique employed is to locate the culvert ends outside of the safe recovery area. Traffic safety standards provide distance from pavement limitations based on speed limits. Culverts should also extend through medians unless safe distances can be maintained.

When culvert ends are not outside the safe recovery area, appropriate inlet and outlet design may reduce the danger they represent. Inlets and outlets can be mitered to conform to the fill slope reducing the obstruction to a vehicle. For culvert ends with headwalls, fill should be warped behind them to limit their exposure (markers should be placed on concealed culvert ends to protect roadside maintenance personnel).

b. Safety Barriers and Grates. Additional traffic safety can be achieved by the installation of safety barriers and grates. Safety barriers should be considered in the form of guardrails along the roadside near a culvert when adequate recovery distance cannot be achieved, or for abnormally steep fill slopes (Figure 2.9). Traversable grates placed over culvert openings will reduce vehicle impact forces and the likelihood of overturning (Figure 2.10).

Safety grates promote debris buildup and the subsequent reduction of hydraulic performance. Thorough analysis of this potential should be undertaken prior to the selection of this safety alternative. Bar grates placed vertically directly against the entrance of the culvert are unacceptable. Good design practice provides an open area between bars of 1.5 to 3.0 times the area of the culvert entrance depending on the anticipated volume and size of debris (Figure 2.11). Research on head loss due to a bar grate resulted in the formula (Davis 1952):

Figure 2.9. Guardrail protection.

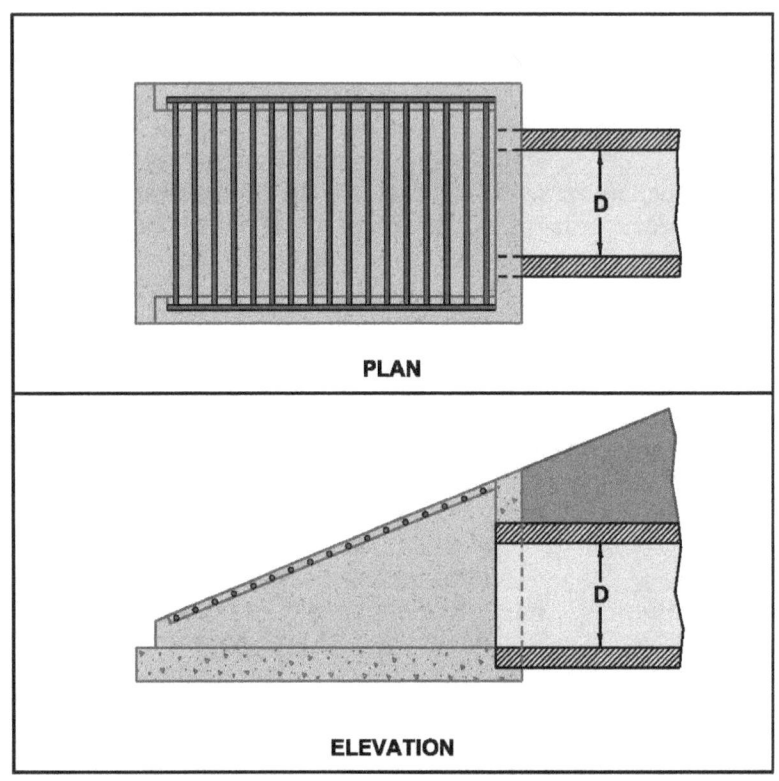

PLAN

ELEVATION

Figure 2.10. Endwall for safety grate.

Figure 2.11. Traversable safety grate with proper open area between bars.

$$H_g = 1.5 \left(\frac{V_g^2 - V_u^2}{2g} \right)$$
(2.1)

H_g = Head loss due to the bar grate, ft (m)
V_g = Velocity between the bars, ft/s (n/s)
V_u = Approach velocity, ft/s (m/s)
g = Acceleration of gravity 32.2 ft/s^2 (9.81 m/s^2)

Another formula for the head loss in bar racks with vertical bars is (Metcalf and Eddy 1972, and Mays et al. 1983)

$$H_g = K_g \left(\frac{W}{X} \right) \left(\frac{V_u^2}{2g} \right) \sin \theta_g$$
(2.2)

K_g is a dimensionless bar shape factor, equal to:

2.42 = Sharp-edged rectangular bars
1.83 = Rectangular bars with semi-circular upstream face
1.79 = Circular bars
1.67 = Rectangular bars with semi-circular upstream and downstream faces
W = Maximum cross-sectional width of the bars facing the flow, ft (m)
X = Minimum clear spacing between bars, ft (m)
θ_g = Angle of the grate with respect to the horizontal, degrees

Both of the above equations are empirical and should be used with caution. In all cases, the head losses are for clean grates and they must be increased to account for debris buildup.

Culverts have always attracted the attention and curiosity of children. In high population areas where hazards could exist, access to culverts should be prevented. Safety grates can serve this function. If clogging by debris is a problem, fencing around the culvert ends is an acceptable alternative to grates.

2.3.4 Culvert Durability

Culvert material longevity is just as important to a successful culvert installation as proper hydraulic and structural design. At most locations, the commonly used culvert materials are very durable. However, there are hostile environmental conditions which will deteriorate all culvert materials. Two problems affecting the longevity of culverts due to adverse environmental conditions are abrasion and corrosion (Figure 2.12). Proper attention must be given to these problems in the design phase. Field inspection of existing culverts on the same or similar streams will prove invaluable in assessing potential problems. Two valuable resources for culvert inspection and assessment that address abrasion and corrosion issues are the Culvert Assessment and Decision-making Procedures Manual (FHWA 2010b), and Chapter 14, Culvert Inspection and Rehabilitation, AASHTO Highway Drainage Guidelines (AASHTO 2007). Research is ongoing at both the Federal and State level to better understand durability issues, particularly related to the increasing range of culvert and lining materials that are available (e.g., Caltrans research as reported by DeCou and Davies 2007).

Figure 2.12a. Abrasion issues in a culvert.

Figure 2.12b. Corrosion issues in a culvert.

The annual cost of a culvert installation is very dependent on its service life. All other conditions being equal, the most durable culvert material should be selected to minimize annual costs. Measures are available to increase the service life of a culvert, such as lining the barrel with a more durable material. When considered, these measures should be included in an economic analysis comparing other culvert materials or other alternatives, including periodic replacement. Periodic replacement of culverts under low fills on secondary roads with light traffic may prove cost effective.

a. Abrasion. Abrasion is defined as the erosion of culvert material due primarily to the natural movement of bedload in the stream. The characteristics of the bedload material and the frequency, velocity, and quantities which can be expected are factors to be considered in the design phase. The resistance of various culvert materials to the expected abrasion is then analyzed. Most materials are subject to abrasion when exposed to high velocity, rock laden flows over a period of time. Performance data on other installations in the vicinity may prove to be the most reliable indicator of abrasion potential and culvert material durability.

When abrasion problems are expected, several options are available to the designer. Debris control structures can often be used to advantage, although they require periodic maintenance. A liner or bottom reinforcement utilizing excess structural material is another option. Concrete or bituminous lining of the invert of corrugated metal pipe is a commonly employed method to minimize abrasion. Concrete culverts may require additional cover over reinforcing bars or high strength concrete mixes. The use of metal or wooden planks attached to the culvert bottom normal to the flow will trap and hold bedload materials, thereby providing invert protection. Oversized culvert barrels which are partially buried accomplish the same purpose.

b. Corrosion. No culvert material exists which is not subject to deterioration when placed in certain corrosive environments. Galvanized steel culverts are generally subject to deterioration when placed in soils or water where the pH falls outside the range of 6 to 10; aluminum deteriorates outside the range of 4 to 9 (NRC 1964). Clay and organic mucks with low electrical resistivity have also proven corrosive to metal culverts. Concrete is adversely affected by alternate wetting and drying with seawater and when exposed to sulfates and certain magnesium salts, and acidic flow with a pH less than 5. Steel deteriorates in saltwater environments. In general, metal culverts are adversely affected by acidic and alkaline conditions in the soil and water, and by high electrical conductivity of the soil. Concrete culverts are sensitive to saltwater environments and to soils containing sulfates and carbonates.

A variety of measures can be taken to prevent the reduction of culvert service life in these hostile environments. These measures are generally categorized as appropriate material selection for the environment or the application of protective coatings. For example, aluminum appears to be resistant to corrosion in salt water installations. Experience has been favorable for fiber-bonded galvanized steel culverts in brackish environments (TRB 1980). Culverts and linings made of vitrified clay, stainless steel, and bituminized fiber perform well in highly acidic conditions. Variations in the concrete mix, such as higher cement content, help to reduce the deterioration of concrete culverts subject to alkaline soils and water. Higher percentages of admixtures (e.g., silica fume) can increase durability as well as the use of water-reducing chemical admixtures. Concrete tends to perform better than metal in clay or organic muck. In areas of severe acidity, such as acid mine drainage, concrete box culverts have been protected by fiberglass linings.

Polymer, bituminous or fiber-bonded coatings on metal culverts may require special consideration. The designer should ascertain that the coating will in fact increase the service life. Delamination is the primary mode of failure and can occur due to sunlight exposure and abrasion. Damage to the coatings during handling and placing is another consideration. Polymer coatings appear to overcome some of these deficiencies. They have excellent corrosion resistance properties and are generally more abrasion-resistant, less subject to damage in handling and placement, and have fewer manufacturing flaws.

(page intentionally left blank)

CHAPTER 3
CULVERT HYDRAULIC DESIGN

3.1 CULVERT FLOW

3.1.1 General

An exact theoretical analysis of culvert flow is extremely complex because the flow is usually nonuniform with regions of both gradually varying and rapidly varying flow. An exact analysis involves backwater and drawdown calculations, energy and momentum balance, and application of the results of hydraulic model studies. Often, hydraulic jumps form inside or downstream of the culvert barrel. The U.S. Geological Survey has defined 18 different culvert flow types based on inlet and outlet submergence, the flow regime in the barrel, and the downstream brink depth (USGS 1968). The flow type can change in a given culvert as the flow rate and tailwater elevations change.

The FHWA has developed a systematic approach to culvert analysis based on the various types of flow and the location of the control section. A control section is a location where there is a unique relationship between the flow rate and the upstream water surface elevation. Many different flow conditions exist over time, but at a given time the flow is either governed by the inlet geometry (inlet control); or by a combination of the culvert inlet geometry, the characteristics of the barrel, and the tailwater (outlet control). Control may oscillate from inlet to outlet; however, in this publication, the concept of "minimum performance" applies. That is, while the culvert may operate more efficiently at times (more flow for a given headwater level), it will never operate at a lower level of performance than calculated.

The culvert design method presented in this publication is based on the use of design charts and nomographs. These charts and nomographs are, in turn, based on data from numerous hydraulic tests and on theoretical calculations. At each step of the process, some error is introduced. For example, there is scatter in the test data and the selection of a best fit design equation involves some error. Also, the correlation between the design equations and the design nomographs is not exact. Reproduction of the design charts introduces additional error. Therefore, it should be assumed that the results of the procedure are accurate to within plus or minus ten percent, in terms of headwater elevation. Additional information on the precision of the design charts is provided in Appendix A.

3.1.2 Types of Control

A general description of the characteristics of inlet and outlet control flow is given below. A culvert flowing in inlet control has shallow, high velocity flow categorized as "supercritical" in the culvert barrel. For supercritical flow, the control section is at the upstream end of the barrel (the inlet). Conversely, a culvert flowing in outlet control will have relatively deep, lower velocity flow termed "subcritical" flow or could be flowing full. For subcritical flow the control is at the downstream end of the culvert (the outlet). The water depth at the culvert outlet is either critical depth or the downstream channel depth, whichever is higher.

Table 1.1 in Chapter 1 provides the factors which must be considered in culvert design for inlet and outlet control. In inlet control, only the inlet area, the inlet configuration, and the shape influence the culvert performance for a given headwater elevation. The headwater elevation is calculated with respect to the inlet invert, and the tailwater elevation has no influence on performance. In outlet control, all of the factors listed in Table 1.1 affect culvert performance. Headwater elevation is calculated with respect to the outlet invert, and the difference between headwater and tailwater elevation represents the energy which conveys the flow through the culvert.

3.1.3 Inlet Control

Figure 3.1 illustrates the types of inlet control flow. The USGS flow type depends on the submergence of the inlet and outlet ends of the culvert. In all of these examples, the control section is at the inlet end of the culvert. Depending on the tailwater, a hydraulic jump may occur downstream of the inlet.

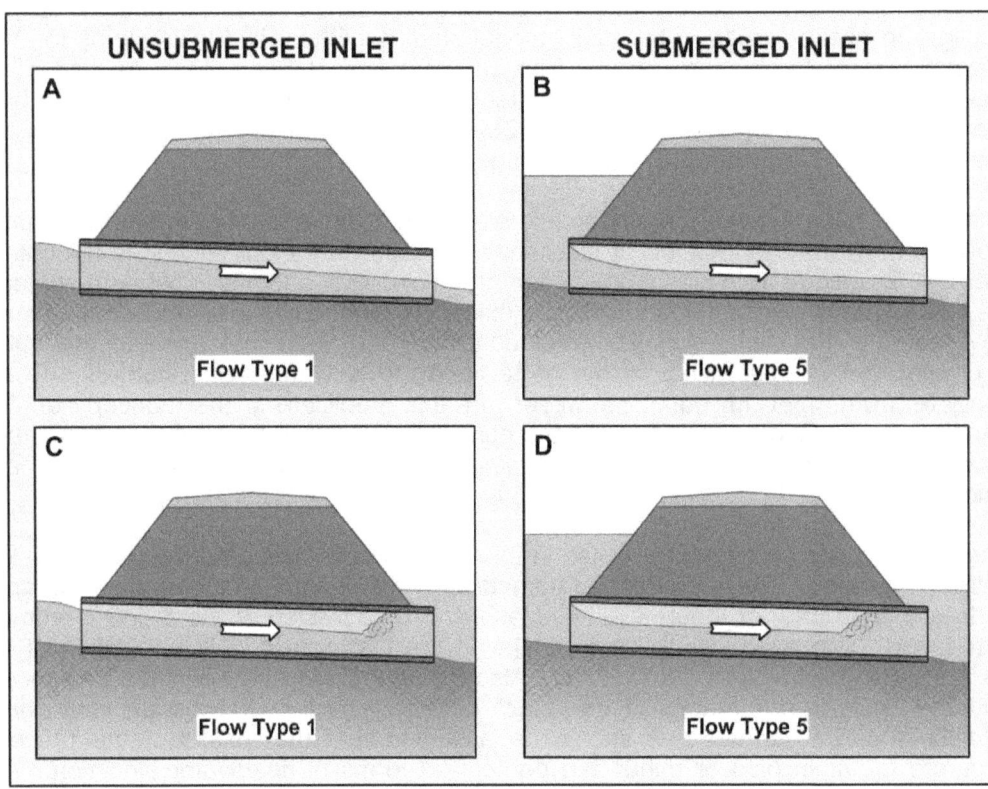

Figure 3.1. Types of inlet control.

Figures 3.1A and C illustrate USGS Flow Type 1 where the inlet is not submerged. The flow passes through critical depth just downstream of the culvert entrance and the flow in the barrel is supercritical. In Figure 3.1A, the barrel flows partly full over its length, and the flow approaches normal depth at the outlet end. In Figure 3.1C, submergence of the outlet end of the culvert does not assure outlet control. In this case, the flow just downstream of the inlet is supercritical and a hydraulic jump forms in the culvert barrel.

Figures 3.1B and D illustrate USGS Flow Type 5 where the inlet is submerged. In Figure 3.1B, the inlet end is submerged and the outlet end flows freely. The flow is supercritical and the barrel flows partly full over its length. Critical depth is located just downstream of the culvert entrance, and the flow is approaching normal depth at the downstream end of the culvert. Figure 3.1D is an unusual condition where submergence of both the inlet and the outlet ends of the culvert does not assure full flow. In this case, a hydraulic jump will form in the barrel. Sub-atmospheric pressures could develop which might create an unstable condition during which the barrel would alternate between full flow and partly full flow.

Factors Influencing Inlet Control. Since the control is at the upstream end, only the headwater and the inlet factors affect the culvert performance (Table 1.1):

- Headwater depth is measured from the invert of the inlet control section to the surface of the upstream pool.
- Inlet area is the cross-sectional area of the face of the culvert. Generally, the inlet face area is the same as the barrel area, but for tapered inlets (Section 3.4) the face area is enlarged, and the control section is at the throat.
- Inlet configuration describes the entrance type. Some typical inlet configurations are thin edge projecting, mitered, square edges in a headwall, and beveled edge as shown in Figures 3.2 and 3.3. Another type of inlet is a tapered inlet that has an enlarged face section, as described in Section 3.4.
- Inlet shape is usually the same as the shape of the culvert barrel; however, it may be enlarged as in the case of a tapered inlet. Typical shapes are rectangular, circular, and elliptical. Whenever the inlet face is a different size or shape than the culvert barrel, the possibility of an additional control section within the barrel exists.
- Barrel slope influences inlet control performance, but the effect is small. Inlet control nomographs assume a slope of 2% for the slope correction term (0.5S for most inlet types). This results in lowering the headwater required by .01D. In the computer program HY-8, the actual slope is used as a variable in the calculation.

The inlet configuration is a major factor in inlet control performance. Typical inlet configurations are shown in Figure 3.2:

- Figure 3.2A is a thin edge projecting inlet, typical of metal pipe.
- Figure 3.2B is a mitered inlet which conforms to the fill slope.
- Figure 3.2C is a square edge in a headwall. A projecting thick-walled inlet gives about the same performance and is typical of concrete pipe without a groove end.
- Figure 3.2D is a groove edge projecting which is typical of a concrete pipe joint (also called a socket end).

A method of increasing inlet performance is the use of beveled edges at the entrance of the culvert. Beveled edges reduce the contraction of the flow by effectively enlarging the face of the culvert. Although any beveling will help the hydraulics, design charts are available for two bevel angles, 45 degrees and 33.7 degrees, as shown in Figure 3.3.

The larger, 33.7-degree bevels (0.083 ft/ft (m/m) or 1.0 in/ft of barrel height) require some structural modification, but they provide slightly better inlet performance than the 45-degree bevels. The smaller, 45-degree bevels (0.042 ft/ft (m/m) or 0.5 in/ft of barrel height) require very minor structural modification of the culvert headwall and increase both inlet and outlet control performances. Therefore, the use of 45-degree bevels is recommended on all culverts that have a headwall, whether in inlet or outlet control. Since the groove end or bell end of a concrete pipe provides about the same performance as a beveled edge, a bevel is not needed if the groove is preserved at the inlet. Other entrance types, such as a stone-faced headwall or an entrance with some type of radius curvature, can also provide performance similar to a beveled edge.

Hydraulics of Inlet Control. Inlet control performance is defined by the three regions of flow shown in Figure 3.4: unsubmerged, transition and submerged. For low headwater conditions, as shown in Figure 3.1A and Figure 3.1C, the entrance of the culvert operates as a weir. A weir is an unsubmerged flow control section where the upstream water surface elevation can be predicted for a given flow rate. The relationship between flow and water surface elevation must be determined by model tests of the weir geometry or by measuring prototype discharges. These tests or measurements are then used to develop equations for unsubmerged inlet control flow. Appendix A contains the equations which were developed from the National Bureau of Standards (NBS) and other model test data.

(A) Thin Edge Projecting – The culvert barrel projects out of the embankment.

(B) Mitered entrance – The culvert barrel is cut so it is flush with the embankment slope.

(C) Square edge in headwall – The end of the culvert barrel is flush with the headwall.

(D) Groove edge projecting – A concrete pipe culvert section extends beyond the fill or headwall.

Figure 3.2. Typical inlet configurations.

For headwaters submerging the culvert entrance, as shown in Figure 3.1B and Figure 3.1D, the entrance of the culvert operates as an orifice. An orifice is an opening, submerged on the upstream side and flowing freely on the downstream side, which functions as a control section. The relationship between flow and headwater can be defined based on results from model tests. Appendix A contains the submerged flow equations which were developed from the NBS and other model test data.

The flow transition zone between the low headwater (weir control) and the high headwater (orifice control) flow conditions is poorly defined. This zone is approximated by plotting the unsubmerged and submerged flow equations and connecting them with a line tangent to both curves, as shown in Figure 3.4.

The inlet control flow versus headwater curves which are established using the above procedure are the basis for constructing the inlet control design nomographs and for developing equations used in software. The original equations for computer software were generally 5th order polynomial curve fitted equations that were developed to be as accurate as the nomograph solution (plus or minus 10%) within the headwater range of 0.5D to 3.0D. These equations are still being used in HY-8, but have been supplemented with a weir equation from 0.0D to 0.5D and an orifice equation above 3.0D.

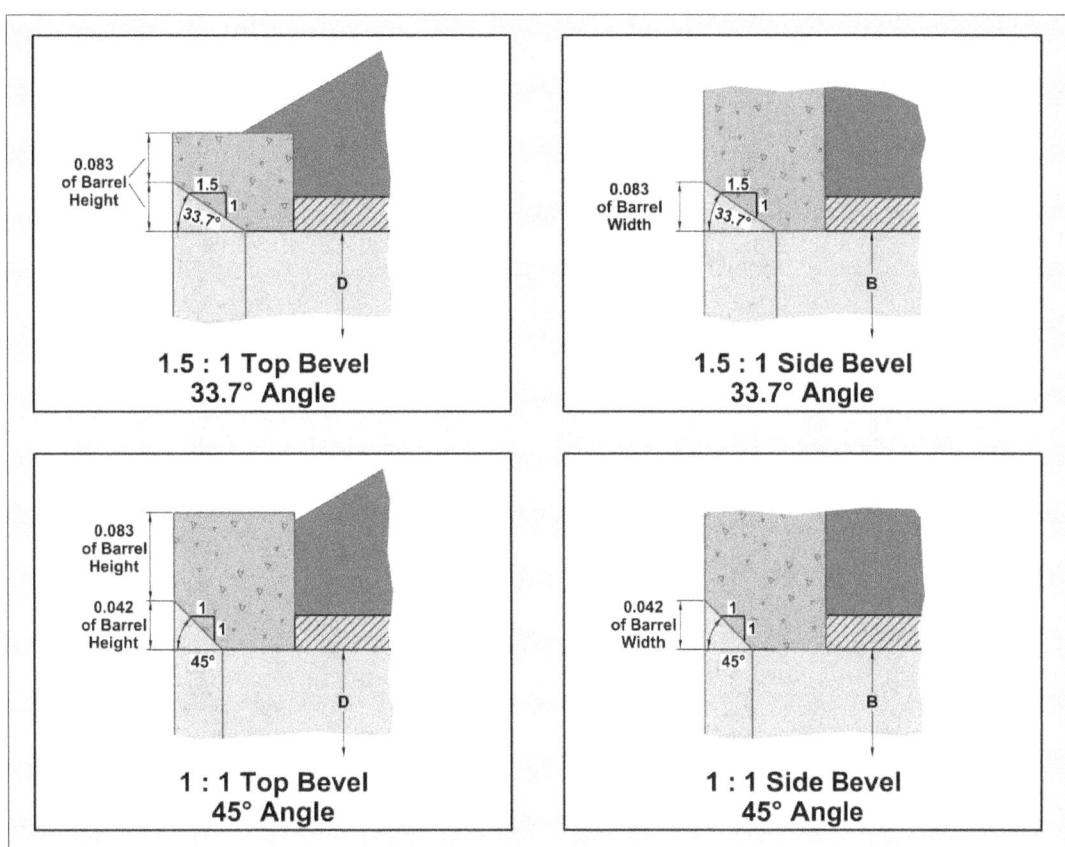

Figure 3.3. Beveled edges.

Inlet Depression. Inlet depression is created by constructing the entrance inlet below the streambed. The amount of inlet depression is defined as the depth from the natural streambed at the face to the inlet invert. The inlet control equations or nomographs provide the depth of headwater above the inlet invert required to convey a given discharge through the inlet. This relationship remains constant regardless of the elevation of the inlet invert. If the entrance end of the culvert is constructed below the streambed, more head can be exerted on the inlet for the same headwater elevation.

Two methods of constructing inlet depression at the entrance end of a culvert are shown in Figures 3.5 and Figure 3.6. Figure 3.5 depicts the use of an approach apron with the fill retained by wingwalls. Paving the apron is desirable. Figure 3.6 shows a sump constructed upstream of the culvert face. Usually the sump is paved, but for small depressions, an unpaved excavation that is lined with riprap to prevent headcutting may be adequate.

3.1.4 Outlet Control

Figure 3.7 illustrates the types of outlet control flow. The USGS flow type depends on the submergence of the inlet and outlet ends of the culvert. In all cases, the control section is at the outlet end of the culvert or further downstream. For the partly full flow situations, the flow in the barrel is subcritical.

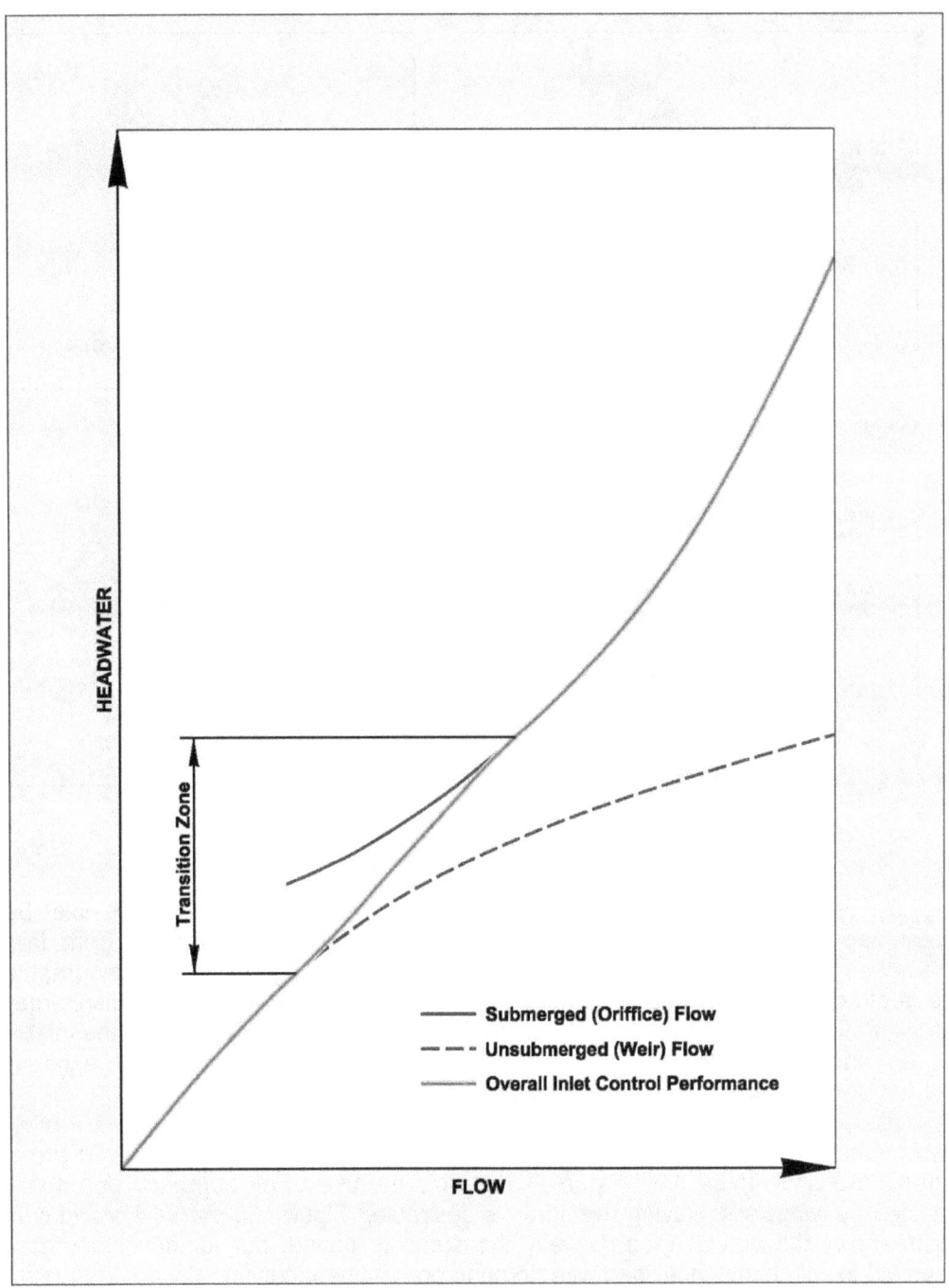

Figure 3.4. Inlet control curves.

<u>Figures 3.7A and C</u> illustrate USGS flow type 2 and 3 where both the inlet and outlet are unsubmerged. The headwater is shallow so that the inlet crown is exposed as the flow contracts into the culvert. The barrel flows partly full over its entire length and is subcritical. For flow type 2 (Figure 3.7A), the flow passes through critical depth at the outlet. For flow type 3 (Figure 3.7C), the tailwater is higher than critical depth and if higher the culvert crown may cause the exit to full flow.

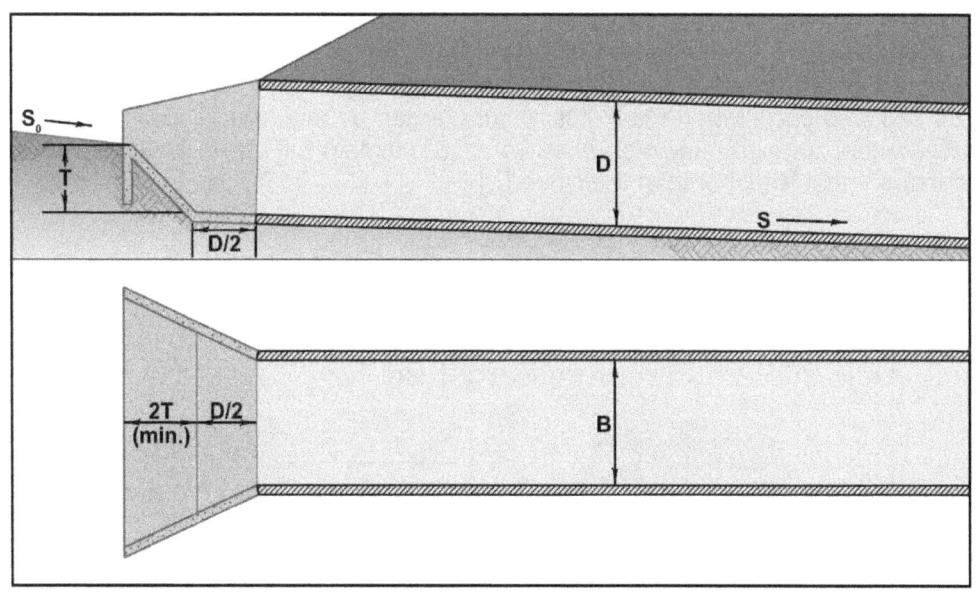

Figure 3.5. Culvert with inlet depression and with apron and wingwalls.

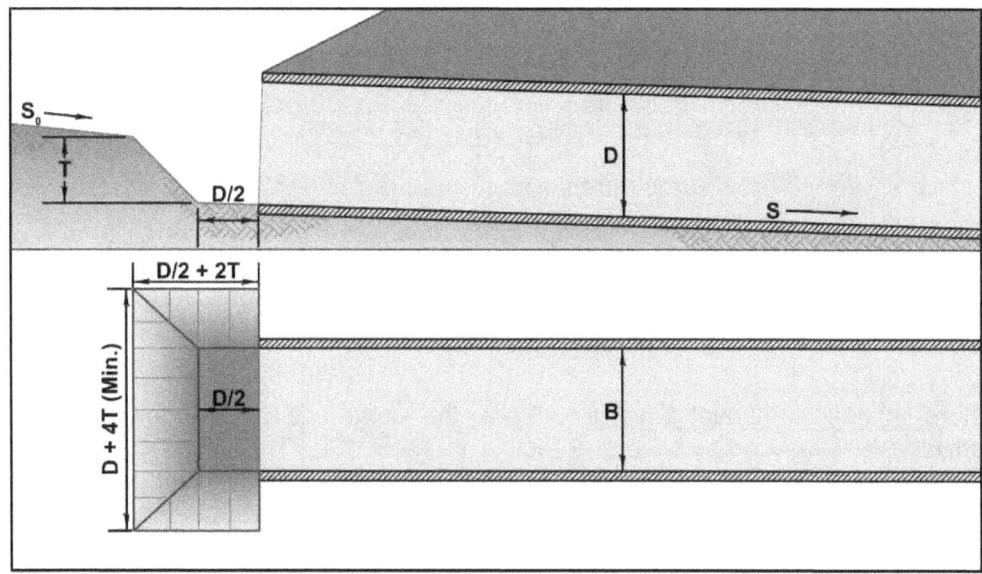

Figure 3.6. Culvert with inlet depression and sump.

Figure 3.7B illustrates USGS Flow Type 6 and 7. The culvert entrance is submerged by the headwater and the outlet end is unsubmerged. For flow type 6, the barrel is assumed to flow full for most of its length (full flow). For flow type 7, the barrel flows partly full over at least part of its length (subcritical flow). For both flow type 6 and 7, the flow passes through either critical depth just upstream of the outlet or the tailwater depth, if higher.

Figure 3.7D illustrates USGS flow type 4 which is the classical full barrel flow where both the inlet and outlet are submerged. The barrel is in pressure flow throughout its length. This condition is often assumed in calculations and was used to construct the nomographs. Flow type 4 can also occur when the exit is unsubmerged by tailwater. This is a rare condition. It requires either an extremely high headwater to maintain full barrel flow with no tailwater or critical depth that is higher than the culvert.

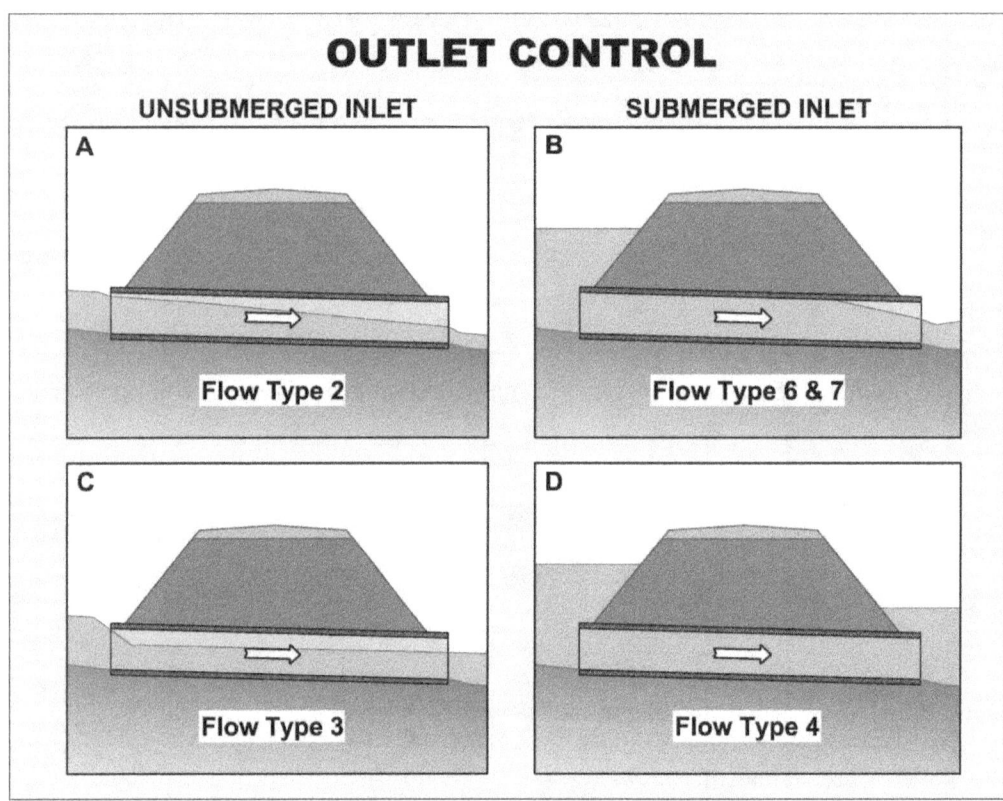

Figure 3.7. Types of outlet control.

Factors Influencing Outlet Control. Since the control is at the downstream end, the headwater is influenced by all of the factors in Table 1.1. The inlet factors influencing the performance of a culvert in inlet control also influence culverts in outlet control (see Section 3.1.3). In addition, the barrel characteristics (roughness, area, shape, length, and slope) and the tailwater elevation affect culvert performance in outlet control:

- Barrel roughness is a function of the material used to fabricate the barrel. Typical materials include concrete, corrugated metal and plastic. The roughness is represented by a hydraulic resistance coefficient such as the Manning's n value. Additional discussion on the sources and derivations of the Manning's n values are contained in Appendix B. Typical Manning's n values used for designing culverts are n = 0.012 for smooth walled culverts and n = 0.024 for rough culverts (corrugated).

- Barrel area is a function of the culvert dimensions. A larger barrel area will convey more flow.

- Barrel shape is function of culvert type and material. Based on the location of the center of gravity for a given area, a box is the most efficient shape, then the arch, followed by the circle.

- Barrel length is the total culvert length from the entrance to the exit of the culvert. Because the design height of the barrel and the slope influence the actual length, an approximation of barrel length is usually necessary to begin the design process.

- Barrel slope is the actual slope of the culvert barrel. The barrel slope is often the same as the natural stream slope. However, when the culvert inlet is raised or lowered, the barrel slope is different from the stream slope. The slope is not a factor in calculating the barrel losses for USGS Flow Types 4, 6, and 7; but is a factor for in calculating USGS Flow Types 2 and 3 when a water surface profile is calculated.

- Tailwater elevation is based on the downstream water surface elevation. Backwater calculations from a downstream control, a normal depth approximation, or field observations are used to define the tailwater elevation.

Hydraulics of Outlet Control (Full Barrel Flow). Full flow in the culvert barrel, as depicted in Figure 3.7D, is the best flow type for describing the hand computation of outlet control hydraulics. Outlet control flow conditions can be calculated based on energy balance. The total energy (H_L) required to pass the flow through the culvert barrel is made up of the entrance loss (H_e), the friction losses through the barrel (H_f), and the exit loss (H_o). Other losses, including bend losses (H_b), losses at junctions (H_j), and loses at grates (H_g) should be included as appropriate. These other losses are discussed in Chapter 5.

$$H_L = H_e + H_f + H_o + H_b + H_j + H_g \tag{3.1}$$

The barrel velocity is calculated as follows:

$$V = \frac{Q}{A} \tag{3.2}$$

V = Average velocity in the culvert barrel, ft/s (m/s)
Q = Flow rate, ft^3/s (m^3/s)
A = Full cross sectional area of the flow, ft^2 (m^2)

The velocity head is:

$$H_V = \frac{V^2}{2g} \tag{3.3}$$

g is the acceleration due to gravity, 32.2 ft/s^2 (9.8 m/s^2)

The entrance loss is a function of the velocity head in the barrel, and can be expressed as a coefficient times the velocity head.

$$H_e = k_e \left(\frac{V^2}{2g} \right) \tag{3.4a}$$

Values of k_e based on various inlet configurations are given in Table C.2, Appendix C.

3.9

The friction loss in the barrel is also a function of the velocity head. Based on the Manning equation, the friction loss is:

$$H_f = \left[\frac{K_U \, n^2 \, L}{R^{1.33}} \right] \frac{V^2}{2g}$$

(3.4b)

$$
\begin{array}{lll}
K_U & = & \text{29 in English Units (19.63 in SI)} \\
n & = & \text{Manning roughness coefficient for a culvert with uniform material on the} \\
& & \text{full perimeter (for composite roughness } (n_c) \text{ see equation 3.8)} \\
L & = & \text{Length of the culvert barrel, ft (m)} \\
R & = & \text{Hydraulic radius of the full culvert barrel} = A/p, \text{ ft (m)} \\
A & = & \text{Cross-sectional area of the barrel, ft}^2 \text{ (m}^2) \\
p & = & \text{Perimeter of the barrel, ft (m)} \\
V & = & \text{Velocity in the barrel, ft/s (m/s)}
\end{array}
$$

The exit loss is a function of the change in velocity at the outlet of the culvert barrel. For a sudden expansion such as an endwall, the exit loss is:

$$H_o = 1.0 \left[\frac{V^2}{2g} - \frac{V_d^2}{2g} \right]$$

(3.4c)

V_d is the channel velocity downstream of the culvert, ft/s (m/s).

Equation (3.4c) may overestimate exit losses, and a multiplier of less than 1.0 can be used (FHWA 2006a) for a transition loss. The downstream velocity is usually neglected, in which case the exit loss is equal to the full flow velocity head in the barrel, as shown in Equation (3.4d).

$$H_o = H_v = \frac{V^2}{2g}$$

(3.4d)

Equation 3.4d is the standard option in HY-8. If the designer chooses the Utah State University (USU) Method (which is the alternate in HY-8), the following equation will be used:

$$H_o = \frac{(V - V_d)^2}{2g}$$

(3.4e)

This equation was formulated for applications like irrigation channels where a small amount of energy is lost in the transition back to the channel.

Bend losses, junction losses, grate losses and other losses are discussed in Chapter 5. These other losses are added to the total losses using Equation (3.1).

Inserting the above relationships for entrance loss, friction loss, and exit loss (Equation 3.4d) into Equation 3.1, the following equation for barrel losses (H) is obtained:

$$H = \left[1 + k_e + \frac{K_U \, n^2 \, L}{R^{1.33}} \right] \frac{V^2}{2g}$$

(3.5)

Figure 3.8 depicts the energy grade line and the hydraulic grade line for full flow in a culvert barrel. The energy grade line represents the total energy at any point along the culvert barrel. HW_o is the depth from the inlet invert to the energy grade line. The hydraulic grade line is the depth to which water would rise in vertical tubes connected to the sides of the culvert barrel. In full flow, the energy grade line and the hydraulic grade line are parallel straight lines separated by the velocity head except in the vicinity of the inlet where the flow passes through a contraction.

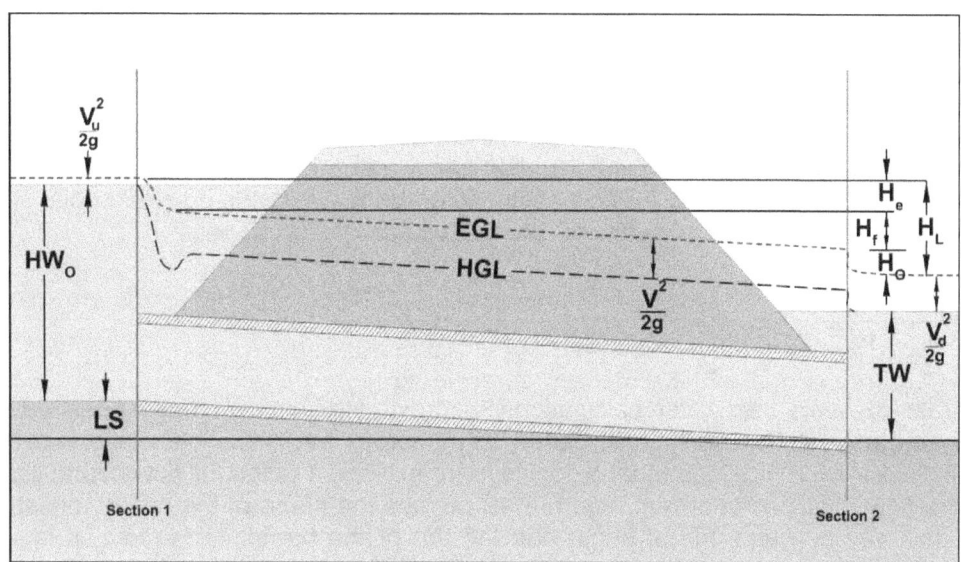

Figure 3.8. Full flow energy and hydraulic grade lines.

The headwater and tailwater conditions as well as the entrance, friction, and exit losses are also shown in Figure 3.8. Equating the total energy at sections 1 and 2, upstream and downstream of the culvert barrel in Figure 3.8, the following relationship results:

$$HW_o + LS + \frac{V_u^2}{2g} = TW + \frac{V_d^2}{2g} + H_L \tag{3.6a}$$

HW_o	=	Headwater depth above the entrance invert in outlet control, ft (m)
V_u	=	Approach velocity, ft/s (m/s)
TW	=	Tailwater depth above the outlet invert, ft (m)
V_d	=	Downstream velocity, ft/s (m/s)
H_L	=	Sum of all losses including entrance (H_e), friction (H_f), exit (H_o) and other losses, (H_b), (H_j), ft (m)
LS	=	Drop through the culvert, ft (m)

In most instances, the approach velocity is low, and the approach velocity head is neglected. However, it can be considered to be a part of the available headwater and used to convey the flow through the culvert.

Likewise, the velocity downstream of the culvert (V_d) is usually neglected. When both approach and downstream velocities are neglected, Equation 3.6a becomes:

$$HW_o = TW + H_L - LS \tag{3.6b}$$

In this case, H_L is the difference in elevation between the water surface elevation at the outlet (tailwater elevation) and the water surface elevation at the inlet (headwater elevation). If it is desired to include the approach and/or downstream velocities, use Equation 3.4c for exit losses and Equation 3.6a instead of Equation 3.6b to calculate the headwater.

<u>Hydraulics of Outlet Control (Unsubmerged Outlet)</u>. Equations 3.1 through 3.6 were developed for full barrel flow (USGS flow type 4), shown in Figure 3.7D. The equations also apply to USGS flow types 6 and 7 shown in Figures 3.7B, which is effectively full flow conditions. Backwater calculations may be required for the partly full flow conditions shown in Figures 3.7A and C. These calculations begin at the water surface at the downstream end of the culvert and proceed upstream to the entrance of the culvert (see Section 3.5). The downstream water surface is based on critical depth at the culvert outlet or on the tailwater depth, whichever is higher. If the calculated backwater profile intersects the top of the barrel, as in Figure 3.7B, a straight, full flow hydraulic grade line extends from that point upstream to the culvert entrance. From Equation 3.4b, the full flow friction slope is:

$$S_f = \frac{H_f}{L} = \frac{K_u n^2}{R^{1.33}} \frac{V^2}{2g} \tag{3.7}$$

In order to avoid backwater calculations, approximate methods have been developed to analyze partly full flow conditions. Based on numerous backwater calculations performed by the FHWA staff, it was found that a downstream extension of the full flow hydraulic grade line for the flow condition shown in Figure 3.9B pierces the plane of the culvert outlet at a point one-half way between critical depth and the top of the barrel. Therefore, it is possible to begin the hydraulic grade line at a depth of $(d_c+D)/2$ above the outlet invert and extend the straight, full flow hydraulic grade line upstream to the inlet of the culvert at a slope of S_f (Figure 3.9D). If the tailwater exceeds $(d_c+D)/2$, the tailwater is used to set the downstream end of the extended full flow hydraulic grade line. The inlet losses and the velocity head are added to the elevation of the hydraulic grade line at the inlet to obtain the headwater elevation.

This approximate method works best when the barrel flows full over at least part of its length (Figure 3.9B). When the barrel is partly full over its entire length (Figure 3.9C), the method becomes increasingly inaccurate as the headwater falls further below the top of the barrel at the inlet. Adequate results are obtained down to a headwater of 0.75D. For lower headwaters, backwater calculations are required to obtain accurate headwater elevations.

The outlet control nomographs in Appendix C provide solutions for Equation 3.5 for entrance, friction, and exit losses in full barrel flow. Using the approximate backwater method, the losses (H) obtained from the nomographs can be applied for the partly full flow conditions shown in Figures 3.7 and 3.9. The losses are added to the elevation of the extended full flow hydraulic grade line at the barrel outlet in order to obtain the headwater elevation. The extended hydraulic grade line is set at the higher of $(d_c+ D)/2$ or the tailwater elevation at the culvert outlet. Again, the approximation works best when the barrel flows full over at least part of its length.

3.12

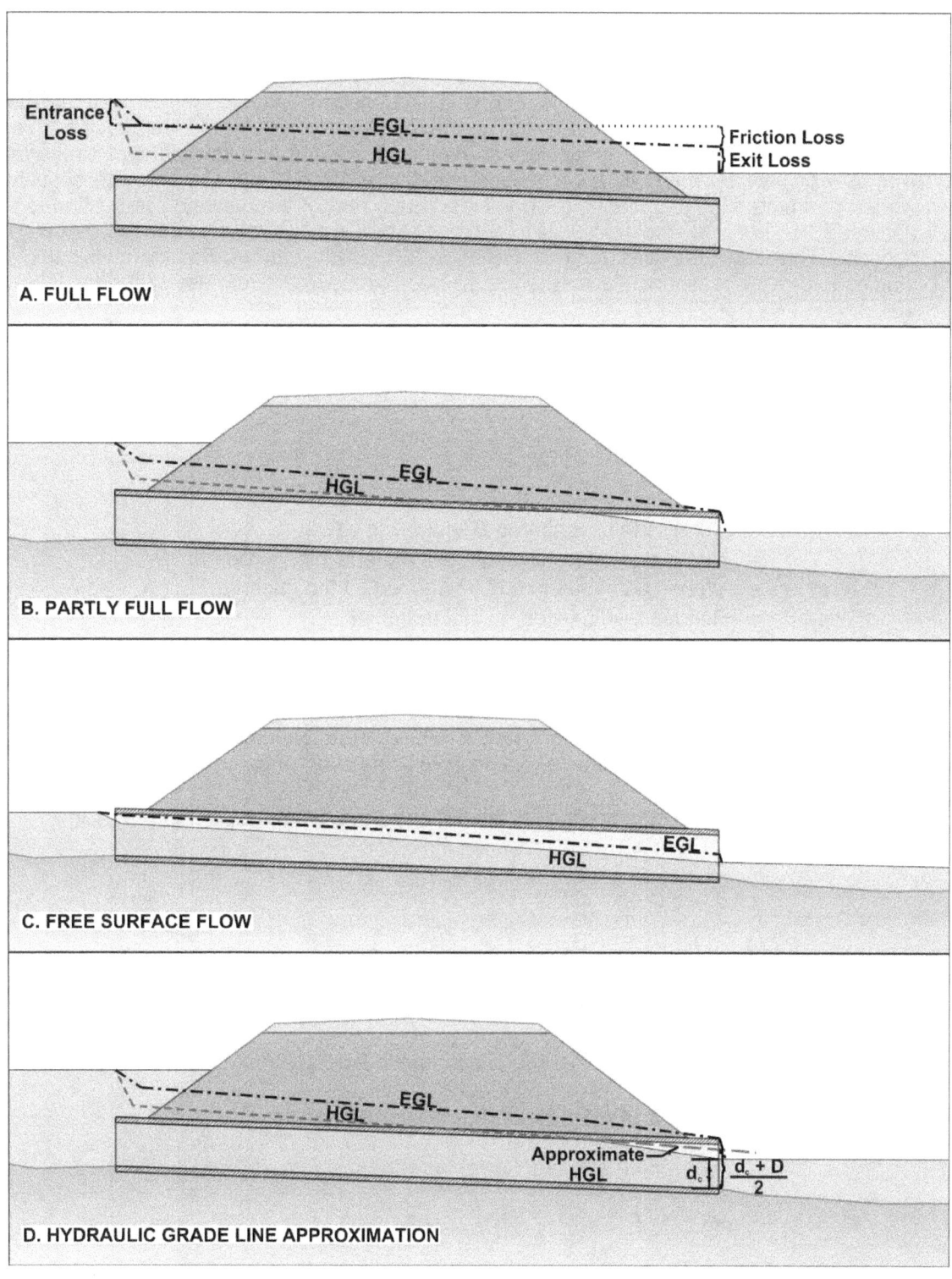

Figure 3.9. Outlet control energy and hydraulic grade lines.

Composite Roughness. Culverts are often fabricated using different materials for portions of the perimeter. Examples include AOP culverts with unlined bottoms or corrugated metal culverts with an invert lining. In order to derive a composite Manning's n value for the above situations, a common practice is to derive a weighted n value based on the estimated Manning's n value for each material and the perimeter of the pipe composed of each material. The method assumes a constant Manning's n value for each material (no variation with size or flow velocity). The conveyance section is broken into G parts with associated wetted perimeters (p) and Manning's n values. Each part of the conveyance section is then assumed to have a mean velocity equal to the mean velocity of the entire flow section. These assumptions lead to Equation 3.8 which was verified by physical model studies to provide reasonably accurate results over a range of roughness scenarios (NCHRP 2011).

$$n_c = \left[\frac{\sum_{i=1}^{G} (p_i n_i^{1.5})}{p} \right]^{0.67}$$

(3.8)

n_c = Composite or weighted Manning's n value
G = Number of different roughness materials in the perimeter
p_1 = Wetted perimeter in feet influenced by the material 1
p_2 = Perimeter influenced by material 2, etc.
n_1 = Manning's n value for material 1, n_2 is for material 2, etc.
p = Total wetted perimeter, ft

Example Problem: Compute the Manning's n value for a 6 ft. diameter corrugated metal pipe with 5 by 1 in annular corrugations, and a smooth lining over 40 percent of the perimeter.

1. Determine the Manning's n for the 6 ft corrugated metal pipe with 5 by 1 in corrugations.

    n = 0.026 (Appendix B)

2. Determine the Manning's n for smooth lining.

    n = 0.013 (assume concrete lining)

3. Determine the relative perimeters composed of each material.

    p = πD = (3.14)(6) = 18.84 ft (total wetted perimeter)

    p_1 (corrugated) = (0.60)(18.84) = 11.30 ft

    p_2 (smooth) = (0.40)(18.84) = 7.54 ft

4. Use Equation 3.8 to calculate the Manning's n_c value

    $$n_c = \left[\frac{(11.30)(0.026)^{1.5} + (7.54)(0.013)^{1.5}}{18.84} \right]^{0.67} = 0.021$$

3.14

3.1.5 Roadway Overtopping

Overtopping will begin when the headwater rises to the elevation of the roadway (Figure 3.10). The overtopping will usually occur at the low point of a sag vertical curve on the roadway. The flow will be similar to flow over a broad crested weir.

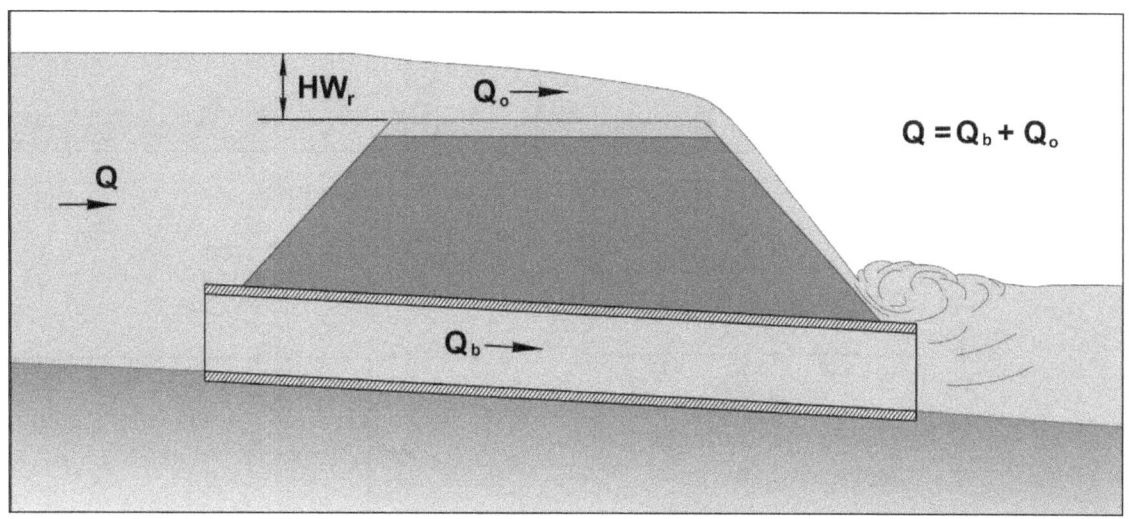

Figure 3.10. Roadway overtopping.

Flow coefficients for flow overtopping roadway embankments are found in HDS 1, Hydraulics of Bridge Waterways (FHWA 1978), as well as in the documentation of Curves from HY-7, Bridge Waterways Analysis Model (FHWA 1986a) are shown in Figure 3.11:

- Figure 3.11A is for deep overtopping
- Figure 3.11B is for shallow overtopping
- Figure 3.11C is a correction factor for downstream submergence. Submergence occurs as the tailwater begins to encroach on the free overfall from the weir.

Equation 3.9 defines the flow across the roadway.

$$Q_o = C_d L \, HW_r^{1.5} \tag{3.9}$$

Q_o = Overtopping flow rate in ft³/s (m³/s)
C_d = Overtopping discharge coefficient = $k_t C_r$ from Figure 3.11
      [C_d (SI) = 0.552(C_d from Figure 3.11)]
L = Length of roadway crest, ft (m)
HW_r = Upstream depth, measured from roadway crest to water surface upstream of weir drawdown, ft (m)

The length and elevation of the roadway crest are difficult to determine when the crest is defined by a roadway sag vertical curve. The sag vertical curve can be broken into a series of horizontal segments as shown in Figure 3.12A. Using Equation 3.9, the flow over each segment is calculated for a given headwater. Then, the incremental flows for each segment are added together, resulting in the total flow across the roadway.

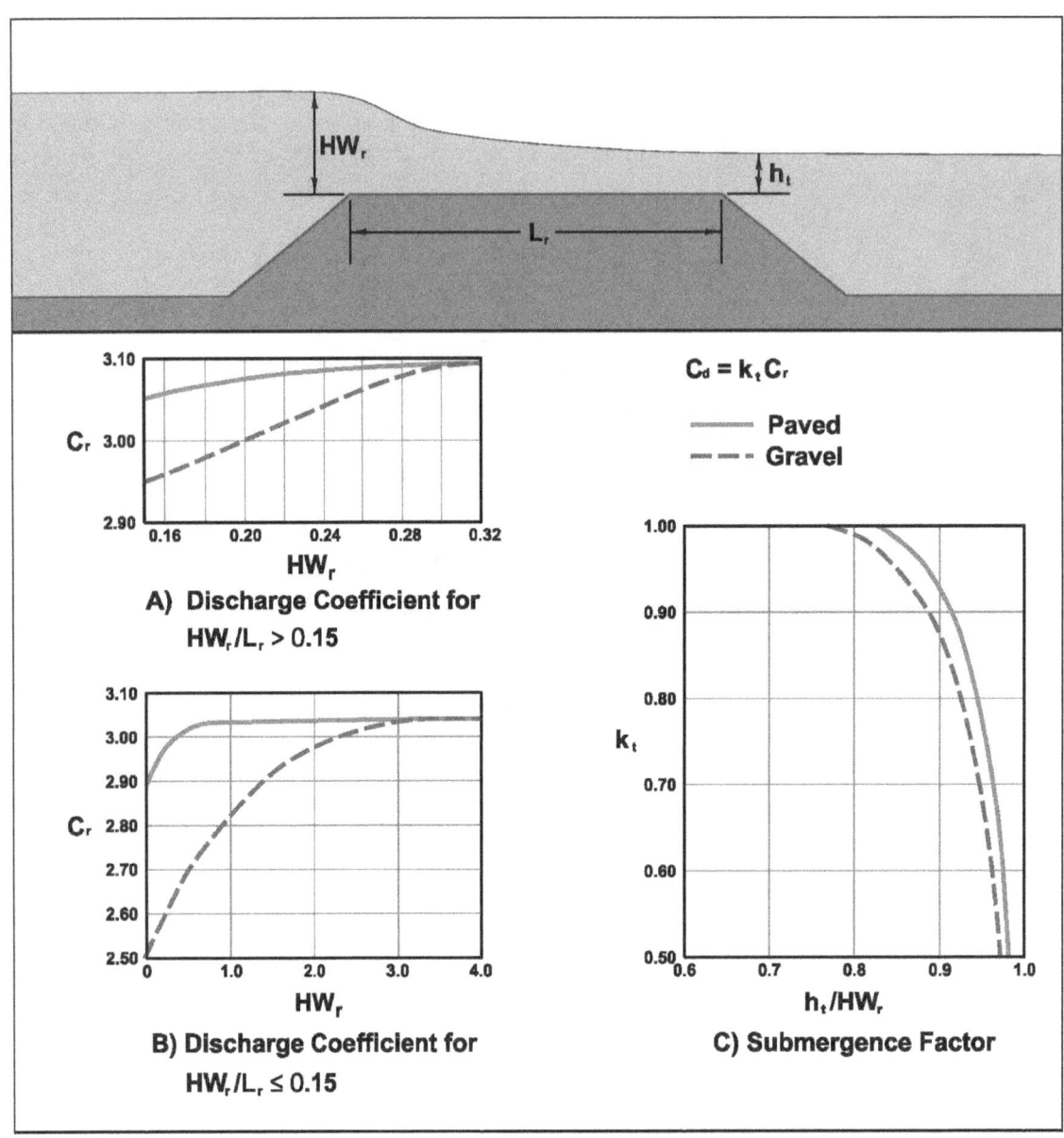

Figure 3.11. Discharge coefficients for roadway overtopping.

Representing the sag vertical curve by a single horizontal line (one segment) is often adequate for culvert design (Figure 3.12B). Using this approach, the length of the weir (L) can be represented by the topwidth of the overflow area in the sag, the upstream depth (HW_r) by the hydraulic depth (overflow area in the sag divided by the top width of flow), and the elevation of the weir crest defined from the lowest point in the sag.

It is a simple matter to calculate the flow across the roadway for a given upstream water surface elevation using Equation 3.9. The problem is that the roadway overflow plus the culvert flow must equal the total design flow. A trial and error process is necessary to determine the amount of the total flow passing through the culvert and the amount flowing across the roadway. Performance curves may also be superimposed for the culvert flow and the road overflow to yield an overall solution as is discussed later in this chapter.

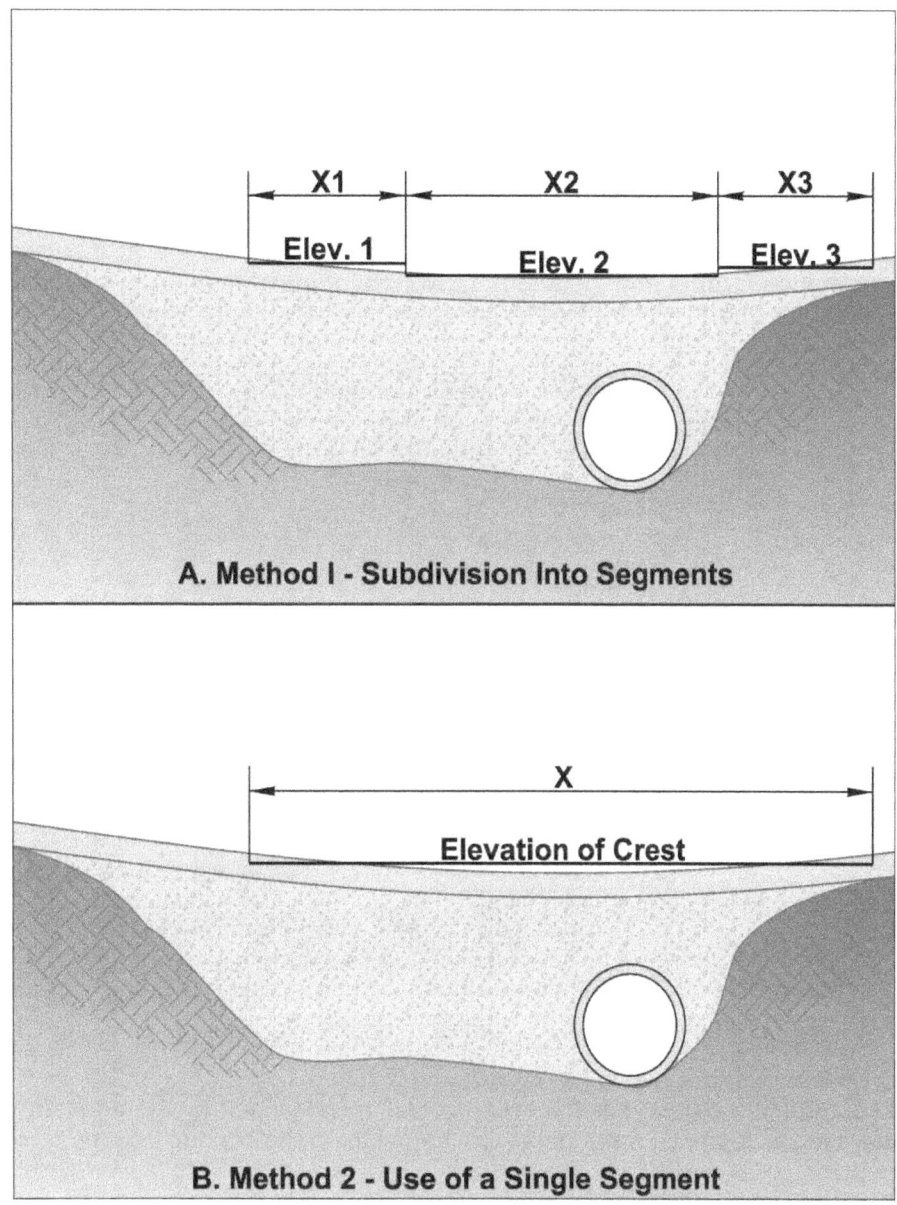

A. Method I - Subdivision Into Segments

B. Method 2 - Use of a Single Segment

Figure 3.12. Weir crest length determinations for roadway overtopping.

3.1.6 Outlet Velocity

Culvert outlet velocities should be calculated to determine the need for erosion protection or an energy dissipater at the culvert exit. Culverts usually result in outlet velocities which are higher than the natural stream velocities. These outlet velocities may require flow readjustment or energy dissipation to prevent downstream erosion.

In inlet control, gradually varied flow calculations may be necessary to determine the outlet velocity. These calculations begin at the culvert entrance and proceed downstream to the exit. The flow velocity is obtained from the flow and the cross-sectional area at the exit (Equation 3.2).

An approximation may be used to avoid backwater calculations in determining the outlet velocity for culverts operating in inlet control. The water surface profile converges toward normal depth as calculations proceed down the culvert barrel. Therefore, if the culvert is of adequate length, normal depth will exist at the culvert outlet. Even in short culverts, normal depth can be assumed and used to define the area of flow at the outlet and obtain the outlet velocity (Figure 3.13). The velocity calculated in this manner may be slightly higher than the actual velocity at the outlet. Normal depth in common culvert shapes may be calculated using a trial and error solution of the Manning equation. The known inputs are flow rate, barrel resistance, slope and geometry. Normal depths may also be calculated using software such as the FHWA Hydraulic Toolbox.

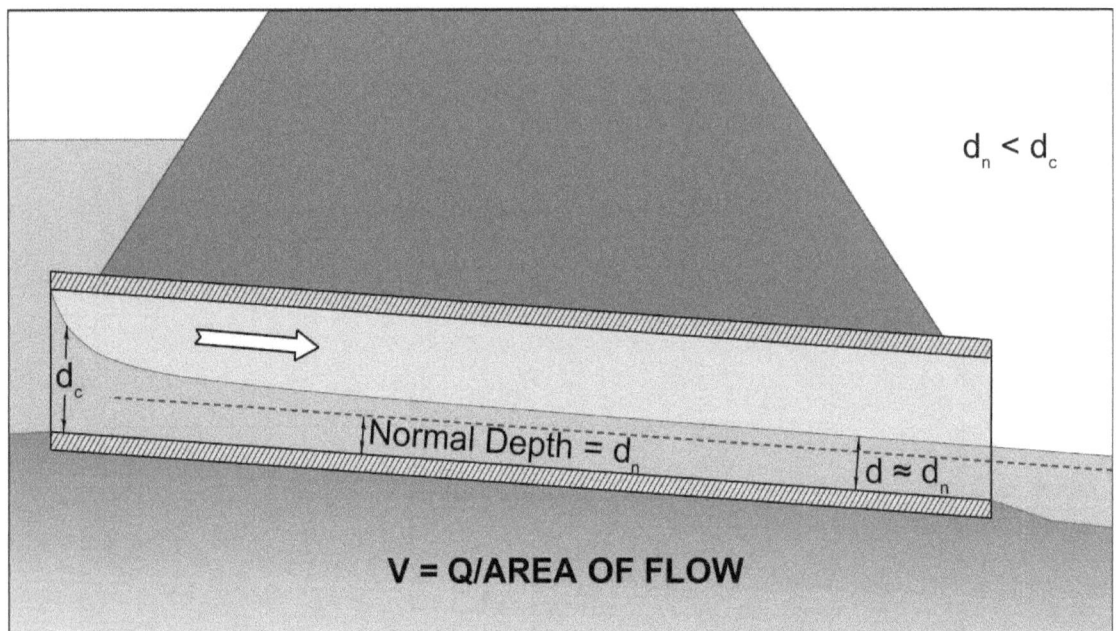

Figure 3.13. Outlet velocity - inlet control.

In outlet control, the cross sectional area of the flow is defined by the geometry of the outlet and either critical depth, tailwater depth, or the height of the conduit (Figure 3.14).

The tailwater depth establishes which depth to use:

- Critical depth is used when the tailwater is less than critical depth
- Tailwater depth is used when tailwater is greater than critical depth but below the top of the barrel
- Total barrel area is used when the tailwater exceeds the top of the barrel

3.2 PERFORMANCE CURVES

Performance curves are representations of flow rate versus headwater depth or elevation for a given flow control device, such as a weir, an orifice, or a culvert. A weir constricts open channel flow so that the flow passes through critical depth. An orifice is a flow control device, fully submerged on the upstream side, through which the flow passes. Performance curves and equations for these two basic types of flow control devices are shown in Figure 3.15.

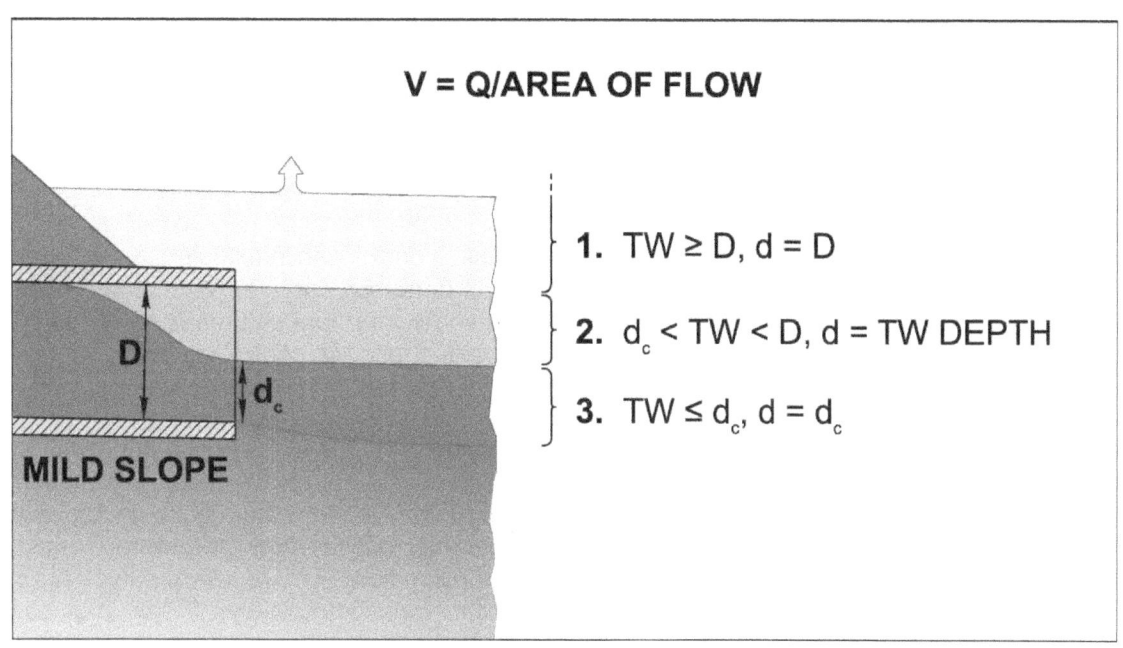

Figure 3.14. Outlet velocity - outlet control.

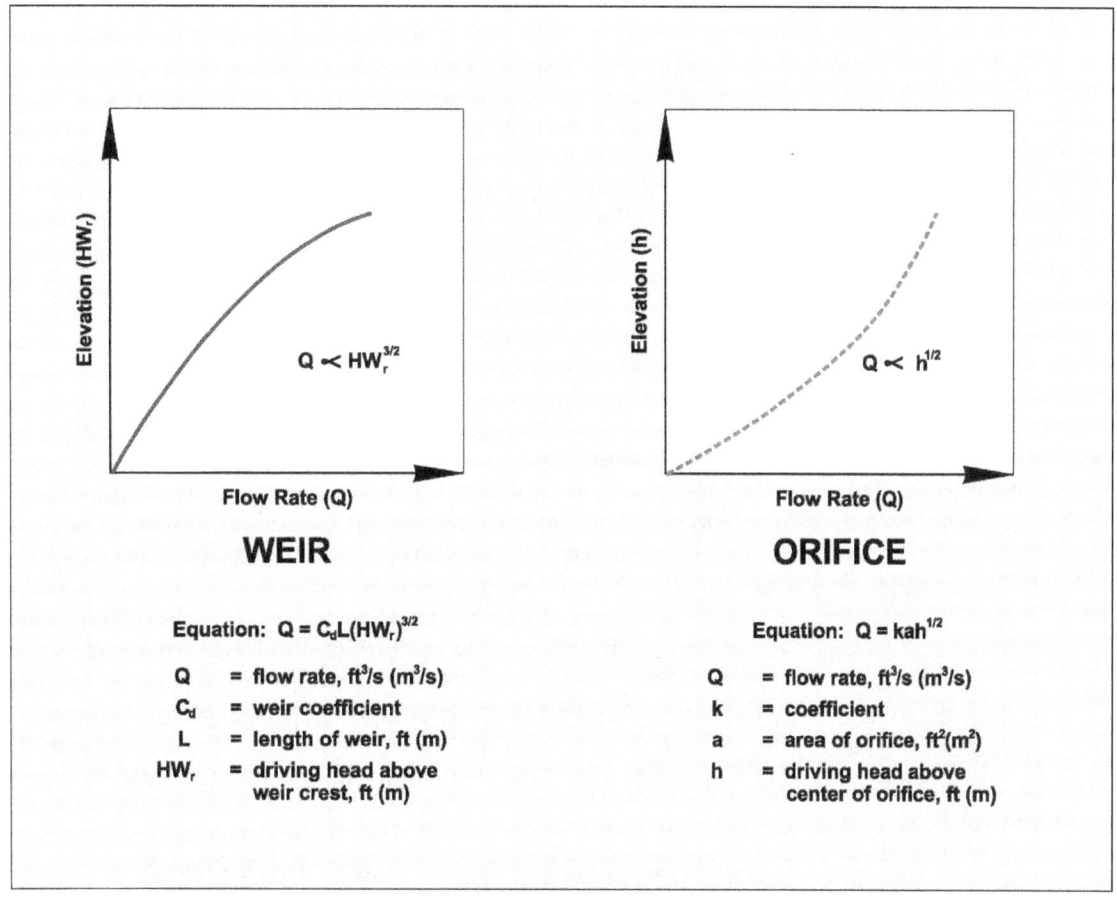

Figure 3.15. Performance curves and equations for weirs and orifices.

When a tailwater exists, the control device may be submerged so that more than one flow-versus-elevation relationship exists. Then, the performance curve is dependent on the variation of both tailwater and headwater. In the case of a weir or orifice, the device is called a submerged weir or a submerged orifice, respectively. For some cases, submergence effects have been analyzed and correction factors have been developed (FHWA 1978, FHWA 1986a, King and Brater 1976).

Culvert performance curves have several possible control sections: inlet, outlet (barrel), roadway. A given culvert installation will have a performance curve for each control section and one for roadway overtopping. The overall culvert performance curve is made up of the controlling portions of the individual performance curves for each control section.

3.2.1 Inlet Control

The inlet control performance curves are developed using either the inlet control equations of Appendix A or the inlet control nomographs of Appendix C. If the equations of Appendix A are used, both unsubmerged (weir) and submerged (orifice) flow headwaters must be calculated for a series of flow rates bracketing the design flow. The resultant curves are then connected with a line tangent to both curves (the transition zone). If the inlet control nomographs are used, the headwaters corresponding to the series of flow rates are determined and then plotted. The transition zone is inherent in the nomographs.

3.2.2 Outlet Control

The outlet control performance curves are developed using Equations 3.1 through 3.6 of this chapter, the outlet control nomographs of Appendix C, or backwater calculations. Flows bracketing the design flow are selected. For these flows, the total losses through the barrel are calculated or read from the outlet control nomographs. The losses are added to the elevation of the hydraulic grade line at the culvert outlet to obtain the headwater.

If backwater calculations are performed beginning at the downstream end of the culvert, friction losses are accounted for in the calculations. The headwater elevation for each flow rate is calculated by adding the inlet loss to the energy grade line in the barrel at the inlet.

3.2.3 Roadway Overtopping

A performance curve showing the culvert flow as well as the flow across the roadway is a useful analysis tool. Rather than using a trial and error procedure to determine the flow division between the overtopping flow and the culvert flow, an overall performance curve can be developed. The performance curve depicts the sum of the flow through the culvert and the flow across the roadway.

The overall performance curve can be determined by performing the following steps.

1. Select a range of flow rates and determine the corresponding headwater elevations for the culvert flow alone. These flow rates should fall above and below the design discharge and cover the entire flow range of interest. Both inlet and outlet control headwaters should be calculated.

2. Combine the inlet and outlet control performance curves to define a single performance curve for the culvert.

3. When the culvert headwater elevations exceed the roadway crest elevation, overtopping will begin. Calculate the equivalent upstream water surface depth above the roadway (crest of weir) for each selected flow rate. Use these water surface depths and Equation 3.9 to calculate flow rates across the roadway.

4. Add the culvert flow and the roadway overtopping flow at the corresponding headwater elevations to obtain the overall culvert performance curve.

Figure 3.16 depicts an overall culvert performance curve with roadway overtopping. The performance curve is used to easily determine the headwater elevation for any flow rate and to visualize the performance of the culvert installation over a range of flow rates. When roadway overtopping begins, the rate of headwater increase will flatten severely. The headwater will rise very slowly from that point on. Design Guidelines in Appendix D illustrate the development of an overall culvert performance curve.

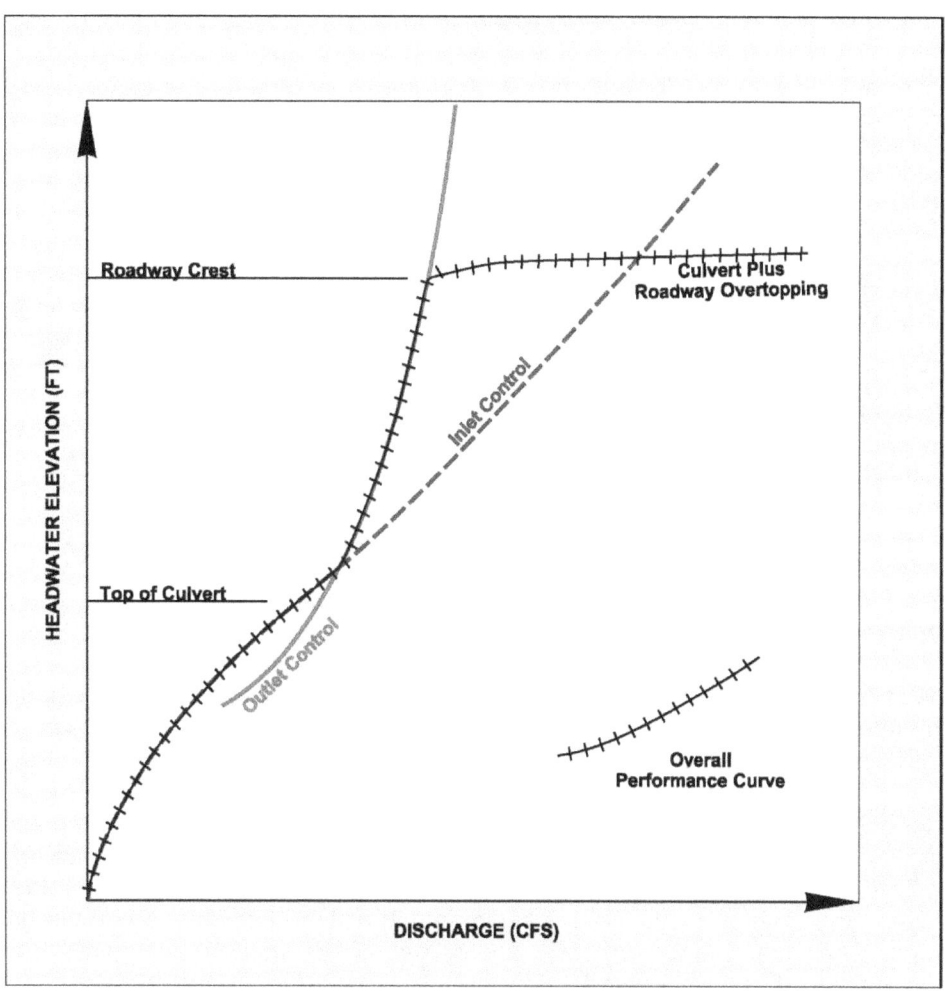

Figure 3.16. Culvert performance curve with roadway overtopping.

3.21

3.3 CULVERT DESIGN USING NOMOGRAPHS

The manual culvert design method using nomographs provides a convenient and organized procedure for designing culverts and for checking software solutions (Section 3.5). While it is possible to follow the manual design method without an understanding of culvert hydraulics, this is not recommended. The result could be an inadequate and possibly unsafe structure. This section provides an overview of culvert design with nomographs. Step by Step design procedures are provided in Design Guideline 1 in the Appendices.

3.3.1 Culvert Design Form

The Culvert Design Form, shown in Figure 3.17, has been formulated to guide the user through the design process. Summary blocks are provided at the top of the form for the project description, and the designer's identification. Summaries of hydrologic data of the form are also included. At the top right is a small sketch of the culvert with blanks for inserting important dimensions and elevations.

The central portion of the design form contains lines for inserting the trial culvert description and calculating the inlet control and outlet control headwater elevations. Space is provided at the lower center for comments and at the lower right for a description of the culvert barrel selected. This form provides adequate documentation for roadway culverts where the site assessments (Section 2.3) do not indicate any unusual conditions.

The first step in the design process is to summarize hydrology data (Section 2.1) and site data (Section 2.2) for the culvert at the top of the Culvert Design Form. This information will have been collected or calculated prior to performing the actual culvert design. The next step is to select a preliminary culvert material, shape, size, and entrance type. The user then enters the design flow rate and proceeds with the inlet control calculations.

3.3.2 Inlet Control

The inlet control calculations determine the headwater elevation required to pass the design flow through the selected culvert configuration in inlet control. The approach velocity head is assumed to be zero for the manual method. If the approach velocity needs to be considered (irrigation structure or AOP design), a software solution should be used (see Section 3.5).

The inlet control nomographs of Appendix C are used to determine the design headwater depth under inlet control (HW_i). If HW_i is greater than the allowable headwater (HW_a) other configurations should be evaluated or an inlet depression can be considered.

An inlet depression is constructing the inlet invert below the streambed while maintaining the outlet invert on the streambed. Essentially, this amounts to rotating the culvert about the outlet invert. While this can allow a given culvert configuration to meet the allowable headwater, this rotation may increase sediment deposition potential both in the barrel and in the sump area created at the entrance, and perhaps more importantly, too much rotation can change the culvert to outlet control.

3.22

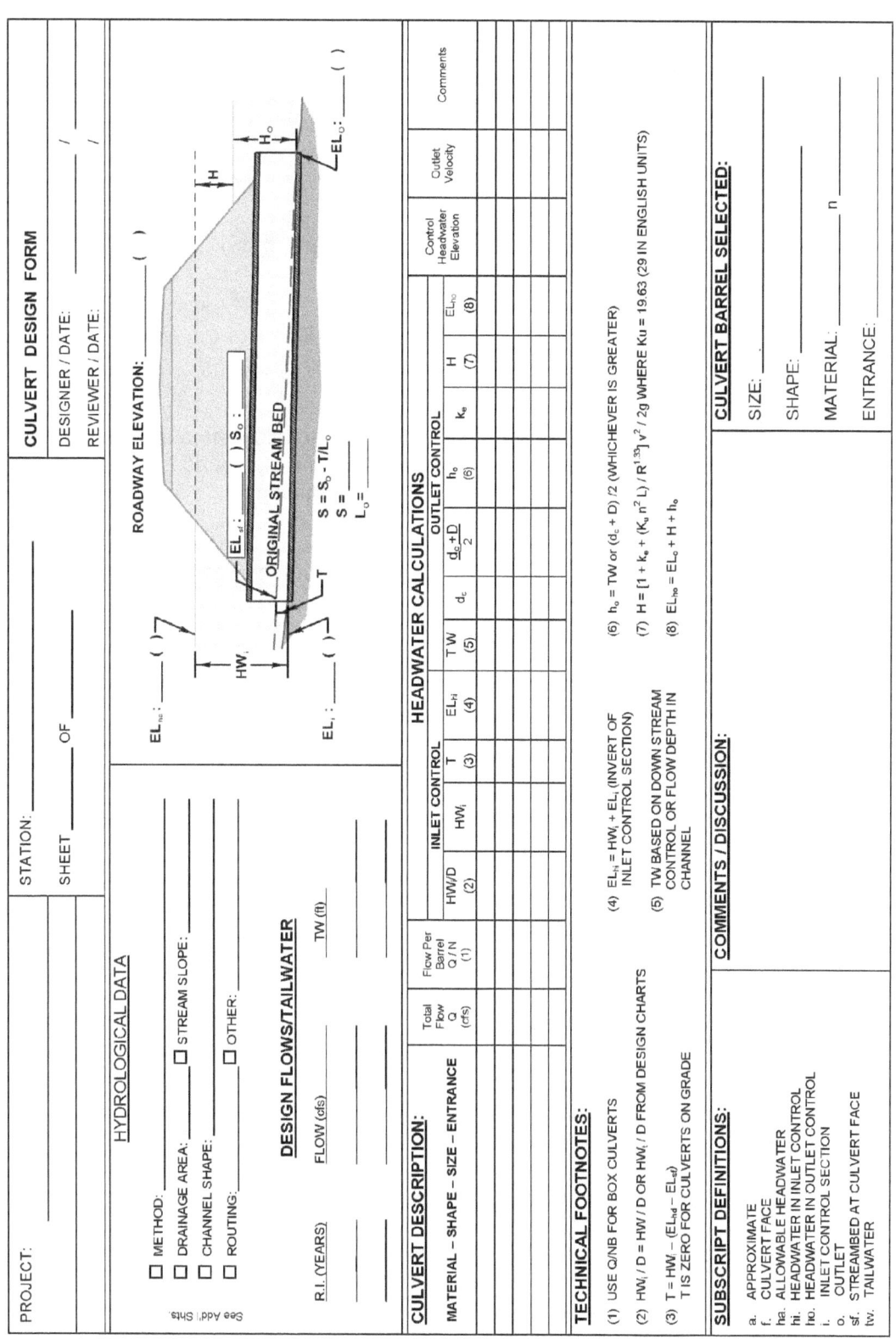

Figure 3.17. Culvert design form.

To calculate the required inlet or face depression (F_D) of the inlet control section below the stream bed the following procedure can be used:

$$HW_a = EL_a - EL_{sf} \tag{3.10}$$

$$F_D = HW_i - HW_a \tag{3.11}$$

HW_a = Allowable headwater depth, ft (m)
EL_a = Allowable headwater elevation, ft (m)
EL_{sf} = Elevation of the streambed at the face, ft (m)
HW_i = Required headwater depth, ft (m)

Possible results and consequences of this calculation are:

(1) If F_D is negative or zero, set F_D equal to zero.

(2) If F_D is positive, the inlet invert should be constructed below the streambed at the face by that amount, assuming that inlet control is maintained. If F_D is not acceptable (e.g., the sump is too large), select another culvert configuration and begin again.

If the controlling headwater is based on inlet control, determine the normal depth and velocity in the culvert barrel. The velocity at normal depth is assumed to be the outlet velocity.

3.3.3 Outlet Control

The outlet control calculations result in the headwater elevation required to convey the design discharge through the selected culvert in outlet control assuming that the barrel is flowing full (USGS flow type 4). The approach and downstream velocities are assumed to be zero for the manual method. If these velocities need to be considered (irrigation structure or AOP design); a software solution should be considered (see Section 3.5). The critical depth charts and outlet control nomographs of Appendix C are used in the design process. The manual method has the assumptions:

- Tailwater - Tailwater depth above the outlet invert (TW) at the design flow rate is obtained from normal depth calculations or from field observations.
- Critical Depth – Appendix C charts are used to read the critical depth (d_c). Critical depth (d_c) cannot exceed D.
- Approximate hydraulic gradeline $h_o = (d_c + D)/2$ can only be used if the barrel flows full for most of its length. It should not be used if the inlet is not submerged.
- Entrance loss coefficient (k_e) - Table C.2, Appendix C provides values which are used in the nomographs. If different k_e values are needed, use Equation 3.5.
- Barrel Losses (H) – use Equation 3.5 if outside the range of the nomograph.

If outlet control governs and the headwater depth (referenced to the inlet invert) is less than 1.2D, it is possible that the barrel flows partly full though its entire length. In this case, caution should be used in applying the approximate method of setting the downstream elevation based on the greater of tailwater or (d_c + D)/2. If a more accurate headwater is necessary, backwater calculations (Section 3.5) should be used to check the result from the approximate method. If the headwater depth falls below 0.75D, the approximate method should not be used.

If the controlling headwater is in outlet control, determine the area of flow at the outlet based on the barrel geometry and the following:

- Critical depth if the tailwater is below critical depth
- Tailwater depth if the tailwater is between critical depth and the top of the barrel
- Height of the barrel if the tailwater is above the top of the barrel

3.3.4 Culverts with Horizontal or Adverse Slopes

The inlet and outlet control procedures that have been described were established assuming that culverts have some positive slope. However, they can be reasonably applied to culverts with no slope (horizontal) or an adverse slope. A horizontal slope is used when flow may occur in either direction. Adverse slopes can occur when entrances settle more than outlet. As a rule of thumb, these adjustments should not be used if the exit invert is more than D/2 higher than the entrance invert.

The inlet control nomograph solution HW/D contains a small slope correction term of -0.5S that was subtracted to account for the control section occurring within the culvert and slightly lower than the inlet invert. For mitered culverts, a correction term of +0.7S is used to account for the control section being outside the culvert barrel and slightly higher. For the nomographs, a slope of 2% was assumed for these adjustments. The inlet control nomograph results can be adjusted as follows:

- Inlets other than mitered - For horizontal slopes, 0.01 should be added to the HW/D obtained from the nomograph. For adverse sloping culverts, 0.01 should be added to the HW/D obtained from the nomograph and an additional (0.5S) added, where S is the adverse slope in ft/ft (for example 3% is .03 ft/ft and 0.5S is 0.015).

- Mitered inlets - For a horizontal slope, 0.014 should be subtracted from the HW/D obtained from the nomograph. For adverse sloping culverts, 0.014 should be subtracted from the HW/D obtained from the nomograph and additional 0.7S subtracted where S is the adverse slope in ft/ft (for example 3% is .03 ft/ft and 0.7S is 0.021).

The outlet control nomograph solution for losses through the barrel (H) for USGS flow type 4 and 6 can be used without adjustment for both a horizontal slope and an adverse slope. The outlet control headwater depth (HW_o) which is the depth above the outlet invert is still equal to the $TW + H_L$.

3.3.5 Evaluation of Results

For the first alternative selected, compare the headwater elevations calculated for inlet and outlet control. The higher of the two is designated the controlling headwater elevation. The culvert can be expected to operate with that higher headwater for at least part of the time.

Repeat the design process until an acceptable culvert configuration is determined. An acceptable culvert based on hydraulic considerations is one where the design HW is less than the allowable HW, and the outlet velocity is not excessive. Once the barrel is selected it must be fitted into the roadway cross section. The culvert barrel must have adequate cover, the length should be close to the approximate length, and the headwalls and wingwalls must be dimensioned.

If the selected culvert will not fit the site, return to the culvert design process and select another culvert. If a multiple barrel configuration is considered the design discharge is typically evenly divided between the barrels in a manual calculation. However, if the barrels are different sizes, different types, or located at different elevations, a software solution (see Section 3.5) should be used as the assumption of a uniform distribution of discharge between the barrels will no longer be valid. Once an acceptable configuration is determined, the selected design should be documented and accompanied by a performance curve which displays culvert behavior over a range of discharges.

A flow chart illustrating the major steps in defining an acceptable culvert alternative based on hydraulic factors is shown in Figure 3.18. Other design variables such as AOP, may also be involved. If the following alternative designs are to be investigated, use the cited guidance:

- Tapered inlets - Section 3.4
- AOP - Chapter 4
- Low Head installations – Section 5.2.2
- Siphons – Section 5.2.5
- Broken-Back Culverts – Section 5.7
- Storage routing - Section 5.8

Special Applications (Section 5.2) should be consulted for the effect on culvert hydraulics of flow control and measurement, junctions, bends, baffles, median drainage and drop inlets.

3.3.6 Example Problems

Design guidelines for the following example problems are found in the Appendices. The guidelines illustrate the use of the design methods and charts for the following culvert configurations and hydraulic conditions:

Design Guideline No. 1a: CMP with standard 2-2/3 by 1/2 in (68 by 13 cm) corrugations with beveled edge, and reinforced concrete pipe with groove end (no inlet depression).

Design Guideline No. 1b: Reinforced cast-in-place concrete box culvert with square edges, and an alternative design with bevels, neither option with an inlet depression.

Design Guideline No. 1c: Elliptical pipe culvert with groove end and with inlet depression.

Design Guideline No. 1d: Analysis of an existing reinforced concrete box culvert with square edges.

3.4 TAPERED INLET DESIGN USING NOMOGRAPHS

3.4.1 Introduction

A tapered inlet is a flared culvert entrance with an enlarged face section and a hydraulically efficient throat section (see Section 1.3.3). Tapered inlets can dramatically improve culvert hydraulic performance for culverts in inlet control. As mentioned in Section 1.3.3 the additional cost of a tapered inlet must be weighed against the savings in barrel cost, and may not be appropriate in some situations such as AOP design. However, when tapered inlets are feasible, the improvement in hydraulic performance can be significant.

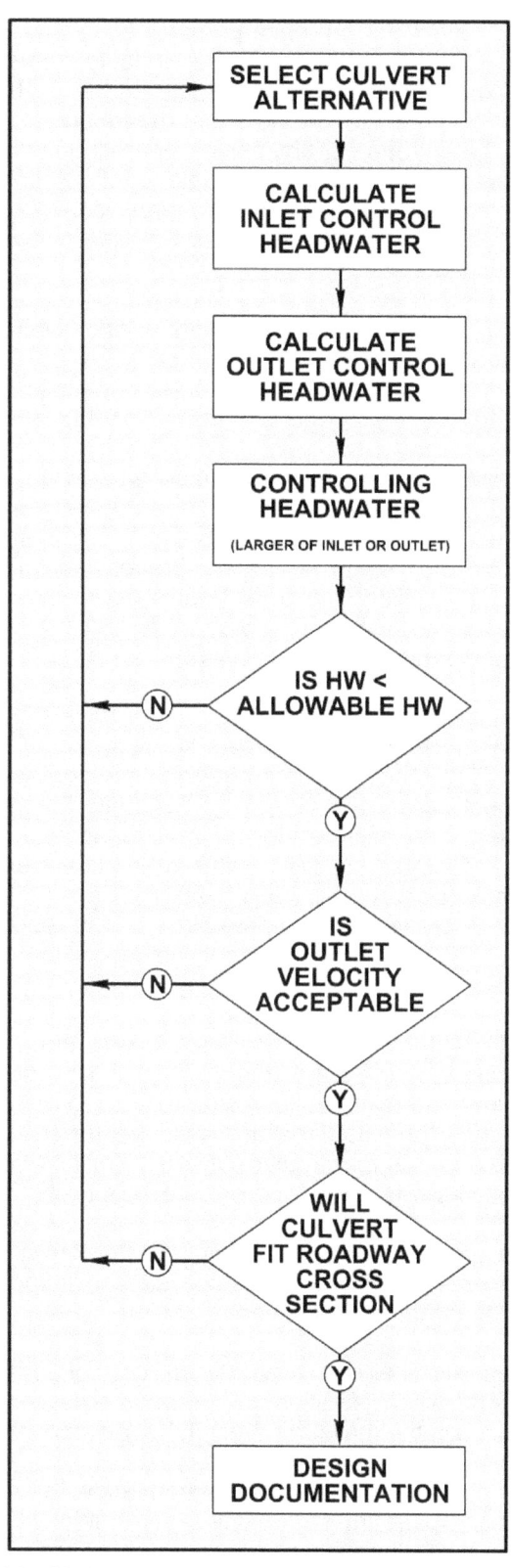

Figure 3.18. Major steps in defining an acceptable alternative.

Tapered inlets improve culvert performance primarily by reducing the contraction at the inlet control section which is located at the throat. Secondarily, some tapered inlet configurations also depress the inlet control section below the stream bed. This depression may be at the face or the throat and is used to create more head on the throat section for a given headwater elevation. The entrance of any culvert operating in inlet control can be depressed to obtain better performance, regardless of the inlet configuration. However, edge conditions are normally improved first and then an inlet depression is applied. If a tapered inlet is selected, the hydraulic performance will be better than the performance of beveled edges for culverts operating in inlet control.

In outlet control the performance of tapered inlets is effectively the same as for inlets with beveled edges. An entrance loss coefficient (k_e) of 0.2 is used for both tapered inlets and beveled edges. Tapered inlets are not recommended for use on culverts flowing in outlet control because the simple beveled edge is of equal benefit.

Design criteria and methods have been developed for two basic tapered inlet designs: the side-tapered inlet and the slope-tapered inlet. Tapered inlet design charts are available for rectangular box culverts and circular pipe culverts. The same principles apply to other culvert barrel shapes, but no design charts are presently available for other shapes. The side-tapered inlet can be installed with or without a depression upstream of the face. The slope-tapered inlet can be designed with a vertical face (illustrated in this chapter) or with a face mitered to the fill slope (discussed in HEC-13).

The inlet configurations presented in this manual are based on research conducted at the National Bureau of Standards (NBS) under the sponsorship of the Bureau of Public Roads (NBS 1961, 1966b, 1967). Many improved inlet configurations were tested; however, only those determined to best satisfy the criteria of hydraulic efficiency, economy of materials, simplicity of construction, and minimization of maintenance problems were selected. For example, while the use of curved surfaces rather than plane surfaces might result in slightly improved hydraulic efficiency at times, the advantages are outweighed by the construction difficulties. Therefore, only plane surfaces are utilized in the recommended designs.

3.4.2 Side-Tapered Inlet

The side-tapered inlet has an enlarged face section with the transition to the culvert barrel accomplished by tapering the side walls (Figure 3.19). The face section is about the same height as the barrel height and the inlet floor is an extension of the barrel invert. The inlet roof may slope upward slightly, provided that the face height (E) does not exceed the barrel height by more than 10% (1.1D) for circular culverts. For box culverts, E should be equal to D. The throat section occurs where the tapered sidewalls meet the barrel.

There are two possible control sections, the face and the throat. HW_f, shown in Figure 3.19, is the headwater depth measured from the face section invert and HW_t is the headwater depth measured from the throat section invert.

The throat of a side-tapered inlet is a very efficient control section. The flow contraction is nearly eliminated at the throat. In addition, the throat is always slightly lower than the face so that more head is exerted on the throat for a given headwater elevation.

The beneficial effect of depressing the throat section below the stream bed can be increased by installing a depression upstream of the side-tapered inlet. Figure 3.20 illustrates a side-tapered inlet with the inlet depression contained between wingwalls. For this type of depression, the floor of the barrel should extend upstream from the face a minimum distance of D/2 before sloping upward more steeply.

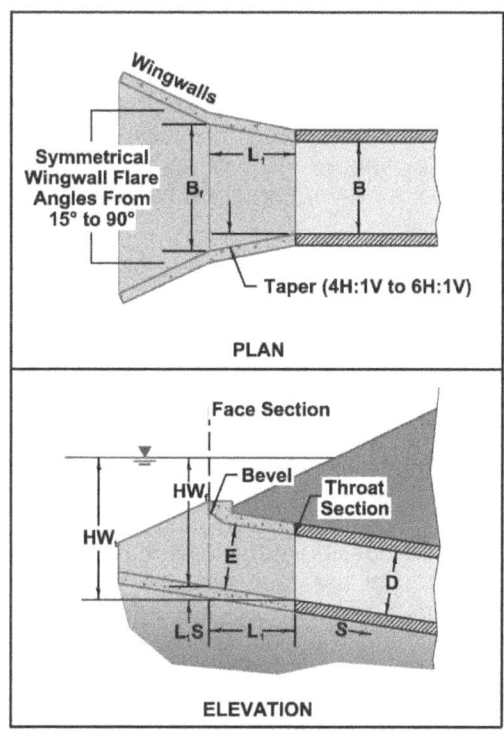

Figure 3.19. Side-tapered inlet.

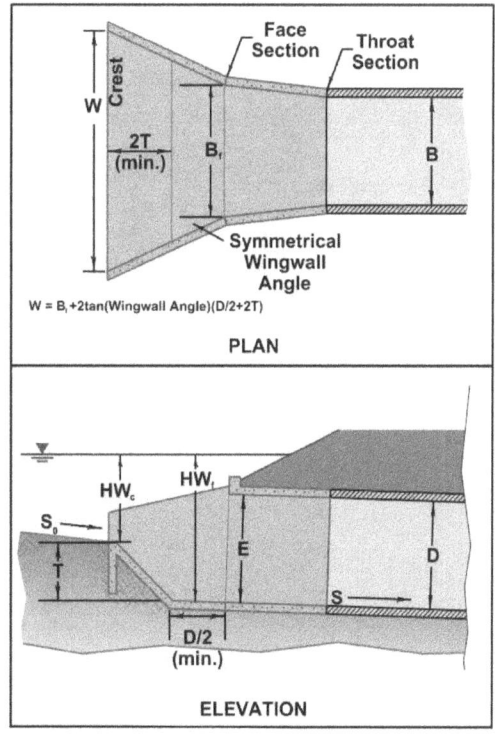

Figure 3.20. Side-tapered inlet with inlet depression.

Dimensional limitations for the designs are shown. The length of the resultant upstream crest where the slope of the inlet depression meets the stream bed should be checked to assure that the crest will not control the flow at the design flow and headwater. If the crest length is too short, the crest may act as a weir control section. For side-tapered inlets with inlet depression, both the face section and the throat section have more head exerted on them for a given headwater elevation. The increased head allows smaller face and throat sections. Beveled edges or other favorable edge conditions also reduce the required size of the face.

3.4.3 Slope-Tapered Inlet

The slope-tapered inlet, like the side-tapered inlet, has an enlarged face section with tapered sidewalls meeting the culvert barrel walls at the throat section (Figure 3.21). In addition, a throat depression is incorporated into the inlet between the face and throat sections. This throat depression creates more head on the throat section. At the location where the steeper slope of the inlet intersects the flatter slope of the barrel, a third control section, designated the bend section, is formed.

A slope-tapered inlet has three possible control sections, the face, the bend, and the throat. Of these, only the dimensions of the face and the throat section are determined by the design procedures of this manual. The size of the bend section is established conservatively so that it will not control by locating it a minimum distance upstream from the throat.

The slope-tapered inlet combines an efficient throat section with additional head on the throat. Since the face section is not depressed, the face sections of these inlets are larger than the face sections of equivalent depressed side-tapered inlets. The required face size can be reduced by the use of bevels or other favorable edge configurations. The vertical face slope-tapered inlet design is shown in Figure 3.21.

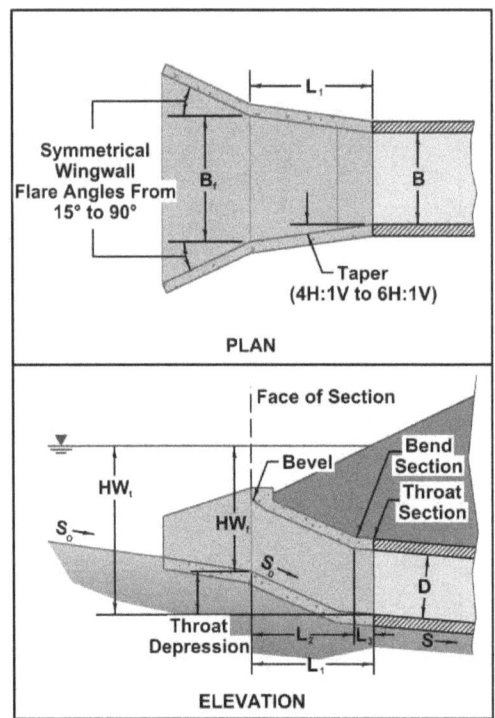

Figure 3.21. Slope-tapered inlet with vertical face.

The slope-tapered inlet has construction difficulties which can increase the cost of the inlet. If the increased cost of inlet cannot be balanced by the benefit in increased performance and/or a reduction in barrel size; this inlet type should not be used.

Slope-tapered inlets can be applied to both box culverts and circular pipe culverts. For the latter application, a square to round transition is normally used to connect the rectangular slope-tapered inlet to the circular pipe.

3.4.4 Inlet Control

Tapered inlets have several possible control sections: the face, the bend (for slope-tapered inlets), and the throat. In addition, a side-tapered inlet with inlet depressed has a possible control section at the crest upstream of the inlet depression. Each of these control sections has a performance curve. The headwater depth for each control section is referenced to the invert of the section. One method of determining the overall inlet control performance curve is to calculate performance curves for each potential control section, and then select the segment of each curve which defines the minimum overall culvert performance (Figure 3.22).

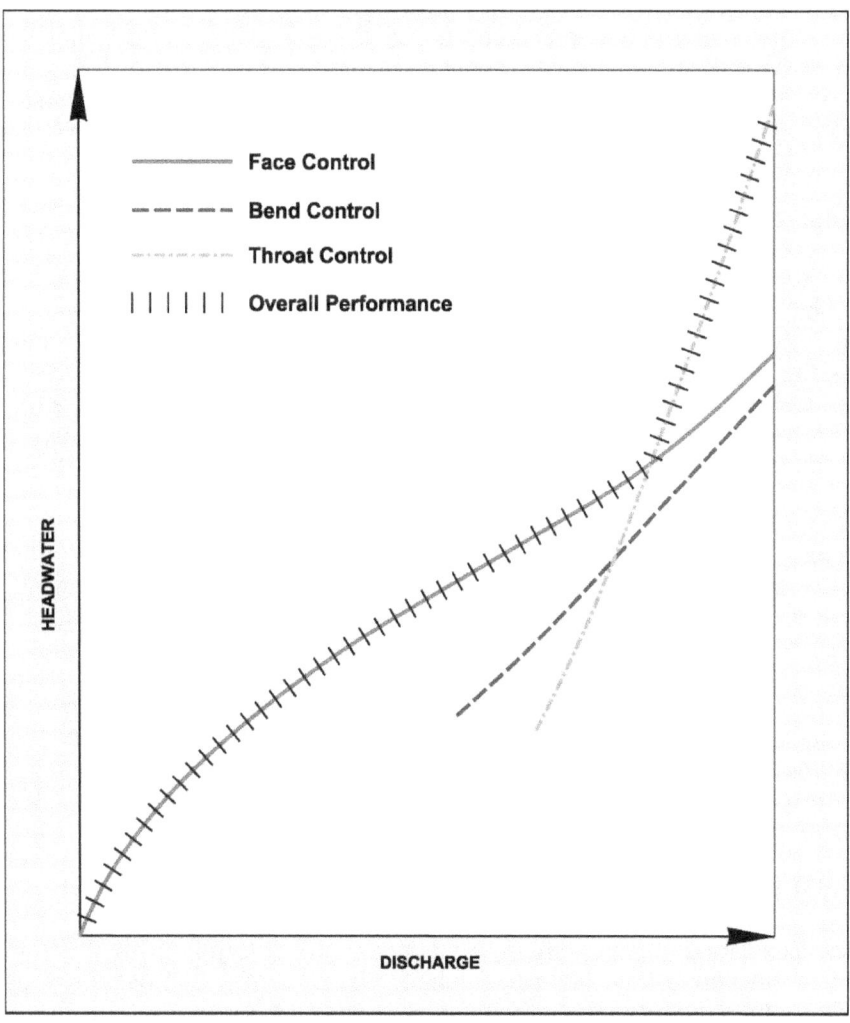

Figure 3.22. Inlet control performance curves.

If the dimensional criteria of this publication are followed, the crest and the bend sections will not function as control sections over the normal range of headwaters and discharges. The crest of the inlet depression may function as a control section for very low flows and headwaters but this is generally not of importance in design. Figure 3.22 depicts performance curves for each of the potential inlet control sections and the overall inlet control performance curves.

The design procedures for tapered inlets include checks on crest lengths for both depressed side-tapered inlets and slope-tapered inlets with mitered faces. As long as the actual crest length exceeds a certain minimum value, there is no need to construct a crest performance curve. Also, if the bend section is located a minimum distance of B/2 upstream of the throat section, the bend will not control and the bend section performance curve does not need to be calculated.

The inlet control equations for tapered inlets are given in Appendix A. The coefficients and exponents for the face and throat control sections were developed based on the NBS hydraulic tests. All of the previously described control sections function in a manner similar to weirs for unsubmerged flow conditions, and in a manner similar to orifices for submerged flow conditions. For each section, there is a transition zone defined by an empirical curve connecting the unsubmerged and submerged curves. The tapered inlet throat should be designed to be the primary control section for the design range of flows and headwaters. The bend section will not act as the control section if the dimensional criteria of this publication are followed. However, the bend will contribute to the inlet losses which are included in the inlet loss coefficient, k_e.

3.4.5 Outlet Control

When a culvert with a tapered inlet performs in outlet control, the computations are the same as described in Section 3.1.4 for all culverts. The factors influencing flow in outlet control are shown in Table 1.1. The inlet area is the area of the face section, the inlet configuration describes the type of tapered inlet as well as the face edge conditions, and the shape is either circular or rectangular. The barrel characteristics refer to the barrel portion of the culvert, downstream of the throat section, except that the barrel length includes the length of the tapered inlet, and the barrel slope may be flatter than the natural stream bed slope.

Equation 3.5 in Section 3.1.4 describes the losses in outlet control. The tapered inlet entrance loss coefficient (k_e) is 0.2 for both side-tapered and slope-tapered inlets. This loss coefficient includes contraction and expansion losses at the face, increased friction losses between the face and the throat, and the minor expansion and contraction losses at the throat.

The headwater depth in outlet control (HW_o) is measured from the invert of the culvert inlet and is calculated using Equation 3.6b (HW_o = TW + H_L -SL). Equation 3.5 or the outlet control nomograph for the appropriate barrel size is used to determine the total losses through the culvert (H_L). The hydraulic gradeline at the culvert exit is either the tailwater (TW) shown in Equation 3.6b or (d_c + D)/2 if larger.

3.4.6 Outlet Velocity

Outlet velocities for culverts with tapered inlets are determined in the same manner as described in Section 3.1.6. Note that when inlet or throat depression is used, the barrel slope is flatter than the stream slope and is calculated as follows.

$$S = \frac{EL_t - EL_o}{L_a - L_1}$$ (3.12)

S	=	Approximate barrel slope, ft/ft (m/m)
EL_t	=	Invert elevation at the throat, ft (m)
EL_o	=	Invert elevation at the outlet, ft (m)
L_a	=	Approximate length of the culvert, ft (m)
L_1	=	Overall length of the tapered-inlet, ft (m)

3.4.7 Performance Curves

Performance curves are important in understanding the operation of a culvert with a tapered inlet. Each potential control section (face, throat, and outlet) has a performance curve, based on the assumption that that particular section controls the flow. Calculating and plotting the various performance curves results in a graph similar to Figure 3.23, containing the face control, throat control and outlet control curves. The overall culvert performance curve is represented by the hatched line. In the range of lower discharges face control governs; in the intermediate range, throat control governs; and in the higher discharge range, outlet control governs. The crest and bend performance curves are not calculated since they do not govern in the design range.

Constructing performance curves for culverts with tapered inlets helps to assure that the designer is aware of how the culvert will perform over a range of discharges. For high discharges, the outlet control curve may have a very steep slope which means that the headwater will increase rapidly with increasing discharge. Since there is a probability that the design discharge will be exceeded over the life of the culvert, the consequences of that event should be considered. This will help to evaluate the potential for damage to the roadway and to adjacent properties.

Performance curves are useful in optimizing the performance of a culvert. By manipulating the depressions of the face and throat sections, it is often possible to achieve a higher flow rate for a given headwater elevation, or to pass the same flow at a lower headwater.

3.4.8 Design Methods

Tapered inlet design begins with the selection of the culvert barrel size, shape, and material. These calculations are performed using the Culvert Design Form shown in Figure 3.17. The tapered-inlet design calculation form (Figure 3.24) and the design nomographs contained in Appendix C are used to design the tapered inlet. The result will be one or more culvert designs, with and without tapered inlets, all of which meet the site design criteria. The designer must select the best design for the site under consideration.

In the design of tapered inlets, the goal is to maintain control at the efficient throat section in the design range of headwater and discharge. This is because the throat section has the same geometry as the barrel, and the barrel is the most costly part of the culvert. The inlet face is then sized large enough to pass the design flow without acting as a control section in the design discharge range. Some slight over sizing of the face is beneficial because the cost of constructing the tapered inlet is usually minor compared with the cost of the barrel.

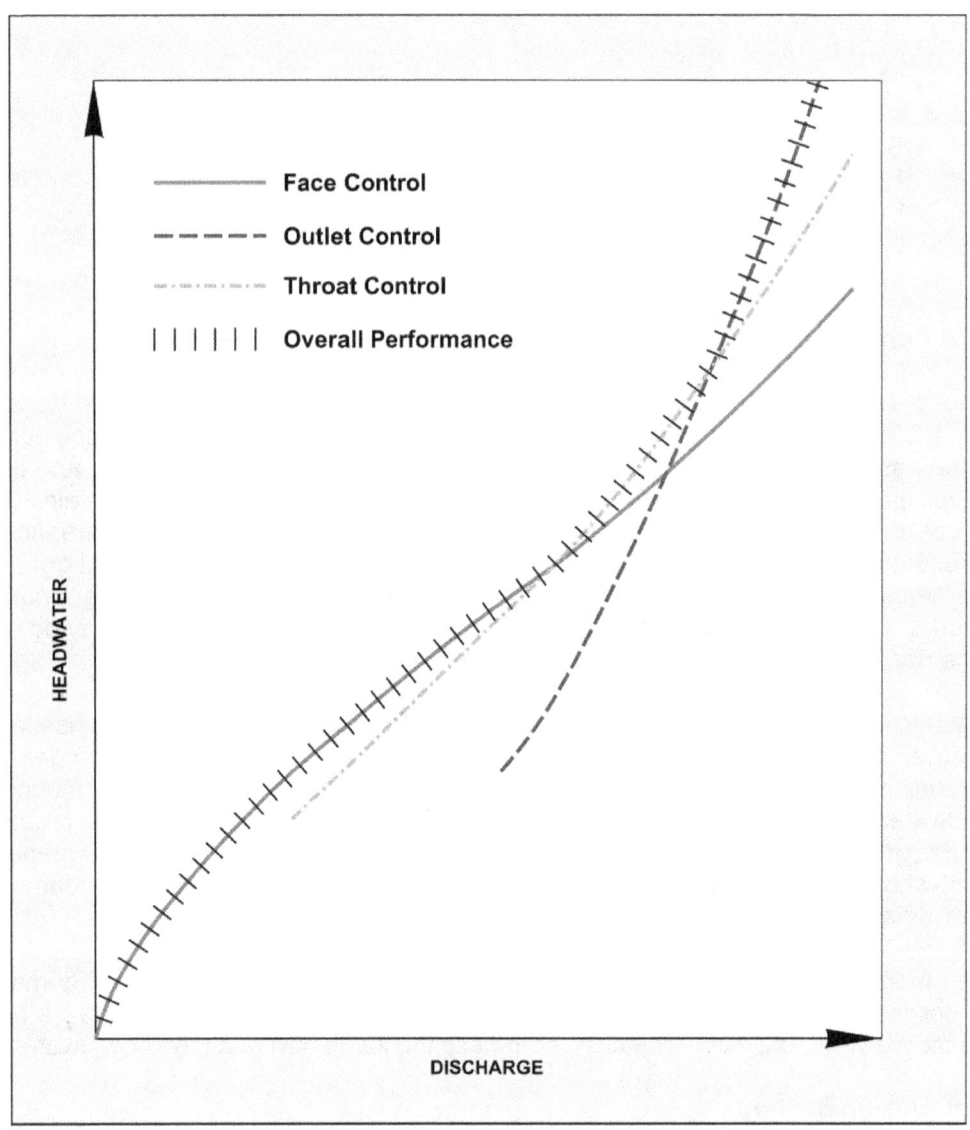

Figure 3.23. Culvert performance curve.

The required size of the face can be reduced by use of favorable edge configurations, such as beveled edges, on the face section. Design nomographs are provided for favorable and less favorable edge conditions.

3.5 CULVERT DESIGN USING SOFTWARE (WATER SURFACE PROFILES)

The manual procedures documented in Chapter 3 were first automated by FHWA in a series of FORTRAN programs for mainframe computers in (HY-1 through 3) in the 1960s. As programmable calculators became available, the procedures were programmed to work with these devices in the early 1980s (CDS 1 through 4). In the 1980s, personal computers became available. The culvert design procedures developed for the HY computer programs and the CDS calculator programs were converted to Apple Basic for an early version of HY-8, to Microsoft Basic for HY-8 Versions 1 (1986) through 6 (1997) and to Microsoft C++ for Version 7 (2005).

Figure 3.24. Tapered inlet design form.

PROJECT: _____

STATION: _____

SHEET _____ OF _____

TAPERED INLET DESIGN FORM

DESIGNER / DATE: ___ / ___

REVIEWER / DATE: ___ / ___

DESIGN DATA:

Q ___ = ___ ( ); EL_{hi} ___ ( )

EL. THROAT INVERT _____ ( )

EL. STREAM BED AT FACE _____ ( )

T ___ TAPER ___ : 1 (4 : 1 TO 6 : 1)

STREAM SLOPE, S_o = ___ ( ) / ( )

SLOPE OF BARREL, S = ___ ( ) / ( )

S_D ___ : 1 (2 : 1 TO 3 : 1)

BARREL SHAPE AND MATERIAL: _____

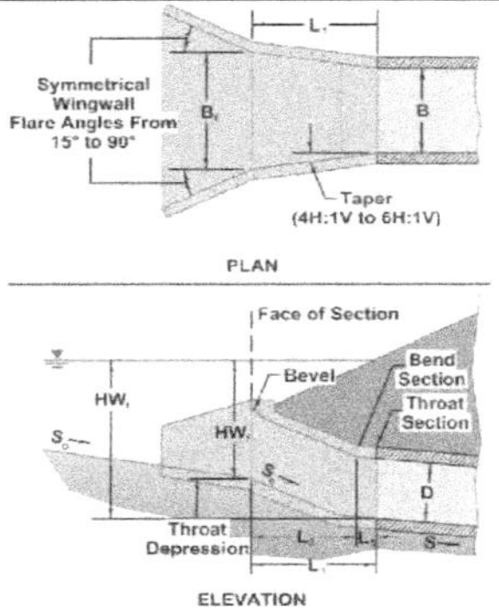

N = ___, B = ___, D = ___

INLET EDGE DESCRIPTION _____

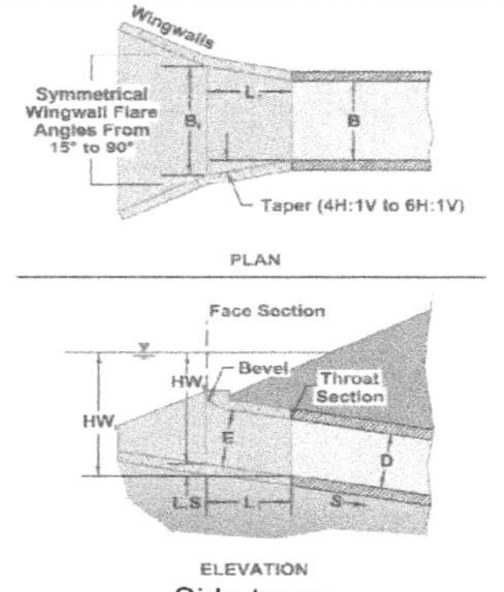

Q ( )	EL_{hi} ( )	EL. Throat Invert	EL. Face Invert (1)	HW_f (2)	$\frac{HW_i}{E}$ (3)	$\frac{Q}{B_f}$ (4)	MIN. B_f (5)	Selected B_f	MIN. L_3 (6)	L_2 (7)	Check L_2 (8)	Adj. L_3 (9)	Adj. Taper (10)	L_1 (11)	EL. Crest Inv. (12)	HW_c (12)	MIN. W (13)

SLOPE-TAPERED ONLY

SIDE-TAPERED w/ depression

COMMENTS

(1) SIDE-TAPERED : EL. FACE INVERT = EL. THROAT INVERT + 1 FT (0.3 M APPROX.)
SLOPE-TAPERED : EL. FACE INVERT = EL. STREAM BED AT FACE

(2) HW = EL_{hi} - EL. FACE INVERT

(3) 1.1 D ≥ E ≥ D; E = D FOR BOX CULVERTS

(4) FROM DESIGN CHARTS

(5) MIN. B_f = Q / (Q / B_f)

(6) MIN. L_3 = 0.5 NB

(7) L_2 = (EL. FACE INVERT- EL. THROAT INVERT) S_D

(8) CHECK $L_2 = \left[\dfrac{B_f - NB}{2}\right] \cdot$ TAPER $- L_3$

(9) If (8)>(7), ADJ. $L_3 = \left[\dfrac{B_f - NB}{2}\right] \cdot$ TAPER - L_2

(10) If (7) >(8), ADJ. TAPER = $(L_2 + L_3)$ / $\left[\dfrac{B_f - NB}{2}\right]$

(11) SIDE-TAPERED : $L = \left[\dfrac{B_f - NB}{2}\right] \cdot$ TAPER

SLOPE-TAPERED : $L_1 = L_2 + L_3$

(12) $HW_C = EL_{hi} =$ EL. CREST INVERT

(13) MIN. $W = K_u$ Q / $HW_c^{1.5}$ Where K_u = 0.35 (0.64 SI)

SELECTED DESIGN

B_f _____

L_1 _____

L_2 _____

L_3 _____

BEVELS ANGLE

b = ___ ( ) ; d = ___ ( )

TAPER ___ : 1

S_D = ___ : 1

Beginning with HY-8, water surface profile computation within the culvert barrel was adopted to refine the computation of flow depth, flow velocity, and length of barrel flowing full. The profile is determined by first establishing if the culvert slope is supercritical (inlet control) or subcritical (outlet control). Next, the tailwater is used to establish which profile to assume and at what depth to start the profile. The parameters and USGS flow type associated with each profile type are illustrated in Section 3.5.1. HY-8 inlet control procedures are discussed in Section 3.5.2. HY-8 outlet control procedures are discussed in Section 3.5.3. Application of HY-8, HEC-RAS and other public domain culvert design software is discussed in Section 3.5.4.

3.5.1 USGS Flow Types and Water Surface Profiles

HY-8 uses the flow type numbers (see Table 3.1) to help define how culvert flow is computed. The flow type numbers are based on USGS flow type numbers 1 to 6 (USGS 1968) supplemented with:

- Simplification of HW = D to indicate submergence at the inlet,
- Flow Type 7 for outlet control with M1 or M2 profile for most of the barrel with HW>D,
- Standard water surface profile indictors (S1, S2, M1, M2, H2 and H3) and
- Tailwater indicator (n, c, t, and f) for normal, critical, tailwater, and full.

Table 3.1. HY-8 Flow Type Numbers.				
Flow Type	Flow Control	Submerged Inlet HW>D	Submerged Outlet TW>D	Length Full
1	Inlet	No	No	None
5	Inlet	Yes	No	None
2	Outlet	No	No	None
3	Outlet	No	No	None
4	Outlet	Yes	Yes	All
6	Outlet	Yes	No	Most
7	Outlet	Yes	No	Part

HY-8 provides a flow type, water surface profile and tailwater indicator for every discharge as a short hand way of identifying how culvert headwater and outlet velocity are calculated. For example, a code of 1-S2n indicates that the culvert is in inlet control with headwater depth less than D, that the water surface is an S2 profile and that the outlet depth will be calculated with the S2 profile since tailwater is less than normal depth. HY-8 includes a complete listing of the culvert flow coding system.

The HY and CDS programs duplicated the manual methods and provided a solution similar to that documented in Section 3.3 Culvert Design Using Nomographs. The HY-8 software provides the option to duplicate the Section 3.3 results (USGS flow type 4), but also provides the option of computing water surface profiles to more accurately definite culvert hydraulics in both inlet and outlet control. This option is very valuable for assessing the location of the maximum shear stress within the culvert, which is required by the AOP design procedure in HEC-26. In inlet control, the S1 and S2 profiles are used to more accurately compute outlet velocity and depth for USGS flow types 1 and 5. In outlet control, the M1 and M2 profiles are used to compute USGS flow types 2, 3, 6 and 7. In all HY-8 Versions through 7.2, hydraulic jumps were only assumed to occur if an S1 curve was calculated or if the tailwater depth was greater than D. For other cases, the jump is assumed to be swept out of the culvert with an S2 profile. In Version 7.3, momentum calculations are used to more accurately determine the location of the hydraulic jump. A profile code of JS1 was added to indicate that a hydraulic jump to an S1 profile occurs in the barrel. In addition, the horizontal profiles of H2 and H3 were added. The general flow types described in Table 3.1 are discussed in the following sections.

USGS Flow Type 1 (Inlet Control). Figure 3.25 shows flow type 1 with tailwater higher than critical depth (S1t). For this case, an S1 curve is computed and used if the S1 curve extends to the face of the culvert. If the S1 curve reaches critical depth, an S2 is calculated and compared to the S1 curve. If the sequent depth for the S2 matches the S1 within the culvert, a JFt code will be shown and outlet velocity will be based on the tailwater. If the sequent depth is not reached within the culvert, the jump is assumed to be swept out and an S2 is used to compute outlet velocity (S2n). If tailwater is higher than D, the barrel flows full at the end, S1f is shown if S1 extends to the entrance. If the S1 reaches critical, JFf code will be shown to indicate that S2 curve jumps to full flow in the barrel.

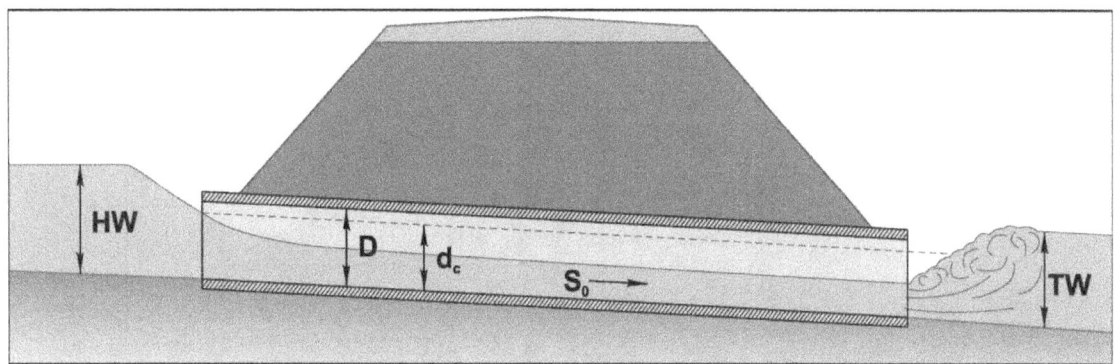

Figure 3.25. USGS flow type 1.

USGS Flow Type 5 (Inlet Control). Figure 3.26 shows flow type 5 with tailwater less than critical depth (S2n). If the Tailwater is higher than critical depth, an S1 curve is computed and used if the S1 curve extends to the face of the culvert. If the S1 curve reaches critical depth, an S2 is calculated and compared to the S1 curve. If the sequent depth for the S2 matches the S1 within the culvert, a JFt code will be shown and outlet velocity will be based on the tailwater. If the sequent depth is not reached within the culvert, the jump is assumed to be swept out and an S2 is used to compute outlet velocity. If tailwater is higher than D, the barrel flows full at the end (S1f) is shown if S1 extends to the entrance. If the S1 reaches critical, JFf code will be shown to indicate that S2 curve jumps to full flow in the barrel.

USGS Flow Type 2 (Outlet Control). Figure 3.27 shows flow type 2 with tailwater lower than critical depth. For this case (M2c), an M2 curve is computed starting at critical depth at the outlet.

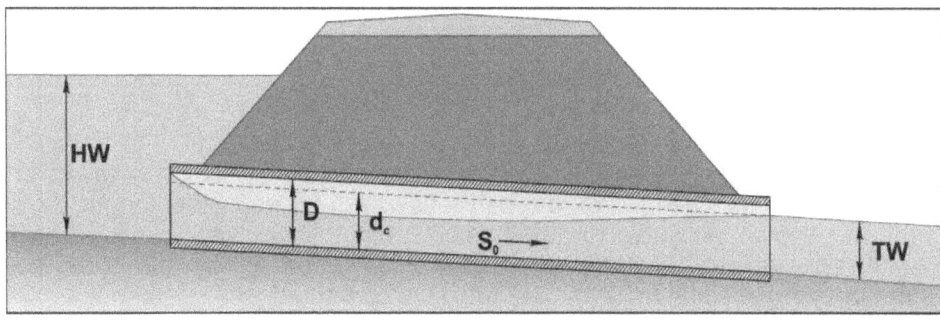

Figure 3.26. USGS flow type 5.

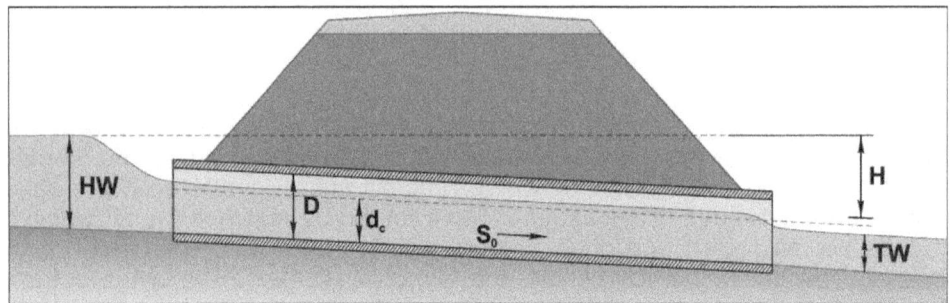

Figure 3.27. USGS flow type 2.

USGS Flow Type 3 (Outlet Control). Figure 3.28 shows flow type 3 with tailwater higher than critical depth. For this case (M1t), an M1 curve is computed starting at the tailwater depth at the outlet.

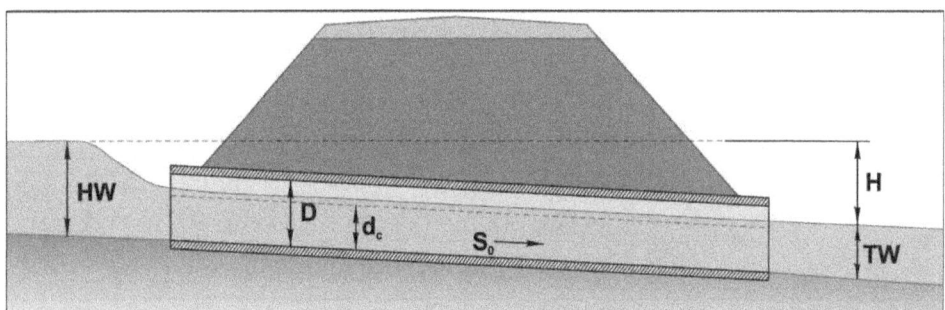

Figure 3.28. USGS flow type 3.

USGS Flow Type 4 and 6 (Outlet Control). Figure 3.29 shows flow type 6 with tailwater higher than critical depth. For this case (FFt), the barrel flows full for most of its length. An M2 curve is computed starting at the outlet to determine the length of culvert that will flow full. Outlet depth is the tailwater for FFt case and is critical depth for the FFc case. For flow type 4, the tailwater is higher than the culvert crown at the exit and the barrel flows full (FFf).

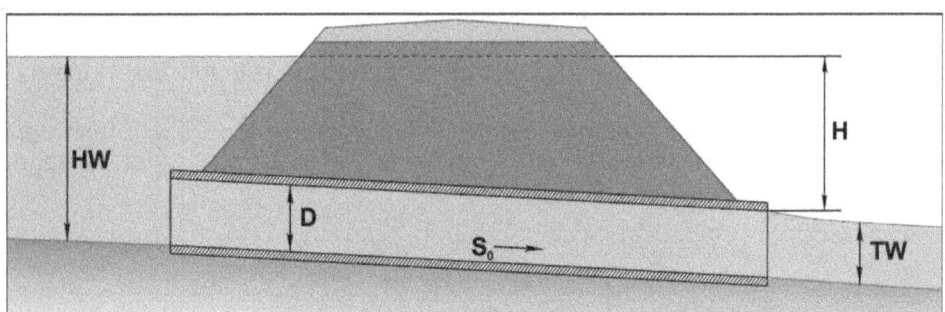

Figure 3.29. USGS flow type 4 and 6.

USGS Flow Type 7 (Outlet Control). Figure 3.30 shows flow type 7 with tailwater lower than critical depth. For this case (M2c), the barrel flows full for part of its length. An M2 curve is computed starting at the outlet to determine the length of culvert that will flow full. Outlet depth is the tailwater depth for M2t case and is critical depth for the M2c case. If the tailwater is higher than normal depth (M1t), an M1 profile is computed.

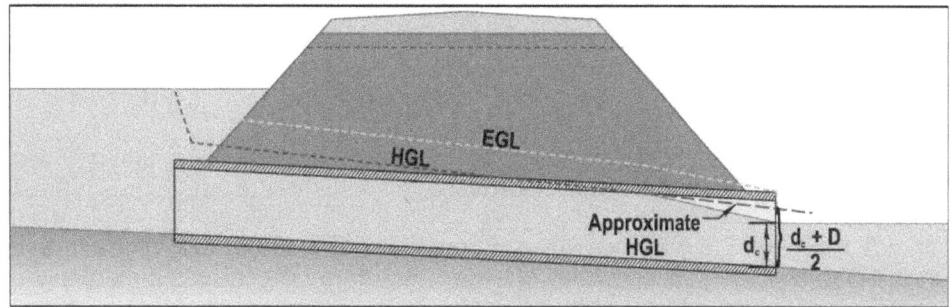

Figure 3.30. USGS flow type 7.

3.5.2 Inlet Control

HY-8 uses the polynomial equations developed for early software to represent the inlet control curves determined in the laboratory (Appendix A). For other shapes, the table of values that represent Chart 52 is used (Section 3.6). These curves are used for HW/D values from 0.5 to 3.0. For HW/D from 0 to 0.5, a general weir equation (Figure 3.15) is fit to the culvert data at HW/D = 0.5. For HW/D above 3.0, a general orifice equation (Figure 3.15) is fit to the culvert data at HW/D = 3.0.

HY-8 versions up to 7.1 used a table of section property values (discharge, area, wetted perimeter, top width, normal depth and critical depth) to interpolate design values between the values that were determined for D/10 increments of depth in the culvert. For irregular shapes, this reduced the time required for trial and error bounded search routines used to balance headwater for multiple culverts and overtopping. Starting with Version 7.2, the section properties are computed directly for each discharge. This improvement permits more accurate depth computation for the low flows used for AOP design.

The polynomial equations and nomographs were developed for a slope of 2%. As shown in Appendix A, the 2% slope means that a slope correction of -0.5S or -0.01 is included in HW/D for all inlets except mitered. For mitered inlets, a +0.7S slope correction term is used to adjust the equations. HY-8 applies an additional slope correction adjustment using the actual slope.

Culvert length is adjusted using equations that are derived in Appendix F of FHWA-IP-83-6, Structural Design Manual for Improved inlets and Culverts (FHWA 1983).

3.5.3 Outlet Control

HY-8 can duplicate the outlet control nomograph solution (USGS flow types 4 and 6) by assuming full flow in the barrel. The nomograph or full flow solution will also apply the approximate hydraulic gradeline if the tailwater is below critical depth (USGS flow type 7).

If water surface profile option is used, normal depth is compared to critical depth to determine if supercritical (inlet control) or subcritical (outlet control) profiles should be computed. Next, the tailwater depth is compared to the culvert height to determine if the culvert will flow full at the exit and to critical depth to determine if downstream control governs or if the control is at the outlet (critical depth). Based on these comparisons, the appropriate USGS flow type and water surface profile is assumed. Section 3.5.1 discusses all the possible combinations of flow type and flow profiles.

HY-8 uses Equation 3.4d to calculate outlet losses. This equation is conservative and assumes that all the velocity head is lost in the downstream tailwater pool. Version 7 provides the option of conserving outlet velocity by selecting the Utah State University (USU) transition loss (Equation 3.4e). This option should only be used if a defined channel exists downstream of the culvert.

3.5.4 Application of Public Domain Software

Section 3.5.1 through 3.5.3 has provided an overview of the hydraulic procedures used in the FHWA HY-8 culvert design software. These procedures are also substantially used in the USACE HEC-RAS software and the Nebraska DOT Broken-Back Culvert Analysis Program (BCAP) which is discussed in Section 5.7. In this section, the capabilities of each software will be discussed.

3.5.5 Application of HY-8

HY-8 is structured to be a culvert design tool. Input data are design discharge range, tailwater channel geometry, a roadway cross section and an embankment template. Any commercially available culvert alternative material and size can easily be selected and a performance curve produced that is compared to design targets. HY-8 should be used if any of the following apply:

- Crossing has only culverts and no nearby upstream or downstream structures
- Crossing is to be designed
- A tapered inlet alternate is to be considered
- An irregular shape culvert has to be considered for joint use
- Embedded culvert is being considered for AOP design
- A broken-back culvert is to be considered
- Energy dissipator design is expected

3.5.6 Application of HEC-RAS

HEC-RAS is structured to be an analysis tool of a stream reach using water surface profiles. Input data are design discharge range, a series of channel cross sections, roadway geometry, bridges and/or culvert descriptions. HEC-RAS has the same commercially available culvert alternative material and size ranges as HY-8. HEC-RAS also uses Appendix A, NBS equations, for inlet control computations. HEC-RAS and HY-8 provide similar solutions for submerged inlets and USGS flow types 4, 5, 6 and 7. For unsubmerged inlets and USGS flow types 1, 2, and 3, HEC-RAS carries the water surface profile through the structure and maintains the approach velocity. HEC-RAS should be used if any of the following apply:

- Crossing has a combination of bridges and culverts
- Crossing has either upstream or downstream structures (bridges or culverts) that affect the crossing
- A water surface profile is needed for the stream reach
- Stream reach is part of an NFIP identified floodplain

3.5.7 Application of BCAP

BCAP is structured to be a design tool for culverts that have combinations of steep and mild slopes within one culvert barrel. Input data is similar to HY-8. BCAP is limited to circular and rectangular shape culverts. BCAP uses HY-8 inlet and outlet control computations, but has the unique ability to analyze hydraulic jumps. Either HY-8 or BCAP can be used if any of the following apply:

- Crossing has a culvert on a long, steep slope
- A hydraulic jump is desirable to reduce outlet velocity

3.6 DESIGN METHODS FOR CULVERTS WITHOUT DESIGN CHARTS

Appendix A lists culvert shapes and materials that have weir and orifice coefficients that have been determined through laboratory testing. Some culvert shapes and materials do not have laboratory determined coefficients. For these culverts, design nomographs and critical depth charts cannot be developed. For example, long span, structural plate, corrugated metal conduits do not have testing and design charts. Developing design charts for all possible conduit shapes and sizes is not practical because they are so numerous and new shapes are constantly being produced. Also, the large size conduits tend to fall outside the nomograph scales. With some modification, usual culvert hydraulic techniques can be used to analyze these culverts.

3.6.1 Inlet Control

Since the inlet has not been modeled, the inlet control equations are necessarily based on hydraulic test results from similar tested conduit shapes. Appendix A contains approximate inlet control equations for nonrectangular conduits with a variety of edge conditions. In order to facilitate the design process, the appropriate inlet control equations of Appendix A have been used to develop dimensionless inlet control design curves for selected conduit shapes and edge configurations. The curves of Figures 3.31 and 3.32 are for nonrectangular, structural plate corrugated metal conduits of two basic shapes and four inlet edge conditions. Figure 3.31 is for circular or elliptical conduits with the long horizontal axis at the mid-point of the barrel. Figure 3.32 is used for high and low profile structural plate arches. Note that these figures are copies of Charts 51b and 52b and are for English Units. For SI Units, see Chart 51a and 51b, respectively in Appendix C.

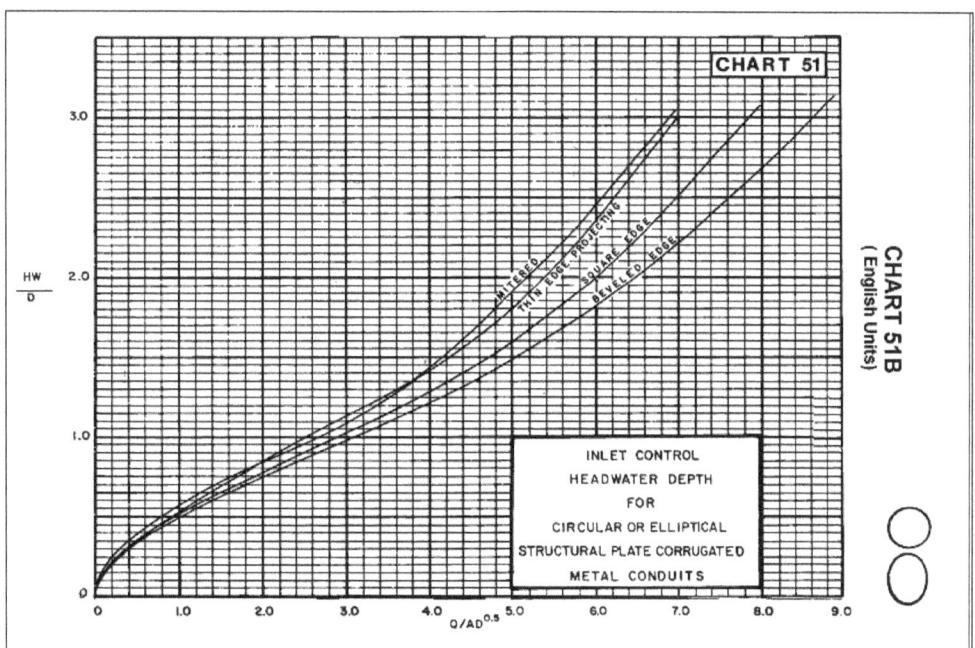

Figure 3.31. Inlet control curves - circular or elliptical structural plate corrugated metal conduits (Chart 51B).

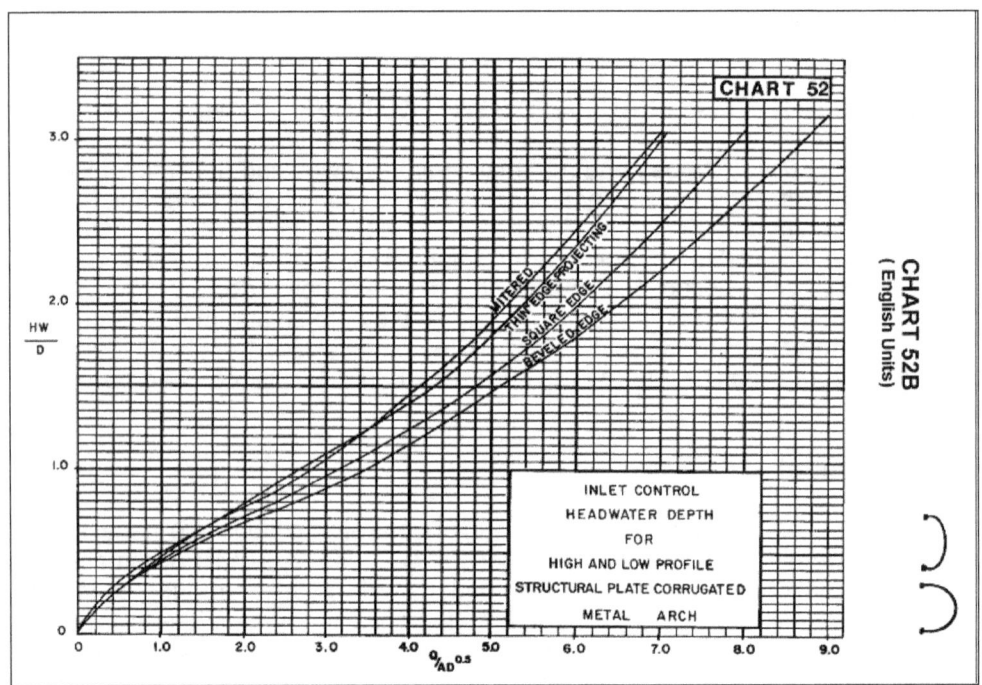

Figure 3.32. Inlet control curves - high and low profile structural plate arches (Chart 52B).

The curves in these figures are for four different inlet edge conditions:

- Thin edge projecting
- Mitered
- Square-edge
- 45-degree bevels

The horizontal axis of the chart is flow rate divided by the area times the square root of barrel height ($Q/AD^{0.5}$) and the vertical axis is headwater depth divided by barrel height (HW/D). Figure 3.31 (Chart 51b) will provide adequate results for any conduit with curved surfaces, including pipe-arches and underpasses. Figure 3.32 (Chart 52b) is used for conduits similar to arches with flat bottoms. The following table of values from Chart 52b is used by HY-8 for interpolating HW/D values for all shapes that do not have laboratory derived weir and orifice coefficients. For $Q/(AD^{.5})$ of 0.5 for a thin edge culvert, HW/D is read under A(1) as 0.3.

Q/A*D^.5 =			0.5	1	2	3	4	5	6	7	8	9
EQ	EDGE	KE	A(1)	A(2)	A(3)	A(4)	A(5)	A(6)	A(7)	A(8)	A(9)	A(10)
1	thin	0.9	0.3	0.47	0.8	1.1	1.41	1.83	2.38	3.02	3.7	4.25
2	mitered	0.7	0.35	0.5	0.78	1.05	1.46	1.92	2.47	3.07	3.7	4.35
3	headwall	0.5	0.3	0.45	0.72	0.95	1.25	1.58	2	2.5	3.07	3.63
4	bevel	0.2	0.3	0.43	0.68	0.88	1.15	1.48	1.8	2.22	2.67	3.17

3.6.2 Outlet Control

a. Partly Full Flow. Large conduits, such as long span culverts, usually flow partly full throughout their lengths. In addition, the invert of the culvert is often natural. In these situations it is advisable to perform backwater calculations to determine the headwater elevation.

The backwater calculations begin at the tailwater level or at critical depth at the culvert exit, whichever is higher. Hydraulic resistance values for the backwater calculations are contained in Hydraulic Flow Resistance Factors for Corrugated Metal Conduit (FHWA 1980). Data from that reference are included in Appendix B. Selected resistance values for natural channels are found in Table C.1 of Appendix C. Note that when the perimeter of the conduit is constructed of two or more materials, a composite resistance value should be used. Methods of calculating composite resistance values are discussed in Section 3.1.4.

b. Full Flow. If the conduit flows full or nearly full throughout its length, Equation 3.6b ($HW_o = TW + H_L - SL$) may be used to calculate the outlet control headwater depth.

H_L is the total loss through the culvert barrel which is calculated using Equation 3.1 or Equation 3.5. TW is either tailwater depth or $(d_c + D)/2$, whichever is larger. Values of critical depth for most conduits are provided in the manufacturers' information. In Equation 3.5 the hydraulic radius and velocity are full flow values. The Manning's n value is a composite value when more than one material is used in the perimeter of the conduit.

3.6.3 Discussion of Results

The inlet control headwater obtained from Figures 3.31 or 3.32 includes the approach velocity head. Therefore, credit may be taken for the approach velocity head in determining the required headwater pool depth.

In outlet control, the same limitations on use of the approximate backwater method apply as for culverts with design charts. That is, if the headwater (referenced to the inlet invert) falls between 1.2D and 0.75D, use the results with caution. For large, expensive installations, check the results using backwater calculations (Section 3.5). If the headwater falls below 0.75D do not use the approximate method. Perform backwater calculations as illustrated in the example problem in Design Guideline 2 in the Appendices.

(page intentionally left blank)

CHAPTER 4

CULVERT DESIGN FOR AQUATIC ORGANISM PASSAGE (AOP)

Chapter 3 of this manual describes the hydraulic design procedure for culverts that is focused on limiting headwater and outlet velocity to meet hydraulic criteria. In many cases, culverts designed with this focus do not provide an environment for aquatic organisms, including fish, to move through the culvert with the same success as they move up and down the unconstricted stream channel. The objective of this chapter is to describe the shift in thinking that is needed to facilitate aquatic organism passage (AOP) through a culvert when such considerations are appropriate and introduce design concepts, criteria, and procedures to facilitate providing that passage.

Most of the material in this chapter is from HEC-26, Culvert Design for Aquatic Organism Passage (FHWA 2010a). AOP culvert design should not be attempted solely based on the material in this chapter; the designer is directed to the more detailed manual as a guide for design.

4.1 AOP CONTEXT

4.1.1 New Installations Versus Retrofit

Although the goals for all AOP culverts are the same, facilitate passage through the culvert, there are differences in the approach and range of options for new culvert installations and culvert retrofits. New installations, including replacement culverts, provide the flexibility to vary the type, size, shape, slope, alignment, and bottom material within the culvert to satisfy AOP goals. Options may include open-bottom culverts, and closed-bottom culverts with the invert embedded below the level of the stream bed. Section 4.2 describes design procedures for new installations.

Conversely, with retrofits the range of options is more limited. The design procedures are different and the design objectives are generally more limited. Cost and disruption to the traveling public are the most common drivers toward retrofits. Section 4.3 briefly describes retrofit options.

4.1.2 Passage Barriers

Culverts designed without consideration of aquatic organisms may create barriers to the movements up and down the stream. Organisms need this mobility to seek food, find shelter, avoid predators, or reproduce. A culvert becomes a barrier to AOP when it poses conditions that exceed the organism's physical capabilities. For example, Figure 4.1 displays a jump barrier and Figure 4.2 displays a jump and velocity barrier. Circumstances that serve as barriers depend on species and life stage (juvenile or adult) of the organism. For fish, the focus of most AOP work to date, common obstructions include:

- Excessive water velocities
- Drops at culvert inlets or outlets
- Physical barriers such as weirs, baffles, or debris caught in the culvert barrel
- Excessive turbulence caused by flow contraction and expansion
- Low flows that provide too little depth for fish to swim

Figure 4.1. Example of a jump barrier.

Figure 4.2. Jump and velocity barrier.

4.1.3 Fish Biology

The physical capabilities of fish depend on two muscle systems intended to accommodate different modes of travel: a red muscle system (aerobic) for low-intensity activities and a white muscle system (anaerobic) for shorter, high-intensity movements. A fish that reaches exhaustion in either low-intensity or high-intensity activity will require a period of rest before continued movement. Extensive use of the white muscle system for high-intensity movements, in particular, causes extreme fatigue, requiring extended periods of rest.

Fish can fail to pass a culvert for a variety of reasons, but among the most common reasons are barriers that require muscle system activity in excess of physical capacity. An outlet drop or high velocity zone will act as a barrier when it exceeds the fish's burst swimming ability, while a long continuous section of culvert with relatively low velocity may require prolonged swimming speeds to be maintained beyond a fish's natural ability.

Life stage is also important. A design to meet the needs of a spawning adult salmon, for example, will not necessarily guarantee that a culvert will allow passage of weaker swimming juvenile salmon. In addition, although fish may be capable of specific swimming energies, it does not mean that fish will choose to expend maximum swimming energy when confronted with specific obstacles. Conversely, fish have also been observed to seek out higher depth lower velocity regions within the complex flow dynamics in a culvert to achieve passage.

4.1.4 Hydrology

Aquatic organism movement requirements vary with species and life stage, but are often related to season and stream flow conditions. For fish, culverts should allow passage for a range of flows corresponding to the timing and extent of fish movement within the channel reach. In a natural stream reach, fish respond to high flow events by seeking out shelter until passable conditions resume. During extreme low flows, shallow depths may cause the channel itself to become impassable. Generally, upper and lower thresholds bound the flow conditions at which passage must be provided and these are referred to as the high and low passage flows.

High passage flow, Q_H, represents the upper bound of discharge at which fish are believed to be moving within the stream, while low passage flow, Q_L, is the lowest discharge for which fish passage is required, generally based on minimum flow depths required for fish passage. High passage and low passage design flows are not defined in the same manner throughout the country. This variation may reflect differences in hydrology and fish species from region to region, but it may also reflect inconsistencies in defining these terms.

4.1.5 Geomorphology and Stream Stability

As a rigid structure in a dynamic environment, all culverts must be designed with channel processes in mind. Effective designs consider the channel and watershed context of the crossing location. Channels are continually evolving, and an understanding of stream adjustment potential must be addressed. Without proper consideration, well-intended plans could detrimentally affect the stream system and related habitat. For example, if a head cut is progressing upstream to the culvert location and it is neither identified nor mitigated, that instability will eventually reach the site. Depending on the type of culvert installation, the head cut will likely result in a drop at the culvert outlet, destabilization of the culvert, or both.

HEC-20 (FHWA 2012a) provides guidelines for identifying stream instability problems at highway stream crossings. This manual covers geomorphic and hydraulic factors that affect stream stability and provides a step-by-step analysis procedure for evaluation of stream stability problems. Stream channel classification, stream reconnaissance techniques, and rapid assessment methods for channel stability are summarized. Quantitative techniques for channel stability analysis, including degradation analysis, are provided, and channel restoration concepts are introduced.

4.2 DESIGN PROCEDURES FOR NEW INSTALLATIONS

New and replacement culvert installations allow for an AOP design to consider a full range of culvert characteristics in the design including the type, size, slope, alignment, and bottom material within the culvert. The state of the art in AOP design is the use of "stream simulation." That is, by applying a stream simulation technique, the designer strives to simulate the conditions of the natural stream within the culvert so that aquatic organisms may move through the culvert as they would be able to move through the stream for those conditions most critical to the organisms. Two stream simulation procedures are briefly described in this section: (1) U.S. Forest Service (USFS) Stream Simulation (USFS 2008) and (2) FHWA Stream Simulation as described in HEC-26 (FHWA 2010a). Both the USFS and FHWA methods apply to embedded closed-bottom culverts and open-bottom culverts. In either case, a natural bed material is provided within the culvert as shown in the schematic of Figure 4.3.

Because of the wide variety of aquatic organisms that may be relevant for a particular site, stream simulation procedures strive to provide passage without the need to analyze any particular species behavior or needs. To accomplish this, surrogate measures have been developed for each stream simulation method to guide the designer. Application of stream simulation design methods is often best accomplished by an interdisciplinary team of aquatic biologists, geomorphologists, and engineers.

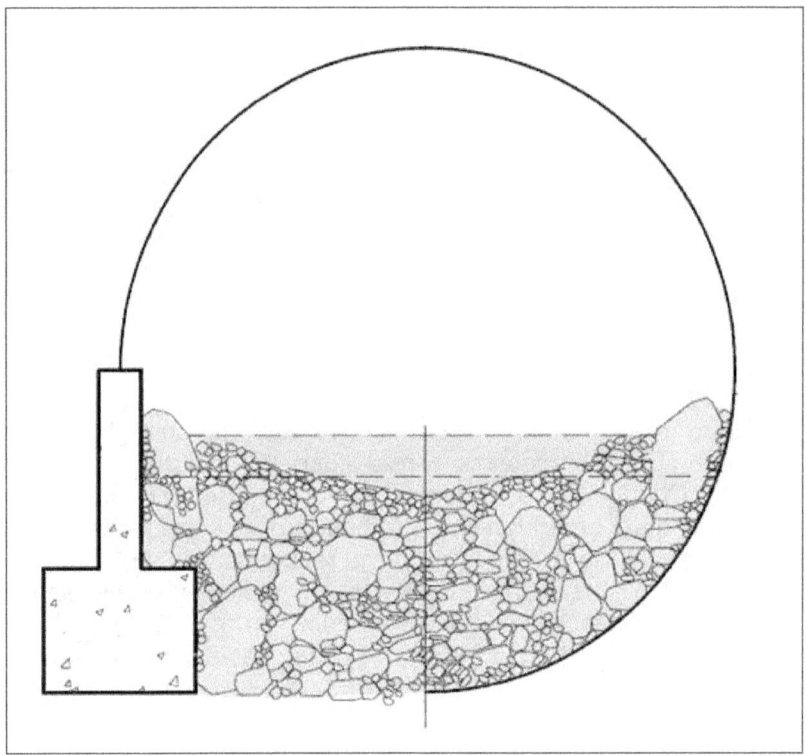

Figure 4.3. Embedment for open-bottom and closed-conduit culverts
(adapted from FSSWG 2008).

4.2.1 Surrogate Measures

Given the diverse behavior and capabilities of fish and other aquatic organisms, design procedures necessarily rely on surrogate parameters and indicators as measures for successful passage design. Some stream simulation procedures, such as the USFS approach, rely on dimensional characteristics of the stream such as bankfull width. A critique of the use of dimensional stream characteristics is that they: (1) can be difficult to identify, (2) can be highly variable within a stream reach, and (3) assume the stream is in dynamic equilibrium.

The HEC-26 procedure uses streambed sediment behavior as its surrogate parameter. The hypothesis of using sediment behavior as a surrogate parameter is that aquatic organisms in the stream are exposed to similar forces and stresses experienced by the streambed material. Therefore, if the culvert design does not alter the forces on the streambed, then it can be presumed not to alter the forces experienced by aquatic organisms. The design goal is to provide a stream crossing that has an equivalent effect, over a range of stream flows, on the streambed material within the culvert compared with the streambed material upstream and downstream of the culvert.

In the case of both surrogates, when the surrogate criteria are satisfactory, the hypothesis is that the conditions through the culvert should present no more of an obstacle to aquatic organisms than conditions in the adjacent natural channel. Although there is no requirement to evaluate specific biological requirements at a given site, such species-specific information can be incorporated in the stream simulation design process when available. A completed stream simulation open-bottom culvert is shown in Figure 4.4.

Figure 4.4. Completed open-bottom installation (from USFS 2008).

4.2.2 USFS Stream Simulation

This section is largely adapted from the USFS stream simulation guidance document (USFS 2008). The goal is to set the stage so that the simulated channel adjusts to accommodate a range of flood discharges and sediment/debris inputs, without compromising aquatic organism passage and without having detrimental effects on upstream or downstream reaches. For the simulated streambed to maintain itself through a broad range of flows, stream processes that control sediment and debris transport and maintain hydraulic diversity must function similarly to the natural channel. In other words, flows that transport sediment and debris and rework the channel bed should not be constrained or accelerated inside the crossing structure. Since bankfull flow is recognized as a good estimator of the channel-forming flow in stable alluvial rivers (Wolman and Miller 1960), the USFS adopted a working criterion that the channel inside the structure be at least as wide as bankfull width in the reference reach. Although this criterion is by no means the only characteristic of a self-maintaining stream-simulation structure, it is an essential one.

The reference reach is the key element of the USFS stream-simulation design. A natural stable reach, preferably upstream and near the project, serves as the design template. The reference reach must satisfy the physical conditions of the crossing site, especially the slope, and it must be self-sustainable inside a confined structure. In other words, flows interacting with the bed and the structure walls will dynamically maintain the streambed within the structure. In high flows, although some features of the simulated bed may be immobile, other streambed materials should mobilize and restructure themselves similarly to the natural channel; sediment transported from upstream should replace eroded material. Simulated channel design means establishing basic characteristics of the reference reach, such as gradient, cross-section shape, bank configuration, and bed material size and arrangement. The reference reach need not reflect the average conditions in the natural channel; however, the condition should not be extreme. It is presumed that if the stream within the culvert is representative of the natural channel (reference reach), passage will be as good as in the natural channel.

Bankfull width and the reference reach are the primary surrogate measures used for the USFS stream simulation procedure. The procedure includes the following primary activities:

1. Initial watershed and reach review
2. Site assessment
3. Design process
4. Construction
5. Post construction monitoring (advised)

The design process addresses the following:

1. Alignment and profile
2. Channel bed cross-section and bed materials
3. Culvert dimensions and type
4. Bed mobility and stability
5. Risk factors

Risk factors are those aspects of the design and channel dynamics that are potential causes for failure. Detailed guidance for the USFS stream simulation method is available (USFS 2008).

4.2.3 HEC-26 Stream Simulation

The HEC-26 stream simulation design procedure was developed based on a streambed stability surrogate measure rather than bankfull width to address the limitations in estimating an appropriate bankfull width in many situations. It was also developed to provide a reproducible approach based on best practices to support design decisions that often result in larger, and therefore, more expensive capital costs to provide for AOP.

The design procedure focuses on five primary variables that must be determined for each site:

1. Peak design flow, Q_P. This flow may be the Q_{25}, Q_{50}, or Q_{100} required for the site to address design flood flows. This is the flow traditionally used to design a culvert. (Q_x represents the x-year recurrence interval flow.)
2. High passage design flow, Q_H. This is the maximum discharge used for passage design. It may apply to the entire year or to a specific season.

3. Low passage design flow, Q_L. This is the minimum discharge used for passage design. It may also apply to the entire year or to a specific season.
4. Bed material characteristics. For noncohesive materials, representative bed quantities including D_{16}, D_{50}, D_{84}, and D_{95} are required. (D_x is the material size for which x percent of the bed material, by weight, is smaller.)
5. Permissible shear stress, τ_p, of the bed material.

Five fundamental tests are applied as part of the procedure. If any test is failed, design adjustments are specified. The tests are:

1. Does the culvert satisfy the peak flow requirements?
2. Is the bed material in the culvert stable (no movement or sediment inflow equals outflow) for the high passage design flow?
3. Is the bed material in the culvert stable for the peak design flow? (An anchoring layer/device below the bed material may be required to satisfy this test.)
4. Is velocity in the culvert for the high passage design flow consistent with upstream and downstream channel velocities?
5. Is depth in the culvert for the low passage design flow consistent with upstream and downstream channel depths?

The first item is the traditional test for hydraulic adequacy. All of the criteria and requirements for hydraulic sufficiency apply when designing for AOP, but will usually not be the limiting factor in the design.

The second test is the initial use of the surrogate – streambed stability. New and replacement AOP culvert installations will have a natural streambed bottom inside the culvert either through the use of an open-bottom culvert or an embedded closed-bottom culvert. It is desirable for the streambed material to be stable at the high passage flow, though this may not be possible for some channel materials, particularly a sand bed channel. The design procedure provides a framework for determining whether a design satisfies this test.

The third test also examines the streambed stability surrogate, but at the peak design flow. This test addresses the long-term performance of the streambed material, especially for an embedded culvert. At the peak flow some material may wash out of the culvert, though in some cases it may be replenished on the receding limb of the flood hydrograph as material is redeposited. In many cases, an oversized armor layer is recommended to avoid exposing the culvert invert and to enhance the redeposition process.

The fourth and fifth tests examine the velocity and depths estimated in the culvert and the upstream and downstream channel reaches. If both are within the range of values observed in the natural stream, it is reasonably concluded that if an aquatic organism can negotiate the natural channel, it can also transit through the culvert. This comparison is between ambient conditions in the stream and the culvert conditions, not with the abilities of any particular fish or other aquatic organism species.

If any of the tests fail, the design procedure offers guidance on how to alter the culvert design to satisfy all five tests. The procedure relies on best practices for estimating streambed properties including permissible shear stress and Manning's roughness. As technical understanding of these processes increases, the same framework is applied, but improved practices may be incorporated in the method.

The five tests are conducted by implementation of a 13-step design procedure that is summarized in Figure 4.5. Step 1 involves determination of the hydrologic requirements for the site for both flood flows and passage flows. The passage flows do not require determination of target species and life stages, though if they are known for a site should be used in defining the passage flows. Step 2 defines the project reach and establishes the representative channel characteristics appropriate for the design. Because it is inadvisable to place a fixed structure, such as a culvert, on an unstable stream, Steps 3 and 4 are to identify whether the stream is stable (Step 3). If not, channel instabilities are analyzed and potentially mitigated (Step 4).

In Step 5, an initial culvert size, alignment, and material are selected based on the flood peak flow. Subsequently, the stability of the bed material is analyzed under the high passage flow (Steps 6 and 7) and flood peak flow (Steps 8 and 9). If any of the criteria are not satisfied, the designer returns to Step 5 to find an alternative culvert configuration, usually larger.

Steps 10, 11, and 12 are focused on the velocity and depth in the culvert. However, these parameters are not compared with species-specific values, but rather are compared with the values upstream and downstream of the culvert insuring that if an organism can pass the upstream and downstream channel, it will also be able to pass through the culvert. If species-specific values are relevant and available for the site, they may also be incorporated into the design. Step 13 is a design review including, but not limited to, assessing design compatibility with project objectives, environmental requirements, and construction/maintenance budgets.

4.3 RETROFIT OPTIONS

Retrofits of existing culverts to facilitate AOP are sometimes required when replacement is not feasible for constructability or cost reasons. In this case the size and material of the culvert are pre-determined as is the horizontal and vertical alignment. The focus in retrofit then generally shifts to comparing the hydraulic conditions – velocity and depth – within the culvert to the movement capabilities of target species, typically fish.

Baffles, sills, and oversized bed materials are techniques that may be used inside the existing culvert barrel to slow velocities, increase depths, and provide more hydraulic diversity so that particular fish at a particular life stage during a particular season is able to negotiate through the culvert. However, the flood flow capacity of a retrofitted culvert must be checked to verify that the culvert capacity is not reduced unacceptably by adding one or more of these features. The designer must recognize that culvert retrofit for AOP is likely to be a series of compromises that attempt to provide acceptable improvements in AOP while maintaining sufficient hydraulic capacity. As with new and replacement AOP culvert design, a multidisciplinary team of biologists and engineers must work together to find the appropriate balance at a particular site.

Another retrofit option for culverts is relining. Generally performed to extend the life of a culvert before full replacement is required, culvert relining may also reduce AOP through the culvert because of increased velocities. Designers should consider whether relining could create or increase passage barriers at a site.

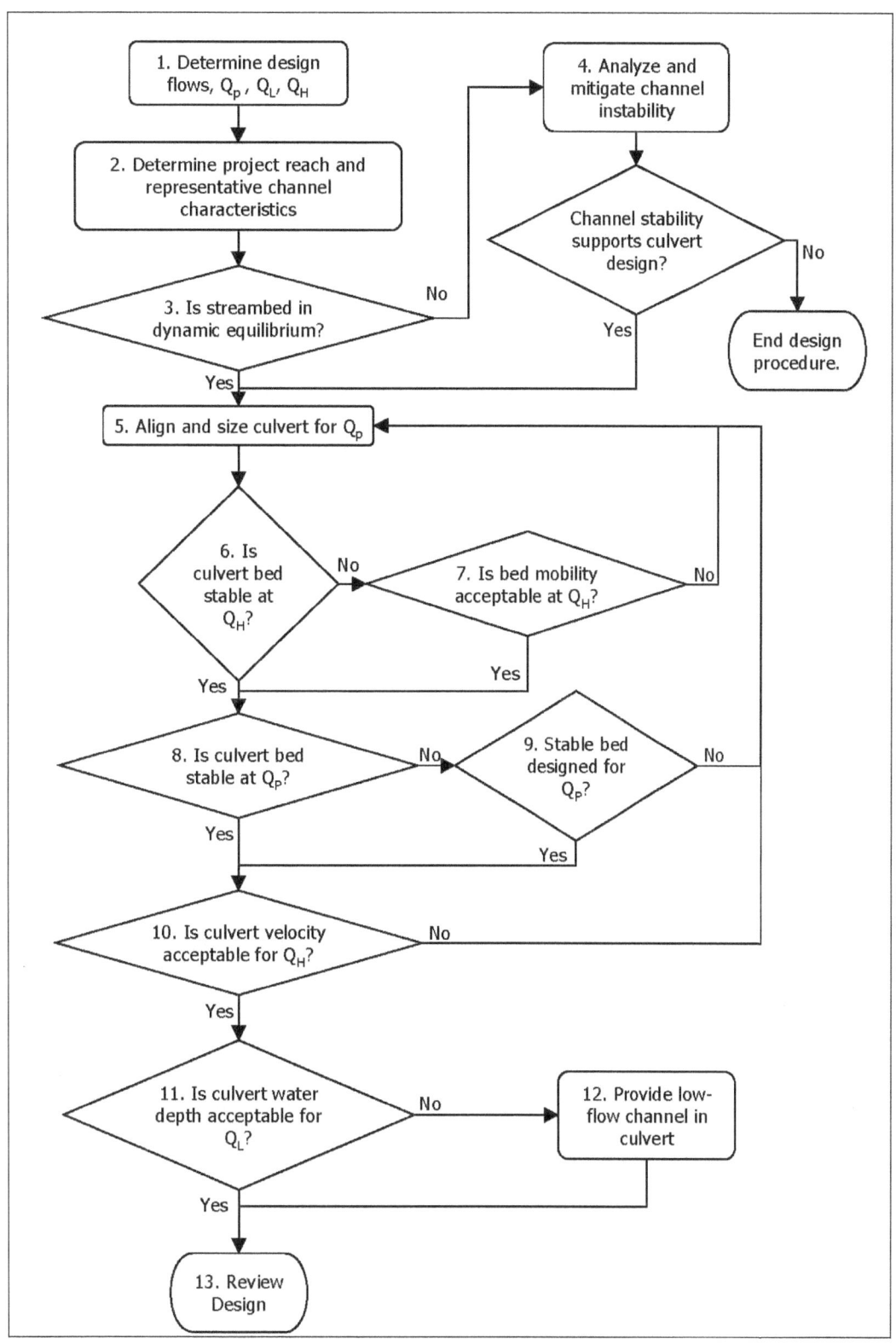

Figure 4.5. Design procedure overview.

4.4 CONSTRUCTION AND POST CONSTRUCTION

Construction of AOP culverts is subject to the same considerations of other culverts at a minimum. The allowable seasons for instream work and protection of the stream banks, vegetation and aquatic resources apply to all culvert construction. In addition, the unique nature of an embedded or open-bottom culvert carries additional considerations for constructability and streambed material placement.

The most significant constructability issue is whether the rise in the culvert provides sufficient clearance to perform work with mechanized equipment or even manual labor. In some cases, small culvert sizes must be increased so that the bed material may be effectively placed inside. In addition, the bed material must be placed so that voids are sealed, preventing interflow (water flowing within the bed rather than on the bed), and that sufficient compaction is obtained. Structural plate culverts may be used to facilitate placement of bed material within the culvert.

After construction is completed, maintenance inspections should be conducted, as with all culverts, but with added attention to the condition of the bed material. Where resources permit, passage monitoring is recommended to determine if the project passage goals are being achieved.

CHAPTER 5

OTHER CONSIDERATIONS

5.1 INTRODUCTION

Culvert design may at times encompass almost every consideration and situation related to hydrologic analysis and hydraulic engineering. While these situations may only occur in a small percentage of culvert designs, or may be specifically related to culvert design in a particular geographic region, they are important and can have a major bearing on the design process. Information in this Chapter includes flow control and measurement (Section 5.2), erosion and sediment control (Section 5.3), site related modifications such as skew (Section 5.4), structural considerations (Section 5.5), broken-back culverts (Section 5.6), storage routing (Section 5.7) and energy dissipators (Section 5.8).

Due to the wide range of topics covered, numerous references are cited to which the culvert designer may refer to for further information. Design guidelines and recommendations are only provided in an abbreviated fashion. It is the designer's responsibility to decide when further study of the specific design situation is necessary.

5.2 SPECIAL APPLICATIONS

Culverts are occasionally designed to fulfill special functions in addition to their primary function as drainage structures. For example, culverts are used as flow control and measurement devices, and can be as effective as weirs and flumes. Culverts can be designed to operate under low heads and minimize energy losses as in roadway crossings for irrigation canals. Often, culverts must be modified in order to fulfill a secondary function. Such is the case with culverts containing bends in plan or profile (broken-back), culverts containing junctions within their barrels, and certain culverts operating as siphons. These special applications are briefly discussed and design guidelines are presented in the following sections.

5.2.1 Flow Control and Measurement

Flow control structures are used to measure and control the rate of discharge in open channels. Culverts are often used as flow control structures due to the in-depth understanding of culvert hydraulics, reliable and accessible design techniques, and the availability of economical construction materials and methods. Discharge measurement and control are required in irrigation canals, stormwater management ponds, and wetland areas (Figure 5.1). In all three applications, a culvert can be used to control water flow rates, water level or flow distribution. The flow rates through the culvert are easily calculated based on the geometry of the structure and coordinated records of headwater and tailwater elevations. The routing procedures of Section 5.7 must be applied to determine the corresponding inflow into the storage pond upstream of the culvert.

Culverts located on small watersheds can be utilized as flow measurement structures to provide streamflow records. Shortly after a flood event, high water marks upstream and downstream of a culvert installation can be measured and documented. Temporary staff gages placed at the site would simplify these efforts. The peak discharge at the culvert site can then be determined. These data help to improve runoff calculation methods and aid in verifying computer models. If discharges for the entire flood event are required, a recording stage gage is required. Techniques and procedures for obtaining peak runoffs using culverts as flow measurement structures are provided by the U.S. Geological Survey (USGS 1968).

Figure 5.1. Culvert as outflow control device.

Flow control structures are also used to prevent tidal or other downstream discharges from flowing back through the culvert. If commercial check values or other devices are used at culvert outlets, the designer should add the manufacturer's recommended headloss values to Equation 3.1.

5.2.2 Low Head Installations

Low head installations are culverts which convey water under a roadway with a minimum headwater buildup and energy loss. These installations are typically found in irrigation systems where the discharge is usually steady, and the available channel freeboard and slope are small. Often the installations flow partly full over the length of the culvert. Energy losses must be minimized to transport the water efficiently. The hydraulic solution imposing the least energy loss would be to bridge the conveyance channel. However, economic considerations may require the use of a low head culvert installation.

Reduction of energy loss and headwater at a culvert installation requires an understanding of the background and theory utilized in the culvert design procedures discussed in Chapter 3. The minimal headwater rise, small barrel slope, and high tailwaters associated with these installations usually result in outlet control. Therefore, minimizing entrance, exit, and friction losses will reduce the required headwater (Equation (3.6a)). Alignment of the culvert barrel with the upstream channel helps to minimize entrance loss and takes advantage of the approach velocity head. Inlet improvements, such as beveled edges, will further reduce entrance loss. However, the hydraulic effects of further entrance improvements, such as side- and slope-tapered inlets are small in outlet control. Thus, the use of these inlets is usually not justified in low head installations. The exit loss can be reduced by smoothly transitioning the flow back into the downstream channel to take advantage of the exit velocity. Friction loss is reduced by the utilization of a smooth culvert barrel.

In analyzing low head installations flowing partly full in outlet control, backwater calculations may be necessary. Beginning at the downstream water surface (tailwater), the hydraulic and energy grade lines are defined. Outlet losses are calculated using Equation (3.4e), considering the downstream velocity. Thus, the calculations proceed upstream through the barrel, until the upstream end of the culvert is reached. At that point, inlet losses are calculated using Equation (3.4a) with the appropriate inlet loss coefficient, k_e. The inlet loss is added to the calculated energy grade line at the inlet to define the upstream energy grade line. Deducting the approach velocity head from the upstream energy grade line results in the upstream water surface elevation (hydraulic grade line).

With minor modifications, the culvert design procedures of this publication are adequate for the design of low head installations. In the usual case of outlet control, the entrance, friction, and exit losses can be obtained from the outlet control nomographs in Appendix C. If the downstream velocity is significant compared with the barrel velocity, the losses should be calculated using Equations (3.4e) instead of the outlet control nomograph. Use of Equation (3.4e) will reduce the exit losses.

It is also advantageous to consider approach and downstream velocities in the design of low head installations. Equation (3.6a) should be used instead of Equation (3.6b) to calculate headwater depth (HW$_o$) in outlet control. In inlet control, the approach velocity head should be considered to be a part of the available headwater when using the inlet control nomographs.

The HY-8 software can be used to size these structures, but currently does not include conservation of approach velocity. These structures can also be designed using the procedures in HEC-14, Section 4.2, which uses transition losses rather than entrance and exit losses. A similar approach can be applied using HEC-RAS.

Sag culverts, called "inverted siphons," are often used to convey irrigation waters under roadways (Figure 5.2). This type of culvert offers the advantage of providing adequate vertical clearance for the pipe under the roadway pavement and subgrade. A possible disadvantage of a sag culvert is clogging due to sediment. The design is not recommended for use on streams. Sag culverts require the use of bends and inclusion of their related energy losses. Losses due to bends are covered in the next section.

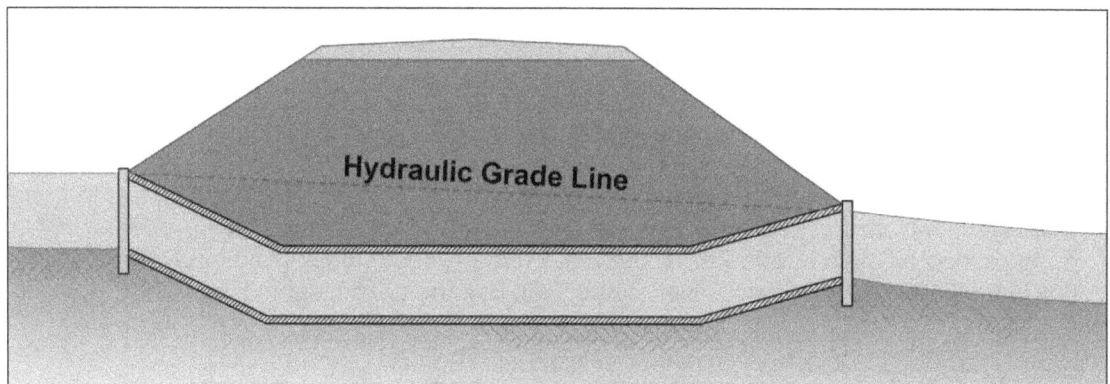

Figure 5.2. Sag culvert.

5.2.3 Bends

A straight culvert alignment is desirable to avoid clogging, increased construction costs, and reduced hydraulic efficiency. However, site conditions may dictate a change of alignment, either in plan or in profile. A change of alignment in profile is generally referred to as a "broken back" culvert and is discussed in Section 5.7. Horizontal bends may also be used to avoid obstacles (e.g. utilities) or realign the flow (Figure 5.3). When considering a nonlinear culvert alignment, particular attention should be given to erosion, sedimentation, and debris control.

Figure 5.3. Culvert with a horizontal bend (Contech).

In designing a nonlinear culvert, the energy losses due to the bends must be considered. If the culvert operates in inlet control, no increase in headwater occurs. If the culvert operates in outlet control, a slight increase in energy losses and headwater will result due to the bend losses. To minimize these losses, the culvert should be curved or have bends not exceeding 15-degrees at intervals of not less than 50 ft (15m) (AASHTO 1975). Under these conditions, bend losses can normally be ignored.

If headwater and flow considerations are critical, accurate hydraulic analysis of bend losses may be required. Bend losses are a function of the velocity head in the culvert barrel. To calculate bend losses, use the following equation

$$H_b = K_b \left(\frac{V^2}{2g} \right)$$
(5.1)

H_b is added to the other outlet losses in Equation (3.1). Bend loss coefficients (K_b) are found in many standard hydraulic references (King and Brater 1976 and Linsley and Franzini 1979). Linsley and Franzini suggest the coefficients in Table 5.1 for bend losses in conduits flowing full which are similar to those found in other publications. These coefficients have been refined by Malone and Parr (2008) in a more recent study for box culverts that are not flowing full. This research also illustrates how to use the coefficient in HEC-RAS.

Table 5.1. Loss Coefficients for Bends.			
	Angle of Bend in Degrees)		
Radius of Bend / Pipe Diameter	90°	45°	22.5°
1	0.50	0.37	0.25
2	0.30	0.22	0.15
4	0.25	0.19	0.12
6	0.15	0.11	0.08
8	0.15	0.11	0.08

5.2.4 Junctions

Flow from two or more separate culverts or storm sewers may be combined at a junction into a single culvert barrel. For example, a tributary and a main stream intersecting at a roadway crossing can be accommodated by a culvert junction (Figure 5.4). A drainage pipe collecting runoff from the overlying roadway surface and discharging into a culvert barrel is an example of a storm sewer/culvert junction. Loss of head may be important in the hydraulic design of a culvert containing a junction. Attention should be given to streamlining the junction to minimize turbulence and head loss in supercritical flow. Also, timing of peak flows from the two branches should be considered in analyzing flow conditions and control. Loss of head due to a junction is not of concern if the culvert operates in inlet control, but the junction must be streamlined to avoid causing a hydraulic jump to occur.

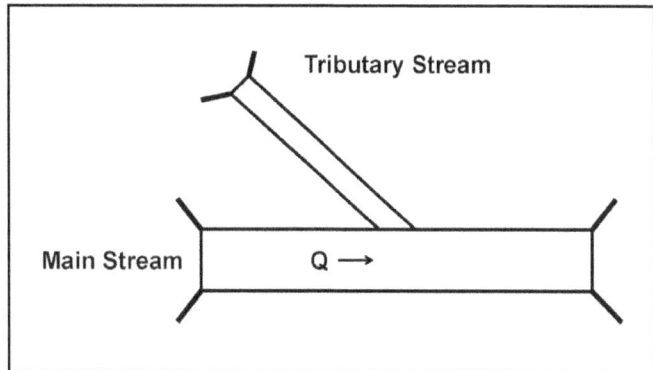

Figure 5.4. Culvert junction.

For a culvert barrel operating in outlet control and flowing full, the junction loss can be calculated recommended in HEC-22 (FHWA 2009a) based on the derivations provided in FHWA (1979). This loss is then added to other outlet control losses in Equation (3.1).

Junctions of culverts with natural bottoms should be avoided if at all possible given erosion concerns. If a junction is used, very gradual alignment, selective invert paving, and strategically placed energy dissipaters within the culvert should be considered.

5.2.5 Siphons

A siphon is a water conveyance conduit which operates at subatmospheric pressure over part of its length. Some culverts act as true siphons under certain headwater and tailwater conditions, but culverts are rarely designed with that intention (Figure 5.5).

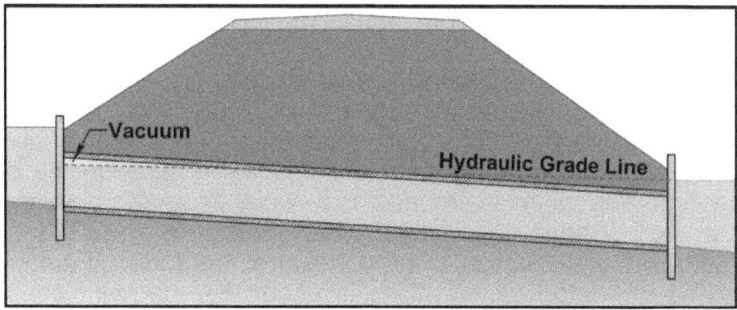

Figure 5.5. Subatmospheric pressure in culverts.

Contrary to general belief, a culvert of constant cross section on a uniform grade may act as a true siphon under certain conditions (Figure 5.5). This was demonstrated by tests at the University of Iowa and later in the NBS, 2nd Progress Report (NBS 1956). However, the additional capacity generated by the siphoning action was sporadic and could be interrupted by any number of changing flow conditions. Such conditions that would permit the admission of air include rapidly declining headwater or tailwater levels, vortices, and entrapment of debris. Since the added capacity was not dependable, it should not be considered in culvert design. In those situations when siphoning does occur culvert performance would be better than design estimates, which is consistent with the overall concept of minimum performance used in the FHWA culvert design procedure (see Section 3.1.1).

Culverts with vortex suppressors may act as siphons under conditions of high headwater. The dependability of such devices under most culvert flow conditions is open to question, and vortex suppressors may be a safety hazard. Therefore, the use of vortex suppression in culvert design applications is not recommended.

Flared-siphon culverts may also act as true siphons. A flared-siphon culvert has an outlet which diverges, much like a side-tapered inlet. The Venturi (expanding tube) principle is used to salvage a large part of the kinetic energy and thereby increase the culvert capacity. The State of California was experimenting with these designs in the early 1950s. Obviously, submergence of the outlet is necessary to achieve the siphoning action. Presumably, the added capacity was not dependable, and their design is rare. However, Cottman (1981) and Apelt (1981) have combined this concept with the slope-tapered inlet concept to produce hydraulically efficient minimum energy culverts and bridges.

Sag culverts are often referred to as "inverted siphons" even though the hydraulic grade line does not intersect the crown of the conduit at any point when the conduit is flowing full. Hence, no portion of the barrel is operating below atmospheric pressure and the name is a misnomer. Sag culverts were discussed in Section 5.2.2.

5.2.6 Baffles

Simulating natural stream bottom conditions in a culvert, typically by constructing open bottom culverts or embedded culverts, is the most desirable design option to accommodate AOP. However, baffles are sometimes used in existing culverts to improve AOP performance when complete culvert replacement is not feasible (Figures 5.6 and 5.7). Baffles have also been used to retain bed material to protect the invert from abrasion damage, particularly in channels that transport large bed material such as cobbles and boulders.

The hydraulic design of culverts with baffles is accomplished by modifying the friction resistance of the barrel in outlet control to account for the resistance of the bed material or the resistance imposed by the baffles if no bed material is retained. HEC-14 provides equations and procedures for estimating the hydraulic loss due to regularly spaced, horizontal baffles. A pair of horizontal baffles that are angled either upstream or downstream should be treated as a single perpendicular baffle when applying the equations and slots should be ignored. The highest composite n value is then used in the outlet control calculations outlined in Chapter 3. The increased resistance equations are also available in HY-8 in the energy dissipator option. For inlet control, the reduced area of the entrance due to the baffles and any edge modifications is used. If a standard configuration has been adopted (e.g. Figure 5.7), a laboratory study such as Thurman and Horner-Devine (2007) is recommended to determine the hydraulic characteristics of the weirs.

Figure 5.6. Alternating baffles.

Figure 5.7. Angled baffles (from FishXing Users Manual).

5.2.7 Median Drainage

Median drainage is sometimes vertically dropped into a cross drainage culvert by including a precast or prefabricated tee section or by providing an opening while forming for a cast-in-place structure. How the median drainage discharge is accommodated depends on whether the culvert is operating in inlet or outlet control and how large the median discharge is in comparison to the culvert discharge. In inlet control, median discharges substantially less than the culvert discharge (e.g. 10% or less) do not have to be considered in sizing the culvert. Larger medium discharges should be added to culvert flow so that the combined discharge normal depth can be compared to the height of the culvert. It may be necessary to construct a divider to merge the flows so that a hydraulic jump is not triggered (see HEC-14, FHWA 2006a). In outlet control, median discharges should be added to the culvert design discharge to determine design headwater and culvert size. If the median discharge is substantial, consider designing the installation as a junction (see Section 5.2.4).

5.2.8 Drop Inlets

A drop inlet is a box that is constructed at the entrance of a culvert because of grade control or right-of-way considerations that limit the ability to place a typical culvert inlet. They are commonly used with ditch relief culverts to discharge runoff to the opposite side of the road in areas of steep relief. Figure 5.8 illustrates one configuration using a curb opening inlet when a curb section and a paved gutter/shoulder are present. Others can include the use of grates in the bottom of a ditch section.

In flat terrain, these structures are designed using the surface drainage procedures of HEC-22 (FHWA 2009a) and the FHWA Hydraulic Toolbox. In steep terrain, the culvert will probably operate in inlet control. The inlet control charts of this publication can be used to size the culvert if a ponded condition can be assumed in the drop inlet box. In the case illustrated, the box is about D by 3D for a minimum 24 inch (600 mm) pipe. If a ponded condition cannot be assumed or a small box that is less than D by 2D has to be constructed, standard weir and orifice coefficients should be used to check the capacity of the outlet culvert. A critical factor in the design of any drop inlet is making sure the inlet configuration is adequate to intercept the design flow. The interception capacity of various inlet types can be evaluated using procedures described in HEC-22 and the FHWA Hydraulic Toolbox.

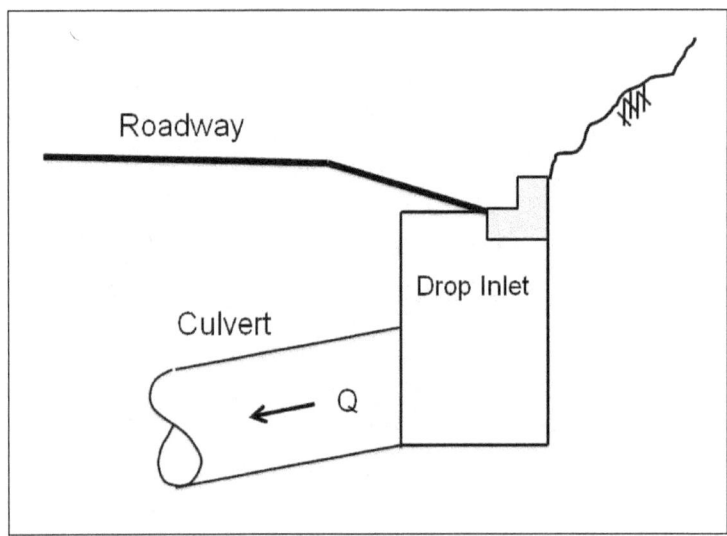

Figure 5.8. Drop inlet to straight culvert.

5.3 EROSION AND SEDIMENTATION

Natural streams and manmade channels are subject to the forces of moving water. Pressure, velocity, and centrifugal forces can be significant depending on the depth of flow, and the slope and sinuosity of the water course. Lateral and vertical migration of the stream is the result with the continuous occurrence and dynamic interplay of erosion and sedimentation, as described in HEC-20 (FHWA 2012a). This process, referred to as fluvial geomorphology, is accelerated during storm events when stream depths and velocities are high. Inserting a culvert into this dynamic environment requires special attention to the effects of these natural phenomena on the culvert and the effects of the culvert on the stream channel. Past experience has shown significant problems, including erosion at the inlet and outlet, and sediment buildup in the barrel.

5.3.1 Scour at Inlets

A culvert barrel normally constricts the natural channel, thereby forcing the flow through a reduced opening. As the flow contracts, vortices and areas of high velocity flow impinge against the upstream slopes of the fill and may tend to scour away the embankment adjacent to the culvert. In many cases, a scour hole also forms upstream of the culvert floor as a result of the acceleration of the flow as it leaves the natural channel and enters the culvert.

Upstream slope paving, channel paving, headwalls, wingwalls, and cutoff walls help to protect the slopes and channel bed at the upstream end of the culvert. Figure 5.9 depicts a culvert with a headwall, wingwalls and riprap protecting the inlet against scour.

Figure 5.9. Inlet protection based on headwall, wingwalls and riprap.

5.3.2 Scour at Outlets

Scour and erosion at culvert outlets is a common occurrence (Figure 5.10). The natural channel flow is usually confined to a lesser width and greater depth as it passes through a culvert barrel. An increased velocity results with potentially erosive capabilities as it exits the barrel. Turbulence and erosive eddies form as the flow expands to conform to the natural channel. However, the velocity and depth of flow at the culvert outlet and the velocity distribution upon reentering the natural channel are not the only factors which need consideration. The characteristics of the channel bed and bank material, velocity and depth of flow in the channel at the culvert outlet, and the amount of sediment and other debris in the flow are all contributing factors to scour potential. Due to the variation in expected flows and the difficulty in evaluating some of these factors, scour prediction is subjective.

Erosion in the vicinity of a culvert outlet can be the result of local scour or stream degradation, or both occurring simultaneously. Local scour produces a scour hole at the culvert outlet that is the result of high exit velocities. The effects of local scour only extend a limited distance downstream (Figure 5.10). Coarse material scoured from the circular or elongated hole is typically deposited immediately downstream, often forming a low bar. Finer material is transported further downstream. The dimensions of the scour hole can change due to sedimentation during low flows and the varying erosive effects of storm events. The scour hole is generally deepest during passage of the peak flow and can partially refill as flows diminish. Methods for estimating local scour hole dimensions are found in HEC-14 (FHWA 2006a).

Figure 5.10. Scour at culvert outlet.

In contrast, stream degradation extends over long distances and time frames as a result of natural and manmade changes in the watershed (HEC 20, FHWA 2012b). Degradation can be caused by a culvert or be completely independent of the culvert. For example, a linear embankment running transverse to overland flow usually concentrates flow at culvert locations, increasing the discharge over natural conditions. This concentration of flow can initiate deep and large scale degradation downstream of the culvert that can extend great distances, particularly in arid regions. Degradation can also occur from changes in the watershed that are unrelated to culverts. Examples include reduced sediment supply, channel straightening and bed lowering in channels downstream of the culvert location. Figure 5.11 illustrates an example of baseline lowering where the downstream main stem channel degraded and initiated a headcut in multiple tributary channels, in this case endangering a culvert crossing on one of those tributaries. The identification of a degrading stream is an essential part of the original site investigation (see Section 2.3.2). Methods for evaluating degradation are provided in HEC-20.

Figure 5.11. Stream degradation downstream of culvert outlet (from Utah DOT).

Since outlet protection is expensive, many state DOTs use riprap aprons (HEC-14, Section 10.2) to provide a minimum amount of protection for small culverts. The riprap apron is intended to withstand damage from small storm events and is typically a standard practice unless field investigation at similar culverts in the vicinity indicates that such protection is not needed. Providing scour protection for large floods or when more serious erosion problems exist requires protection measures designed based on HEC-14.

5.3.3 Sedimentation

The companion problem to erosion is sedimentation. Most streams carry a sediment load and deposit this load when their velocities decrease. Therefore, barrel slope less than the natural channel and roughness greater than the channel are key indicators of potential problems at culvert sites. Other important factors in sedimentation processes are the magnitude of the discharge and the characteristics of the channel material.

Culverts which are located on and aligned with the natural channel generally do not have a sedimentation problem. A stable channel is expected to balance erosion and sedimentation over time; a culvert resting on such a channel bed behaves in a similar manner. In a degrading channel, erosion, not sedimentation, is a potential problem. However, a culvert located in an aggrading channel may encounter some sediment accumulation (Figure 5.12).

Stream channel aggradation and degradation, and characteristics of each type of stream are discussed in Section 2.3.2 based on information from HEC-20 (FHWA 2012a). Fortunately, storm events tend to cleanse culverts of sediment when increased velocities are experienced. Helical corrugations tend to promote this cleansing effect if the culvert is flowing full.

Figure 5.12. Excessive sediment deposition in culvert.

Certain culvert situations are more prone to sedimentation problems. For example, roadway alignment often follows natural benches in the terrain, resulting in a grade break to a flatter slope at the inlet of the culvert. Sediment deposition problems are also common in multibarrel installations and culverts built with depressions at the entrance. Culverts with more than one barrel may be necessary for wide shallow streams and for low fills. It is well documented that one or more of the barrels will accumulate sediment, particularly the inner barrel in a curved stream alignment. It is desirable for these installations to be straight and aligned with the upstream channel. Culverts built with an upstream depression possess a barrel slope which is less than that of the natural channel. Sedimentation is the likely result, especially during times of low flow. However, self-cleaning usually occurs during periods of high discharge. Both design situations should be approached cautiously with an increased effort in the field investigation stage to obtain a thorough knowledge of stream characteristics and bed-bank materials.

5.4 SITE RELATED MODIFICATIONS

A good culvert design is one that limits the hydraulic and environmental stress placed on the existing natural water course. This stress can be minimized by utilizing a culvert which closely conforms to the natural stream in alignment, grade, and width. Often the culvert barrel must be skewed with respect to the roadway centerline to accomplish these goals (Section 5.4.1). Skewed inlet alignment is also quite common (Section 5.4.2). Multiple barrels are used in wide, shallow streams to accommodate the natural width of the stream (Section 5.4.3). The roadway embankment may be designed to handle the effects of overtopping (Section 5.4.5) or may have relief openings (Section 5.4.6).

5.4.1 Skewed Barrels

The alignment of a culvert barrel with respect to the roadway centerline is referred to as the barrel skew angle. A culvert aligned normal to the roadway centerline has a zero barrel skew angle. For any other alignment, the barrel skew angle is the angle from a line normal to the highway to the culvert centerline. Right or left skew indicates where the culvert ends are in reference to the barrel zero skew angle (Figure 5.13). Left and right is applied from the roadway centerline looking either upstream or downstream.

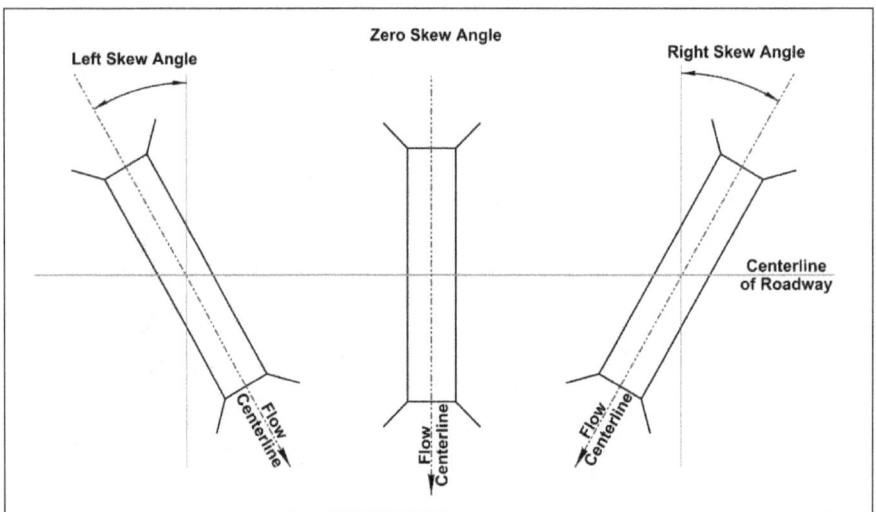

Figure 5.13. Barrel skew angle.

The culvert barrel is commonly designed to conform to the existing stream channel's alignment and grade. The advantages of this design practice include a reduction of entrance loss, equal depths of scour at the footings, less sedimentation in multi-barrel culverts, and less excavation. The disadvantages of this design procedure are that the inlet may be skewed with respect to the culvert barrel and the culvert will be longer.

5.12

Modifications to reduce the barrel skew angle and shorten the culvert barrel may produce a more economical solution in some situations. Chapter 2 contains a discussion of alternative culvert location procedures as related to culvert length.

5.4.2 Skewed Inlets

The angle from the culvert face to a line normal to the culvert barrel is referred to as the inlet skew angle (Figure 5.14). The inlet skew angle varies from 0-degrees to a practical maximum of about 45-degrees. The upper limit is dictated by the difficulty in transitioning the flow from the stream into the culvert.

Culverts which have a barrel skew angle often have an inlet skew angle as well. This is because headwalls are generally constructed parallel to a roadway centerline to avoid warping of the embankment fill (Figure 5.15).

Skewed inlets slightly reduce the hydraulic performance of the culvert under inlet control conditions. The differences are minor and are incorporated into the inlet control nomographs for box culverts (Charts 11 and 12). As an illustration of the minor effects of inlet skew, comparisons of flow capacity were made on a single barrel 6 ft by 6 ft (1829 mm by 1829 mm) box culvert with various inlet skew angles operating in inlet control (Table 5.2a, b).

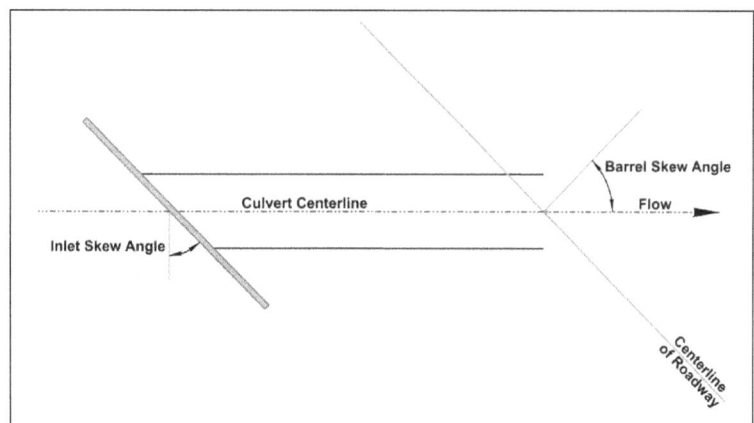

Figure 5.14. Inlet skew angle.

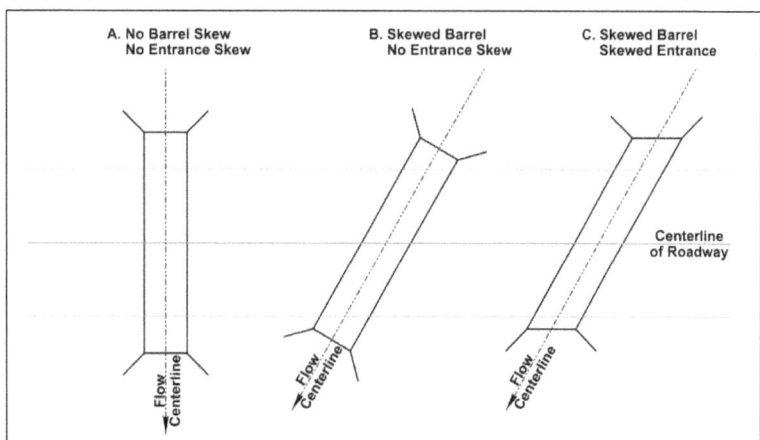

Figure 5.15. Barrel and inlet skew.

Table 5.2a. Effect of Inlet Skew Angle on Flow Capacity of (6 ft by 6 ft) Box Culvert.[1] (Flow in ft³/s, Skew Angle in Degrees)				
	SKEW	SKEW	SKEW	SKEW
HEADWATER	0°	15°	30°	45°
3 ft	85 ft³/s	85 ft³/s	82 ft³/s	80 ft³/s
6 ft	240 ft³/s	234 ft³/s	228 ft³/s	222 ft³/s
9 ft	396 ft³/s	396 ft³/s	390 ft³/s	384 ft³/s
[1]Values from Chart 11B, Appendix C				

Table 5.2b. Effect of Inlet Skew Angle on Flow Capacity of 1829 mm x 1829 mm Box Culvert.[1] (Flow in m³/s, Angle in Degrees)				
	SKEW	SKEW	SKEW	SKEW
HEADWATER	0°	15°	30°	45°
0.91 m	2.40 m³/s	2.407 m³/s	2.322 m³/s	2.265 m³/s
1.83 m	6.796 m³/s	6.626 m³/ s	6.456 m³/s	6.286 m³/s
2.74 m	11.214 m³/s	8.382 m³/s	11.044 m³/s	10.874 m³/s
[1]Values from Chart 11B, Appendix C				

Inlet skew should be avoided for culverts with tapered inlets and for multiple barrel culverts. Structural design complications result when a tapered inlet is skewed (FHWA 1983). Both tapered inlets and multiple barrel culverts perform better with the inlet face oriented normal to the barrel. The interior walls of multiple barrel culverts may promote sedimentation and unequal flow in some barrels when the inlet is skewed. The embankment fill should be warped to fit the culvert when avoiding inlet skew (Figure 5.16).

Figure 5.16. Fill warped to avoid inlet skew.

5.4.3 Multiple Barrels

Multiple barrel culverts may be necessary due to site conditions, stream characteristics, or economic considerations (Figure 5.17). Roadway profiles with low fills often dictate the use of a series of small culverts. Multiple barrel culverts are also used in wide, shallow channels to limit the flow constriction. To accommodate overbank flood flows, relief culverts with inverts at the flood plain elevation are occasionally used. Multiple barrel box culverts are more economical than a single wide span because the structural requirements for the roof of the long span are costly.

Figure 5.17. Multiple barrel box culvert.

The most significant problems associated with the use of multiple barrel culverts are sedimentation and debris. In alluvial channels, normal flows tend to pass through one of the barrels, while sediment and debris collect in the others. To reduce this problem, one barrel is installed at the flow line of the stream and the others at higher elevations (Figure 5.18). This will encourage the flow and sediment to follow the lower barrel. Sediment and debris accumulation in the other barrels will be reduced since the barrels will only be used to convey higher than normal flows.

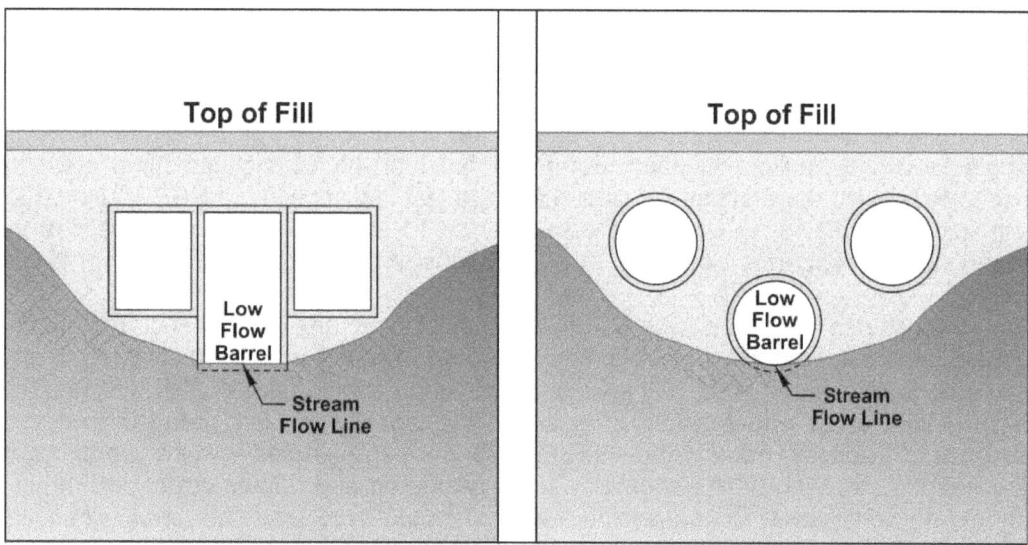

Figure 5.18. Multiple barrel culverts with one low flow barrel.

The nomographs of this publication can be used to determine the capacity of multiple barrel culverts with only minor alterations. The nomographs provide the culvert discharge rate per barrel for pipes or the flow per foot (meter) of span width for box culverts. For multiple barrels with identical hydraulic characteristics, the total discharge is assumed to be divided equally among the barrels. An iterative procedure or development of a combined performance curve is required for culverts with dissimilar barrels or invert elevations (see

Section 3.5). The discharge of the component barrels must add up to the total peak design flow at a common headwater elevation. For multiple barrel installations with bevel-edged inlets, the bevels are sized on the basis of the total clear width of the barrels. No more than two barrels may be used with tapered inlets using the design procedures of this manual.

5.4.4 Barrels on Steep Slopes

Based on open channel hydraulics and the Froude number, hydraulically mild and steep slopes are related to the occurrence of subcritical and supercritical flow, respectively. However, culverts on extremely steep slopes can create additional complications beyond that typically encountered in supercritical flow. Such conditions are similar to spillways as described by the USBR (1987):

> "When culvert spillways placed on steep slopes flow full, reduced or negative pressures prevail along the boundaries of the conduit. Large negative pressures may cause cavitation to the surfaces of the conduit or even its collapse. Where cracks or joints occur along the low-pressure regions, there is also the possibility of drawing in the soil surrounding the conduit. Culvert spillways, therefore, should not be used for high-head installations where large negative pressures can develop. Furthermore, the transition flow phenomenon, when the flow changes from partial flow to full stage, is accompanied by severe pulsations and vibrations that increase in magnitude with increased culvert fall. For these reasons, culvert spillways should not be used for hydraulic drops exceeding 25 feet."

Drops of 25 ft (7.6 m) and more are not uncommon for culverts in steep terrain. Applying the USBR guidance, the designer should be conservative and make sure that the culvert will operate in inlet control at least through the check discharge to avoid the barrel flowing full. In addition, any dissipators that are considered should be external and have no chance of causing the barrel to flow full.

Most culverts are constructed on slopes that are less than 10% and the effects of slope are neglected in hydraulic equations. For these culverts, no special structural measures are used. However, some states do require that the last few sections of RCP be bolted together with pipe ties (tie rods) which are also referred to as u-bolts. These guidelines are also reasonable for moderate slopes (10% to 20%) if they do not exceed 25 ft (7.6 m) of drop.

For steep slopes (over 20%), the USBR hydraulic caution and additional structural measures should be considered. Cast-in-place concrete boxes are the best alternative for steep slopes since no joints are required. Otherwise, construction should start at the downstream end to assure tight joints and the use of gaskets is encouraged. For metal pipe delivered in sections, the longest transportable sections should be requested to minimize joints. If joints are needed, coupling bands should be fully corrugated and double width, and thrust blocks should be considered. Concrete pipe sections should have pipe ties for every section, and the use of keys cast into the pipe are also recommended.

5.4.5 Embankment Overtopping

Embankment overtopping is an accepted result of highway crossings of floodplains. The overtopping can be infrequent as in the case of highways on the National Highway System (NHS) which are designed not to overtop for the 50-year flood event, or can be frequent for highways that are designed as low water crossings (Section 1.2.6). If the overtopping is infrequent, overtopping mitigation measures are not used and flood damage is repaired after flood events. If the overtopping is frequent, the potential damage of overtopping can be mitigated using the countermeasure design guidelines in HEC-23 (FHWA 2009b):

- Downstream fill slope can be protected using Design Guideline 5 – Riprap Design for Embankment Overtopping, or

- Embankment itself can be constructed of erosion resistant material using Design Guideline 7 – Soil Cement.

5.4.6 Relief Opening

Relief openings provide additional capacity to supplement the main channel crossing. Culverts are often used as relief opening structures. For wide floodplains, relief culverts may be necessary at side channels, including those that flow on a regular basis as well as those that only see runoff during large, infrequent flood events. Sometimes relief culverts may not even be located at a defined channel, but are necessary to pass overbank flow during a large flood. Relief culverts may also be provided in a high fill as an emergency spillway to provide an alternative path for extreme flood events. Note that high fill slopes should also be evaluated for slope failure by a geotechnical engineer if flood routing indicates a pond height or duration that will cause saturation of the fill. Hydraulic analysis and design of a floodplain crossing with relief openings is based on summing the performance curves of the separate culverts or by using HY-8 software.

5.5 STRUCTURAL CONSIDERATIONS

Proper structural design is critical to the performance and service life of a culvert. The structural design of a highway culvert begins with the analysis of moments, thrusts, and shears caused by embankment and traffic loads, and by hydrostatic and hydrodynamic forces. The culvert barrel, acting in harmony with the bedding and fill, must be able to resist these sizeable forces (Section 5.5.1). Anchorage is sometimes needed at inlets to prevent flotation (Section 5.5.2). Headwalls, wingwalls and cutoff walls are often required to maintain the structural integrity of a culvert barrel and the embankment. Piping and seepage must be prevented or minimized to prevent culvert material failure from hydrostatic pressure and embankment failure from loss of material (Section 5.5.4).

5.5.1 Structural Analysis

Most highway culverts do not have to be structurally designed because standard specifications exist in all State DOTs that cover commonly used spans, rises and materials. These design aids have been developed to be consistent with AASHTO and ASTM specifications. If a proposed installation requires a special design, a structural engineer should be consulted. Tables, charts, and formulas are available from manufacturers to streamline the process of structural design. The "Structural Design Manual for Improved Inlets and Culverts" (FHWA 1983) provides:

- Methods for hand calculation for reinforced concrete box (RCB), pipe (RCP), and ellipse (RCPE) and for corrugated metal pipe (CMP) and ellipse (CMPE)
- Computer solutions for RCB, RCP, RCPE, CMP, and CMPE
- Structural design of tapered inlets
- Typical details for headwalls, wingwalls, side-tapered and slope-tapered culverts

5.5.2 Flotation and Anchorage

Flotation is the term used to describe the failure of a culvert due to the tremendous uplift forces caused by buoyancy. The buoyant force is produced when the pressure outside the culvert is less than the pressure in the barrel. This occurs in a culvert in inlet control with a submerged upstream end. The phenomenon can also be caused by debris blocking the culvert end or by damage to the inlet. The resulting uplift may cause the outlet or inlet ends of the barrel to rise and bend. Occasionally, the uplift force is great enough to dislodge the embankment. Generally, only flexible barrel materials are vulnerable to failure of this type because of their light weight and lack of resistance to longitudinal bending. Large, projecting or mitered corrugated metal culverts are the most susceptible (Figure 5.19). In some instances, high entrance velocities will pull the unanchored inlet edges into the culvert barrel, causing blockage and additional damage. Events have been recorded in which the culvert barrel has been turned inside out by the forces of the flow.

Figure 5.19. Unanchored thin edge projecting.

A number of precautions can be taken by the designer to guard against flotation and damages due to high inlet velocities. Steep fill slopes which are protected against erosion by slope paving help inlet and outlet stability (Figure 5.20). Large skews under shallow fills should be avoided. Rigid pipe susceptible to separation at the joints can be protected with commercially available tie bars.

Figure 5.20. Slope paving around a mitered inlet.

When these precautions are not practical or sufficient, anchorage at the culvert ends may be the only recourse. Anchorage is a means of increasing the dead load at the end of a culvert to protect against floatation. Concrete and sheet pile cutoff walls and headwalls are common forms of anchorage. The culvert barrel end must be securely attached to the anchorage device to be effective. Protection against inlet bending, inlet warping and erosion to fill slopes represent additional benefits of some anchorage techniques.

5.5.3 Headwalls, Wingwalls, and Cutoff Walls

Culvert barrels are commonly constructed with headwalls and wingwalls. These appurtenances are often made of cast-in-place concrete but can also be constructed of precast concrete, corrugated metal, timber, steel sheet piling, or gabions. Headwalls are used to shorten the culvert length, maintain the fill material, and reduce erosion of the embankment slope. Headwalls also provide structural protection to inlets and outlets and act as a counterweight to offset buoyant forces. Headwalls tend to inhibit flow of water along the outside surface of the conduit (piping).

Wingwalls can be used to hydraulic advantage for box culverts by maintaining the approach velocity and alignment, and improving the inlet configuration. However, their major advantage is structural in eliminating erosion around a headwall. Additional protection against flotation is provided by the weight of the wingwalls.

Cutoff walls can provide protection from erosion, either at the inlet or outlet of a culvert. They can also be the first step in controlling piping or seepage problems, prior to considering more extensive anti-seep collars (see below).

5.5.4 Piping and Seepage

Piping is a phenomenon that begins with seepage along the outside of a culvert barrel, which progressively removes fill embankment material, forming a hollow similar to a pipe, hence the term "piping." Fine soil particles are washed out freely along the hollow, and the erosion inside the fill may ultimately cause failure of the culvert or the embankment. Piping may also result from the exfiltration of flow through open joints in the culvert barrel. The possibility of piping can be reduced by decreasing the velocity of the seepage or by decreasing the quantity of seepage flow. Methods of achieving these objectives are discussed in the following paragraphs.

A. Joints

To decrease the velocity of the seepage flow, it is necessary to increase the length of the flow path and thus decrease the hydraulic gradient. The most direct flow path for seepage and thus the highest hydraulic gradient is through open pipe joints. Therefore, it is important that culvert joints be as watertight as practical. If piping through joints could become a problem, flexible, long-lasting gaskets should be specified as opposed to mortar joints.

B. Anti-seep Collars

Piping should be anticipated along the entire length of the culvert when ponding above the culvert is planned, and especially when fine backfill material must be used. Anti-seep or cutoff collars increase the length of the flow path, decrease the hydraulic gradient and the velocity of flow, and thus the probability of pipe formation. Anti-seep collars usually consist of bulkhead type plates or blocks around the entire perimeter of the culvert. They may be of metal or of reinforced concrete and, if practical, dimensions should be sufficient to key into impervious material. Design guidance on longitudinal spacing and dimensions is available from the USBR (1987). Figure 5.21 shows anti-seep collars installed on a culvert under construction.

Figure 5.21. Anti-seep collars (FEMA 2005).

C. Weep Holes

Weep holes are sometimes used to relieve pressure that can occur on wingwalls and aprons. They have also been used along the barrel, however, this may be better addressed by a separate underdrain system. Weep hole design should include filter materials, designed similar to an underdrain filter, to prevent clogging and eliminate piping through the pervious material and weep hole.

5.6 BROKEN-BACK CULVERTS

An alternative to installing a steeply sloped culvert is to break the slope into a steeper portion near the inlet followed by a horizontal runout section. This configuration is referred to as a broken-back culvert. Broken-back culverts can be considered an internal (integrated) energy dissipater if designed so that a hydraulic jump occurs in the runout section to dissipate energy (FHWA 2006a). The broken-back culvert design procedure presented below is based on material from the *TxDOT Hydraulic Design Manual*, (2011).

5.6.1 Broken-Back Culvert Guidelines

One potential mechanism for creating a hydraulic jump is a broken-back culvert. Two common configurations are shown in Figures 5.22 and 5.23. When used appropriately, a broken-back culvert can influence and contain a hydraulic jump. However, there must be sufficient tailwater, and there should be sufficient friction and length in Unit 3 (Figures 5.22 and 5.23) of the culvert. In ordinary circumstances for broken-back culverts, the designer should employ one or more devices, such as roughness baffles, to create tailwater that is high enough to force a hydraulic jump.

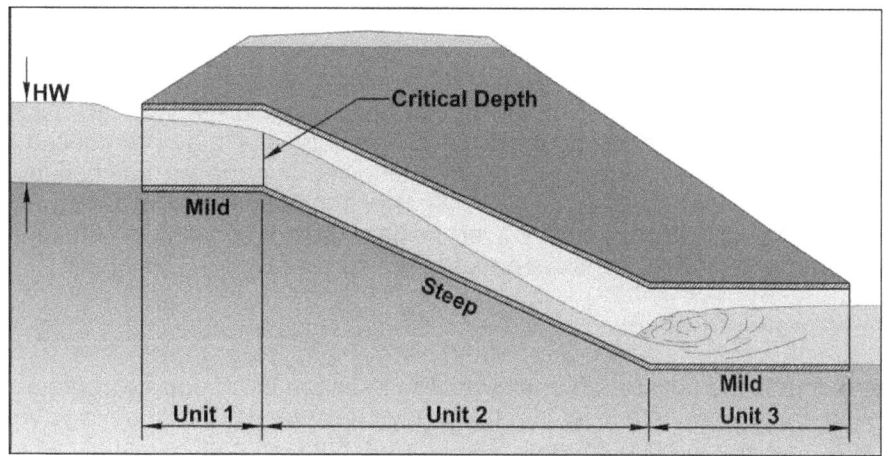

Figure 5.22. Three-unit broken-back culvert.

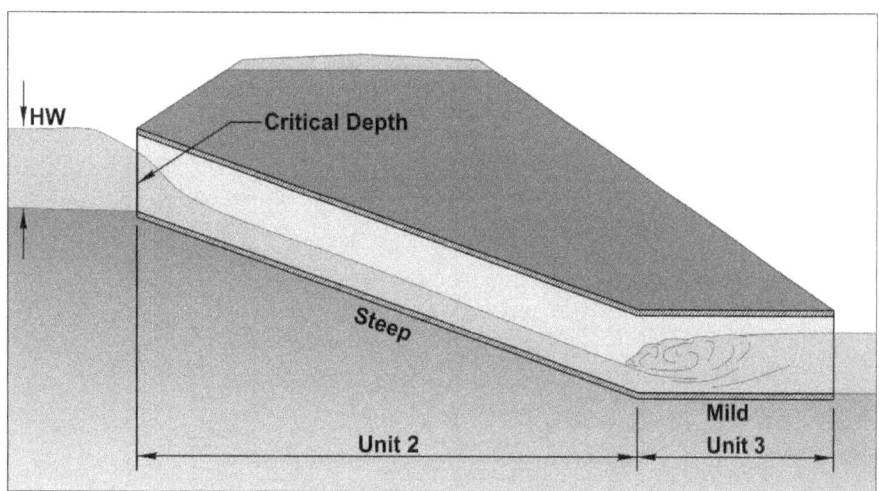

Figure 5.23. Two-unit broken-back culvert.

5.6.2 Broken-Back Design Procedure

The design of a broken-back culvert is not difficult, but provisions must be made so that the primary intent of reducing velocity at the outlet is realized. Table 5.3 outlines the broken-back culvert design process. The hydraulics of circular and rectangular culverts can be determined using the FHWA HY-8 software or the Broken-back Culvert Analysis Program (BCAP) software from the Nebraska Department of Roads (see Section 3.5). The design of associated energy dissipators is contained in HEC-14, Chapter 7.

| Table 5.3. Broken-back Culvert Design Steps. ||
Step	Action
1	Establish a flow-line profile
2	Size the culvert
3	Begin to calculate a supercritical profile
4	Complete profile calculations
5	Consider hydraulic jump cautions

5.7 STORAGE ROUTING

The significant storage capacity behind a highway embankment can attenuate a flood hydrograph. Because of the reduction of the peak discharge associated with this attenuation, the required capacity of the culvert, and its size, may be reduced. This section outlines how to complete hydrologic routing. Detailed information on routing is provided in HEC-22 (FHWA 2009a). While the calculation is not difficult and is readily completed with the FHWA Hydraulic Toolbox, most culvert designs do not consider attenuation upstream of the embankment, but rather consider it part of the safety factor in the design.

5.7.1 The Routing Concept

Storage routing is the calculation of the change in shape of a flood hydrograph over time. A pronounced shape change occurs in a flood wave when a significant storage volume such as a pond or a reservoir is encountered. The storage concept can be visualized by means of a hypothetical situation. In this situation, a spigot discharges water into an empty barrel which has an orifice (hole) at the bottom (Figure 5.24). A plot of the inflow and the outflow reveals some important characteristics of the storage routing process.

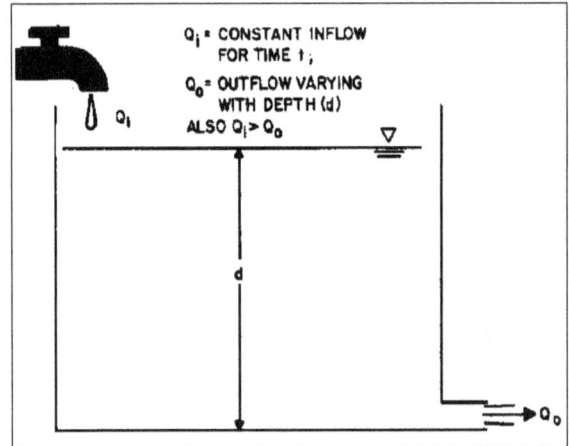

Figure 5.24. Hole in the barrel analogy.

The spigot is turned on at t=0 and discharges a constant flow rate, Q_i, until $t=t_i$, at which time the spigot is turned off. The flow rate entering the barrel exceeds the discharge capacity of the hole. This results in the storage of water in the barrel. As the depth increases, the discharge of water through the hole increases due to the rising head on the hole. The maximum outflow is reached at a time when the depth is at a maximum. This peak outflow occurs when the spigot is turned off since there is no additional inflow after that time. Figure 5.25 is a schematic representation of the inflow and outflow hydrographs.

Additional information about the storage routing concept may be obtained by examining Figure 5.25 more closely. An area on a graph of discharge versus time represents a volume; that is, a discharge increment multiplied by a time increment. The area under the inflow hydrograph depicts the volume of water entering the barrel. The area under the outflow hydrograph depicts the volume of water leaving the barrel. The area between the two curves is the volume stored in the barrel. This volume (area) reaches a maximum when the spigot is closed. From that point on, the area under the outflow hydrograph represents the discharge of the volume stored in the barrel. This equals the maximum storage area previously defined. The total area under the inflow and outflow curves should be equal since the volume of water entering and the volume of water leaving the barrel are the same.

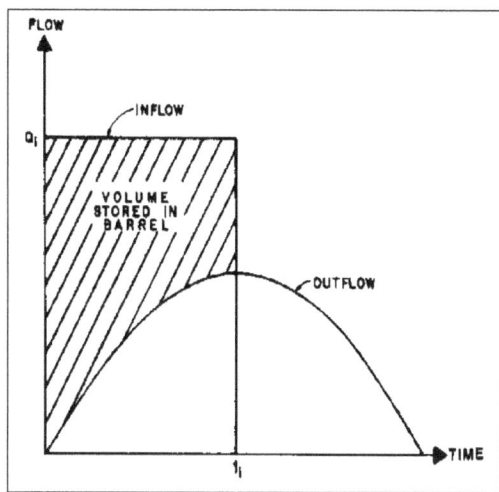

Figure 5.25. Inflow and outflow hydrographs.

5.7.2 Methodology

The mathematical solution of the preceding situation is referred to as a storage routing problem. Conservation of mass, as defined in the Continuity Equation, is essential in formulating the solution. Simply stated, the rate of change in storage is equal to the inflow minus the outflow. In differential form, the equation may be expressed as follows:

$$ds/dt = I - O \qquad (5.2)$$

ds/dt is the rate of change of storage
I is the rate of inflow
O is the rate of outflow

An acceptable solution may be formulated using discrete time steps (Δt). Equation (5.2) may be restated in this manner:

$$(\Delta s / \Delta t)_{ij} = I - O \qquad (5.3)$$

I and O equal the average rates of inflow and outflow for the time step Δt from time $_i$ to time $_j$.

By assuming linearity of flow across a small time increment, the change of storage is expressed as:

$$\left[\left(\frac{I_i + I_j}{2} \right) - \left(\frac{O_i + O_j}{2} \right) \right] \Delta t = \Delta s \qquad (5.4)$$

"$_i$" and "$_j$" represent the time at the beginning and end of the time increment Δt.

There are two unknowns represented in Equation (5.4); therefore, the equation cannot be solved directly. The two unknowns are the increment of storage, Δs, and the outflow at the end of the time increment, O_j. Given a design inflow hydrograph, the known values include

5.23

each inflow value, the time step which is selected, and the outflow at the beginning of the time step solved for during the previous time step. Equation (5.4) can be rewritten as:

$$I_i + I_j + (2s/\Delta t - O)_i = (2s/\Delta t + O)_j \qquad (5.5)$$

where the two unknowns are grouped together on the right side of the equality. Because an equation cannot be solved with two unknowns, it is desirable to devise another equation with the same two unknowns. In this case, a relationship between storage and outflow is required. Since both storage and outflow can be related to water surface elevation, they can be related to one another. This second relationship provides a means for solving the routing equation. The method of solution is referred to as the storage-indication method. A design guideline using the method is included in Appendix D.

5.7.3 Data Requirements

All reservoir routing procedures require three basic data inputs: an inflow hydrograph, an elevation versus storage relationship, and an elevation versus discharge relationship. A complete inflow hydrograph, not just the peak discharge, must be generated. Elevation, often denoted as stage, is the parameter which relates storage to discharge providing the key to the storage routing solution. Elevation versus storage data can be obtained from a topographic map of the culvert site. The volume needed is the accumulated volume at each elevation. Elevation versus discharge data can be computed from culvert data and the roadway geometry. Discharge values for the selected culvert and overtopping flows are tabulated with reference to elevation and the combined discharge is utilized in the elevation versus discharge relationship.

5.7.4 Storage-Indication Method

The storage-indication routing method is outlined in the following steps. A Design Guideline and worked example is found in Appendix D. The FHWA Hydraulic Toolbox applies this method and completes all the necessary calculations.

Step 1. Generate an inflow hydrograph by an appropriate hydrologic procedure.

Step 2. Select a time interval for routing (Δt). Remember that linearity over the time interval is assumed. Generally, a routing interval of one-tenth the time-to-peak is adequate.

Step 3. Determine the elevation-discharge and elevation-storage relationships for the site and outlet device(s) selected.

Step 4. For convenience in solving the routing equation, tabulate the storage-outflow relationship.

Step 5. Plot the ($2s/\Delta t + O$) versus (O) relationship from step 4.

Step 6. Using Equation 5.5, perform the routing. A tabular format may prove convenient in performing the storage routing calculations.

5.8 ENERGY DISSIPATORS

Energy dissipators are devices designed to protect downstream areas from erosion by reducing the velocity of flow to acceptable limits. The failure or damage of many culverts and detention basin outlet structures can be traced to unchecked erosion. Erosive forces, which are at work in the natural drainage network, are often increased by the construction of a highway or by urban development (see Section 5.3). The interception and concentration of overland flow and constriction of natural waterways inevitably results in an increased erosion potential. To protect the culvert and adjacent areas, it is sometimes necessary to employ an energy dissipater.

5.8.1 HEC-14

FHWA guidance for energy dissipators is provided in HEC-14 (FHWA 2006a). HEC-14 includes procedures for designing dissipators that are both internal and external to the culvert and that are located on or below the streambed. This section only provides a brief overview of energy dissipators. For detailed design information and guidance, refer to HEC-14. The design standards provided in HEC-14 are automated in the FHWA software HY-8.

5.8.2 Selection Guidelines

The following general guidelines, with a reference to the applicable Chapter in HEC-14, can be used to limit the number of alternative types of dissipators to consider. The terms "internal" and "external" are used to indicate the location of the dissipator in relationship to the culvert. An external dissipator is located outside of the culvert, and an internal dissipator is located within the culvert barrel. For more detailed information, including the recommended energy dissipators in each category, see HEC-14.

Internal Dissipators (HEC-14, Chapter 7). Internal dissipators are used where:

- Estimated outlet scour hole is not acceptable
- Right-of-way is limited
- Debris is not a problem
- Moderate velocity reduction is needed

Natural Scour Holes (HEC-14, Chapter 5). Natural scour holes are used where undermining of the culvert outlet will not occur or it is practical to be checked by a cutoff wall, and:

- Estimated scour hole will not cause costly property damage, or create a public nuisance

External Dissipators (HEC-14, Chapters 9, 10, and 11). External dissipators are used where:

- Estimated outlet scour hole is not acceptable
- A moderate amount of debris is present
- Culvert outlet velocity (V_o) is moderate (Fr≤3)

Stilling Basins (HEC-14, Chapter 8). Stilling basins are used where:

- Estimated outlet scour hole is not acceptable
- Debris is present
- Culvert outlet velocity (V_o) is high (Fr>3)

Drop Structures (HEC-14, Chapter 11). Drop structures are used where:

- Downstream channel is degrading, or channel headcutting is present.

5.8.3 Design Considerations

A number of design considerations are involved in selecting and designing appropriate energy dissipators. The flood frequency used in the design of the energy dissipator device should be the same flood frequency used for the culvert design. The use of a design flood of less magnitude may be permitted if justified by low risk of failure of the crossing, substantial cost savings, limited or no adverse effect on the downstream channel, and limited or no adverse effect on downstream development.

The review flood frequency should also be evaluated. For most external dissipators, the review flood check will indicate that the dissipator will have a higher outlet velocity than the design flood. If this higher velocity causes concern, it should be mitigated. Internal dissipators and some external dissipators may cause the culvert to flow full for the review flood. If this is likely and if the higher headwater causes concern, a different dissipator should be evaluated.

If ice buildup is a factor, it can be mitigated by sizing the structure to not obstruct the winter low flow, or using external dissipators. Debris control (based HEC-9, FHWA 2005a)) should be considered where clean-out access is limited or if the dissipator type selected cannot pass debris.

CHAPTER 6

CULVERT REPAIR AND REHABILITATION

6.1 ASSESSMENT OF EXISTING CULVERT CONDITIONS

The assessment of existing culvert conditions is an important first step in the design process for repair and rehabilitation efforts. All the condition factors and indicators regarding an existing culvert and channel must be assessed and considered in the selection of appropriate repair and rehabilitation methods and technologies. The selection must also consider the effect of rehabilitative techniques on hydraulic performance of the structure and channel.

Useful references for culvert assessments include the FHWA Bridge Inspector's Manual (FHWA 2006b), the FHWA Culvert Inspection Manual – Supplement to the Bridge Inspector's Training Manual (FHWA 1986b), and the FHWA Federal Lands Highway (FLH) Culvert Assessment and Decision-Making Procedures Manual (FHWA 2010b). Information in this chapter is based largely on the FLH document.

6.1.1 FLH Culvert Assessment and Decision-Making Procedures

The FLH culvert assessment procedure applies to culverts with a span of less than 20 feet, and is readily adaptable for programmatic and inventory use. It was developed to provide a quick visual assessment of culverts on a planned roadway project. The assessment procedure identifies the minimum set of parameters necessary to effectively and efficiently evaluate existing condition and performance for a broad range of culvert structure types, materials, and applications that may be encountered. The procedure describes the defining criteria for each parameter, provides a rating system, and suggests methods and tools for measuring and recording the parameters. The assessment procedure also describes the evaluation of channel stability and performance as it relates to the culvert.

The FLH culvert assessment tool, commonly referred to as a Level 1 assessment, is intended for rapid assessment of a culvert's condition and performance. Culvert condition refers to the level of physical deterioration of the culvert barrel and appurtenances. Culvert performance refers to the functionality of the structure as a water conveyance device. With the increasing interest in AOP, culvert performance in that regard might also need to be considered. The Level 1 assessment procedure may identify the need for a more in-depth investigation, termed a Level 2 assessment. Level 2 assessments require the involvement of technical discipline specialists in hydraulic, geotechnical, structural or materials engineering in order to facilitate further investigation, and may also require special equipment for access and inspection.

The Level 1 assessment procedure typically leads to one of the following recommendations, for each culvert assessed:

1. Condition and performance appear to be acceptable, and no further action is needed with respect to the project being undertaken;
2. Maintenance (e.g., cleaning/clearing) is needed to remedy an observed performance problem and/or facilitate completing the Level 1 assessment;
3. Repair or replace the culvert or appurtenances, with assistance from the decision-making portion of this procedure; or
4. Conduct a Level 2 assessment due to indicators identified by the Level 1 assessment.

The FLH procedure also describes recommended inspection and safety equipment, including guidance on culvert entry by personnel based on OSHA regulations concerning confined space entry, as contained in 29 CFR 1910.146. Alternative assessment methods to personnel entry are described, such as culvert end-only assessments and the use of probing rods, divers and a remotely operated vehicle (ROV). Note that while end-only assessments are widely used, internal conditions cannot always be dependably inferred solely by conditions at the end of the pipe.

The FLH assessment form (Figure 6.1) contains data entry fields for general project information and the overall rating for the culvert, which is generally governed by the lowest rating of the individual elements. This form illustrates the range of problems that can exist. A comprehensive assessment is the only way to identify necessary repair and rehabilitation actions.

6.1.2 Structural Condition

The FLH culvert assessment procedure includes a Culvert Assessment Guide to assist in assigning the appropriate condition rating codes to the various culvert material types based on structural deterioration levels. The guide consists of eleven tables, the first of which describes the five possible rating codes and their general meanings (Table 6.1). The remaining tables describe each major culvert material type and common appurtenances, with typical modes of deterioration for that material type and rating codes. Important considerations and special conditions that might trigger in-depth Level 2 investigations above and beyond this initial Level 1 assessment are discussed, and a photographic guide to assist in assigning rating codes is provided.

Table 6.1. FLH Culvert Assessment Guide Condition Ratings.	
Rating Code	Meaning of Rating Code
Good	Like new, with little or no deterioration, structurally sound and functionally adequate
Fair	Some deterioration, but structurally sound and functionally adequate
Poor	Significant deterioration and/or functional inadequacy, requiring repair action that should, if possible, be incorporated into the planned roadway project
Critical	Very poor conditions that indicate possible imminent failure that could threaten public safety, requiring immediate repair action.
Unknown	All or part of the culvert is inaccessible for assessment or a rating cannot be assigned.

The common modes of culvert deterioration associated with structural or material/physical condition are often related and influence one another, and eventually the culvert's performance. For example, abrasion and corrosion work in tandem to remove material section from the invert of a metal culvert (see Section 2.3.4). Progressive section loss will lead to perforation of the invert and resultant backfill loss, which may lead to loss of structural integrity, cross-section deformation and reduced capacity. In contrast, what appears to be significant physical deterioration may not necessarily translate to performance issues, and may not even warrant repair.

FLH CULVERT ASSESSMENT FORM

Notes by:_____ Date: _____ Project: _____

Measurements by:_____ Time: _____

Site Information:

Facility Location: _____Lat/Long _____

Milepost:_____ Project Station: _____ GPS Road CL Waypoint No. _____

Named waterway:_____ Direction of Flow:_____

Culvert Information:

No. of Barrels: _____ Barrel Length (approx):_____ Barrel Slope: Mild / Steep / _____

Skew (0 degrees = perpendicular to road): _____ Approx Cover: Upstream _____ Downstream_____

Barrel Shape (circle one)   Circular   Box   Elliptical   Pipe Arch   Arch

Diameter: _____ / Span _____ x Rise _____

Pipe Material (circle one):   Metal - Concrete / RCP - Corrugated Plastic - Smooth Plastic - Timber – Masonry

Appurtenances (circle one):

Upstream : Projecting / Mitered / Headwall / Headwall & Wingwalls / Flared End Section / _____

Downstream : Projecting / Mitered / Headwall / Headwall & Wingwalls / Flared End Section / _____

Flowing or standing water? N / Y Depth:____(ft) Est. Flow Velocity:_____(ft/s) Possible AOP/fish passage? Y / N

Utilities Present (list)? Y / N_____ Possible historic features? Y / N   Open Bottom? Y / N

Culvert Condition and Performance (circle / check all that apply and provide appropriate explanations below)

Category	Rating
Invert deterioration	Good Fair Poor Crit Unk N/A
Joints & Seams	Good Fair Poor Crit Unk N/A
Corrosion / Chemical	Good Fair Poor Crit Unk N/A
Cross-Section Deform	Good Fair Poor Crit Unk N/A
Cracking	Good Fair Poor Crit Unk N/A
Liner / Wall	Good Fair Poor Crit Unk N/A
Mortar and Masonry	Good Fair Poor Crit Unk N/A
Rot and Marine Borers	Good Fair Poor Crit Unk N/A
Headwall/Wingwall	Good Fair Poor Crit Unk N/A
Apron	Good Fair Poor Crit Unk N/A
Flared End Section	Good Fair Poor Crit Unk N/A
Pipe End	Good Fair Poor Crit Unk N/A
Scour Protection	Good Fair Poor Crit Unk N/A

Performance Problems Requiring Level 1 Action	
Debris/Veg Blockage > 1/3 of rise at inlet or outlet	❏
Sediment Blockage 1/3 to 3/4 of rise at inlet/outlet	❏
Buoyancy or Crushing-Related Inlet Failure	❏
Poor Channel Alignment	❏
Previous and/or Frequent Overtopping	❏
Local Outlet Scour	❏

Performance Problems Requiring Level 2 Action	
Embankment Piping	❏
Channel Degradation / Headcut (circle one)	❏
Embankment Slope Instability	❏
Sediment Blockage > 3/4 Rise at Inlet or Outlet	❏
Sediment Blockage > 1/3 Rise Throughout Barrel	❏

Other Problems Requiring Level 2 Action	
No Access / Ends Totally Buried / Submerged	❏
Aggressive Abrasion/Corrosion/Chemical (circle)	❏
Exposed Footing (Open-Bottom Culvert Only)	❏

Overall Rating
Good
Fair
Poor
Critical
Unknown
Performance Problems

Photos (number):   ___ Inlet ___ Outlet ___ Roadway (ahead) ___ Roadway (back) ___ View downstream

___ View upstream Others: _____

Notes / Recommendations:

_____

_____

❏ Additional notes / Sketches on back of form   A.2

Figure 6.1. FLH culvert assessment form.

The most important feature of the rating system for the FLH culvert assessment procedure is the link of a Poor condition to repair actions needed. Significant indicators are embedded in the guide that trigger a Poor rating and repair recommendations, depending upon the material type and mode of deterioration. For concrete boxes and arches, cracks larger than ¼ in. (6 mm) wide with significant infiltration or exfiltration and voids, or cracking of any size covering more than 50 percent of the culvert surface area, results in a Poor rating. Reinforced concrete pipe (RCP) receives a Poor rating if there are random cracks greater than 1/8 in. (3 mm) in width, or cracks of any size along the crown, haunches or covering more than 25 percent of the pipe surface area. Concrete culverts with exposed steel reinforcement, open joints with significant infiltration or exfiltration of soil and/or water and voids visible, or perceptible cross-section deformation are also rated Poor and repairs recommended.

The primary modes of deterioration for corrugated metal pipe (CMP) are corrosion, abrasion, deformation and displacement. A significant trigger for assigning a Poor rating and recommending repairs is perforation, wherein holes in the metal are either visible or easily made by the inspector's pick hammer. The underlying assumption of this trigger is that once perforation occurs in CMP, the rate of deterioration accelerates as water exfiltrates the pipe, creating voids outside the pipe and exposing the exterior metal surface to increased corrosion. A Poor rating for CMP is also triggered by joints and seams with connection hardware failure, or that are open or displaced with significant infiltration and exfiltration of soil and/or water and visible voids.

The primary modes of deterioration for plastic pipe are wear from abrasion, deformation and displacement. Significant wear and perforations in the invert, or splits, cracks and tears in the walls in excess of 6 in. (150 mm) long, trigger a Poor rating and repair recommendations. Significant local buckling or perceptible cross-section deformation also triggers a Poor rating for plastic pipe.

Timber culverts are rated Poor and repairs recommended if there is significant section loss or perforations in the invert, or open joints and seam, with accompanying infiltration and exfiltration and visible voids. Failure of connection hardware, significant warping or breakage of members, or significant crushing, cracking or section loss due to rot or insect borer attack also triggers a Poor rating for timber culverts.

Masonry culverts are rated Poor and repair recommendations made if there is perceptible cross-section deformation and cracks in the crown, invert and/or spring line areas. Holes in the invert, displaced or missing mortar or blocks, and infiltration or exfiltration and visible voids also triggers a Poor rating for masonry culverts.

The condition assessment of appurtenances is also important. Failure of a culvert appurtenance, such as crushing of a pipe end treatment, can quickly lead to degradation and performance or structural failure of the culvert. A flared end section or pipe end with significant cracking over more than 50 percent of its area, or that is crushed or separated from the barrel, is assigned a Poor condition rating and repair recommendations. A headwall, wingwall or apron with over 50 percent of its surface area cracked and spalling, and/or rebar exposure or an exposed or undermined foundation, is rated Poor. Scour protection such as riprap is rated Poor if there is significant displacement, undermining or deterioration affecting its performance or that of the culvert itself.

6.1.3 Soil Support

Proper soil support is important for flexible culverts, and ultimately the roadway above the culvert. Soil support can be degraded through voids in, and settlement of, the embankment through piping, erosion and scour. These types of problems can result from poor backfill material, inadequate compaction, not using a headwall or cutoff walls in granular soil types and insufficient surface drainage design. Multiple barrel culverts that are placed too close together for good compaction can also lead to problems. For all culvert types, the FLH culvert assessment guide triggers a Poor rating and repair recommendations when soil support problems are indicated, typically by significant infiltration or exfiltration of soil and/or water through open joints and/or deteriorated inverts. A Critical rating is assigned to all culvert types requiring immediate repair action due to threatened public safety, typically resulting from holes and depressions in the roadway related to underlying embankment issues.

Voids in the embankment under and around the barrel are commonly caused by soil infiltration through open joints and seams, or post-construction consolidation due to inadequate soil preparation (backfill and compaction). Water overtopping the culvert and roadway may lead to embankment piping and erosion, respectively. Embankment piping is flow outside of the culvert and through the embankment, which can be caused and exacerbated by open joints. Inspection of the roadway is an important part of culvert assessment, in particular end-only assessments, since problems with voids in the embankment are often manifested in roadway settlement and pavement failures (Figure 6.2).

Figure 6.2. Roadway settlement caused by voids around a culvert.

6.1.4 Culvert and Channel Performance

The performance of a culvert and associated channel must be assessed before culvert repair design. The first consideration is to determine if these structures are hydraulically adequate. If they are not, replacement should be considered. If they are adequate, typical problems might include debris or sediment blockage, buoyancy problems, poor channel alignment, overtopping and scour. Depending on the severity of the problem a Level 1 maintenance or repair action may be required, or a more extensive Level 2 investigation and/or repair.

The culvert will fail to perform as designed if the entrance is blocked by a combination of vegetation, trash, sediment and other debris (Figure 6.3).

Figure 6.3. Severe debris blockage.

Sediment accumulation in the barrel can be a problem related to the culvert or the channel and watershed conditions (see Section 5.3.3). An accumulation of sediment, generally devoid of vegetation debris, that is local to either the inlet or outlet and greater than or equal to 1/3, but less than or equal to 3/4 of the rise of the barrel may be considered a Level 1 maintenance issue. The localized blockage should not extend more than a few feet into the barrel from the culvert end, which would be indicative of greater channel aggradation problems and trigger Level 2 action (Figure 6.4). In most cases, a minor accumulation is due to embankment sloughing around the pipe end, or settling out of sediment loads conveyed by the flow. In cases where the blockage is less than 1/3 of the rise, with sufficient invert slope and periodic high flows, the culvert will likely clear out the blockage as a self-cleaning mechanism. If the blockage exceeds 3/4 of the rise, self-cleaning may not occur and the culvert should be a candidate for maintenance to clear the sediment.

Figure 6.4. Barrel filled with sediment up to half its rise.

Buoyancy problems typically occur on corrugated metal culverts, more often on large diameter barrels (greater than 48 in. (1200 mm)) projecting from the fill, but can also occur on small diameter culverts (Figure 6.5). The problem is typically related to a submerged thin-edge projecting inlet operating in inlet control (see section 5.5.2). If the buoyancy force from the entrapped air is greater than the dead weight of the culvert and the small amount of fill on the entrance, and the bending resistance of the pipe material, the metal pipe material is compressed at the crown and the inlet is bent up by the buoyancy force. The typical repair is to anchor the entrance with a headwall, slope pavement anchor, or a properly installed terminal end section. The buoyancy force can also be reduced by using a bevel on the headwall, which increases the amount of water and decreases the amount of air in the barrel.

Figure 6.5. Buoyancy uplift.

Poor channel alignment exists if the channel approaching the culvert from upstream or exiting the culvert downstream is highly skewed (more than roughly 45 degrees) from the axis of the culvert barrel, and there is scour at the outside channel bank that is causing damage to the culvert, headwall, wing walls or road embankment. Figure 6.6 is an idealized example sketch of poor channel alignment.

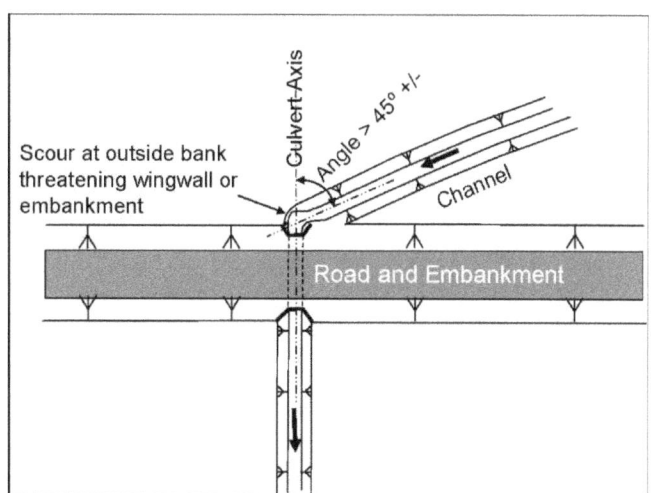

Figure 6.6. Idealized example sketch of poor channel alignment.

Embankment damage at the culvert site may be present because of previous overtopping, potentially due to inadequate hydraulic capacity. Indicators of overtopping include, but are not limited to, drift on the guardrail above the culvert and/or extensive erosion of the downstream embankment, often accompanied by loss of the pavement section along the downstream edge (Figure 6.7). The most likely location of overtopping is at the low point in the road profile, which may be offset from the culvert crossing location. Overtopping indicators typically lead to a recommendation for maintenance to repair any related erosion damage, and potentially a recommendation to add erosion protection to accommodate future overtopping (see Section 5.4.5). If overtopping is known to be frequent at the culvert and if the condition rating is Poor or Critical, consideration should be given to replacement with an adequately sized structure.

Figure 6.7. Erosion damage to downstream embankment slope from overtopping.

Scour at the outlet (see Section 5.3.2) is a concern if a large and noticeable scour hole occurs that endangers the culvert (Figure 6.8). In addition to the repair of the damage to the culvert itself, such problems would result in a recommendation for installation or repair of outlet protection, such as to line the existing scour hole with riprap.

Figure 6.8. Scour hole at outlet of a RCP culvert.

Open-bottom culverts often have shallow foundations that can be undermined by scour (Figure 6.9). If an open-bottom culvert is to be rehabilitated because of a condition rating of Poor or Critical, care must be taken to ensure that the rehabilitation does not increase the risk of undermining the foundations.

Figure 6.9. Exposed spread footing in an open-bottom culvert.

6.2 CULVERT REHABILITATION VERSUS REPLACEMENT

6.2.1 FLH Culvert Decision-Making Procedure

The FLH decision-making procedure provides guidance for post-assessment actions to be taken for existing roadway culverts. Post-assessment actions could include maintenance, repair, replacement, and further investigation and/or analysis. Information is provided to assist with repair or replacement technique selection. A set of eight flowcharts outlines the logical decision-making process for the various culvert types and scenarios. A repair liner selection comparison matrix is included, which provides rough cost information, capabilities and limitations for each commonly-used liner option. The procedure also includes matrices for considering and comparing culvert person-entry repairs and replacement techniques, as well as culvert-related construction activity options based on the FLH bid history database (Table 6.2). Also included in the procedure is a photographic guide to culvert rehabilitation, which illustrates some of the more common techniques.

6.2.2 Excavation Requirements

Depth of excavation or cover is an important consideration in deciding what design option is best for a deteriorated culvert. If the depth of cover is less than or equal to 4 ft (1200 mm), it is viewed as a "shallow" culvert. All other factors such as traffic disruptions aside, if a shallow culvert is a "small" pipe (less than 3 ft (900 mm) rise) with no headwalls, it is assumed to be most cost-effective to address it with open-trench replacement. Open-trench excavation remains a viable replacement option for culverts where the excavation depth is up to 20 ft (6 m) to the bottom of the invert, there are no other culverts being replaced with trenchless methods or in-situ repairs within the project, and traffic closures are allowed. Shoring is sometimes required in open-trench excavation (Figure 6.10).

Repair Type	Cost Estimates Based on FLH Bid History	Rough Cost Estimates from Other Agencies	Maximum Size Limits	Other Limitations
		Table 6.2. Example Person-entry Repair Selection Matrix (based on FLH culvert decision-making procedure).		
Grouted Repair Sleeves or Short Cured-in-Place Sleeves (CIPP)	No Estimate Available	Low cost; $2000 to $5000 per CIPP repair for 18 inch diameter	Up to 54 inch diameter for CIPP; Up to 54 inch for stainless steel; Up to 108 inch for PVC	Mechanical seals work poorly with helical and small diameter CMP; may fail if separated or offset joints present; CIPP available in 36 in. connectible lengths; can be used on deformed flexible pipes
Grouting voids	Medium Cost; $330/cu.yd.	Low cost; $10/lin.ft. for small voids; $100-$150/cu.yd. for large voids	N/A	Difficult to judge completeness of repair; toxicity with manned-entry
Crack Epoxy Injection/Mortar	No Estimate Available	Low cost	N/A	Toxicity with manned-entry; not recommended for cracks greater than 0.1 inch wide
Crack/Spall Patching and Rebar Coating with epoxy grout	High Cost; General repair of concrete $860/sq.yd. or $2020/cu.yd.; epoxy coated rebar $1.30/lb	Low cost	N/A	Toxicity with manned-entry; hand-applied above or underwater via man-entry; repair may only slow deterioration or be cosmetic
Joint Sealing with Expansion Gasket Seal Ring	No Estimate Available	Low cost	Up to 216 inch diameter	No more than 10% displacement tolerated; more applicable to RCP than flexible pipe
Invert Lining	No Estimate Available	Medium Cost	N/A	Difficulties tying into host pipe; cement is subject to breakdown if runoff is acidic; modified high-strength concrete mix required; steel plating is best for CMP and RCP
Repoint Masonry	Low Cost; $55/sq.ft.	No Estimates Available	N/A	N/A

Figure 6.10. Open-trench excavation with shoring.

6.2.3 Traffic Disruption

The issue of traffic disruption will play a major part in the selection of rehabilitation design options for a deteriorated culvert. Whether or not temporary road or lane closures will be allowed will usually dictate if either an open-trench or trenchless replacement is used. If closures are not permitted the only remaining design option is typically trenchless replacement techniques. If closures are permitted, but in-situ repair of the existing culvert is either impossible or trenchless replacement methods are already being used elsewhere in the project, the best design option may still be trenchless replacement for cost efficiency.

In some cases where significant embankment and/or roadway damage has already occurred, traffic disruptions may already exist and repair-related closures unavoidable. If culvert repairs require or are accompanied by surface excavation and rebuilding of the embankment and roadway, the cost-effective assumption is that culvert should be replaced via open-trench replacement. The most cost-effective and least intrusive option for "deep" excavations in excess of 20 ft (6 m) to the bottom of the culvert is also assumed to be trenchless replacement.

6.2.4 Condition and Remaining Service Life

Culvert condition and estimated remaining service life influence the decision of whether to rehabilitate the structure, as well as selection of the best design option. A determination that the culvert in its current state will not last another 20 years or until the next roadway project, is followed by the design question of what rehabilitative method will facilitate reaching this life expectancy. Factors such as an aggressive chemical and/or corrosive environment must be considered in the repair and rehabilitation design to ensure the culvert reaches its intended service life.

Concrete culverts with Poor or Critical cross section deformations or cracking are assumed to have lost most of their structural capacity and are recommended for replacement. Repair options for this level of deterioration of this particular culvert type are assumed to fall short of the needed capacity and extension of service life. Similarly, if most of the culvert is affected by Poor or Critical conditions of one type or another, the assumption is that the host pipe has neared the end of its service life and spot repairs on 50 percent or more of a pipe is not cost efficient when compared to lining or replacement.

6.2.5 Construction and Installation Issues

If cover is deeper than 4 ft (1.2 m) and traffic disruptions are discouraged, a liner repair is generally called for with small culverts, provided there is sufficient access. Access for lining repair refers to available right of way, means of ingress/egress, and work space for the lining equipment, machinery and crew at the ends of the culvert (Figure 6.11). If there is not access for a liner repair, the recommendation for a small culvert with Poor conditions generally becomes replacement; however, the feasibility and costs of creating access at the culvert ends should be considered.

Figure 6.11. Example of a spiral wound liner installation with restricted access at culvert end.

Joint repair by person-entry is feasible and desirable if the pipe is large enough and the number of joints needing repair is reasonably small; however, if many joints need repair, a liner or replacement may be more cost effective. Similarly, the most cost-effective solution for a deteriorated appurtenance on a culvert where the barrel is to be replaced is likely to replace the appurtenance as well, possible historic issues aside.

Special permitting issues may exist that will affect the culvert repair and rehabilitation design process. For example, if aquatic organism passage (AOP) issues exist (see Chapter 4) environmental permitting requirements may restrict the time of year for repair or replacement, the duration of the construction, and/or the type of materials used (such as a liner product that might introduce water quality concerns). Similarly, culverts in historical sites or classified as historic structures may have cultural or historic permit requirements with special restrictions.

6.3 CULVERT REHABILITATION TECHNIQUES

6.3.1 Liner Repairs

The FLH culvert decision making procedure includes a one-page culvert Repair Liner Selection Matrix, which summarizes properties, advantages and disadvantages of some of the liners commonly used in full-length, full-circumference repairs. More options and considerations for liner selection are also presented in the Culvert Pipe Liner Guide and

Specifications (FHWA 2005b), including typical construction specifications/details, installation requirements, manufacturers and cost estimates. Note that not all liner types provide structural repair and may not be appropriate for use if the original pipe has become structurally inadequate. The following paragraphs summarize information related to different liner types.

a. Slip-Lining. A slip liner is essentially a smaller-size conduit that is slipped inside a host pipe with the annulus between the two conduits typically grouted (Figure 6.12). The pipe used for slip lining may be a continuous length, or may be segmental. Slip lining is common in round pipes, but other liner shapes such as pipe arch are also available. Note that deformations or discontinuities in the host pipe can block insertion and limit the size of the liner. Abrasion and corrosion resistance of the rehabilitated pipe is good, with structural restoration possible depending upon the liner and annulus composition. Installation space requirements vary, and flow bypass is sometimes required. There is low to moderate safety concern for installers and the environment in the installation process, in particular with low-density grout. The fusion-welding process for segmental liners can be labor intensive.

b. Spiral Wound. Spiral wound lining uses interlocking profile strips, most commonly made from PVC, to line a deteriorated culvert. Coiled, interlocking profile strips are fed through a winding machine that mechanically forces the strips to interlock and form a smooth, continuous, spirally wound liner (Figure 6.11). When expandable profile strips are used the liner is pressed against the host pipe creating a close fit liner that eliminates the grouted annulus (Figure 6.13). Spiral wound lining is generally limited to circular pipes, with less tolerance for bulges and deformations then other methods. Abrasion and corrosion resistance is very good, with structural restoration also possible depending upon the liner and annulus composition. Installation space requirements are small, and flow bypass is sometimes required. There is low to moderate safety concern for installers and the environment in the installation process. Specialized equipment and trained personnel needed, with some larger manual systems requiring person-entry. Spiral wound liners may become brittle in freezing temperature.

Figure 6.12. Slip lining a 30 inch (750 mm) CMP with a 24 inch (600 mm) PVC pipe.

Figure 6.13. Close-fit spiral wound liner installation in a masonry pipe.

c. <u>Cured-in-Place</u>. Cured-in-place lining installations involve insertion of a flexible tube coated with a thermosetting resin into an existing culvert. Once installed, the resin is cured under ambient conditions or through applied heat provided by circulating stream or hot water. The resulting liner is close-fitting and introduces minimal diameter reduction. This technique is not limited to circular pipes and has a higher tolerance for bulges, deformations and discontinuities then other methods. Abrasion and corrosion resistance is very good, with structural restoration possible depending upon the liner wall thickness. Installation space requirements are small to moderate, but flow bypass is most always required. Given the resin products involved and use of steam or hot water, there are concerns for the environment and installers during installation. Specialized equipment and trained personnel are needed for this type of liner installation. Figure 6.14 shows an example of a cured-in-place liner installation.

Figure 6.14. Cured-in-place liner installation.

d. Close-fit Lining Methods. Sometimes referred to as modified slip lining, close-fit lining involves the insertion of a thermoplastic pipe with an outside diameter equal to or slightly larger than the inside diameter of the host culvert. As a result, the liner must be modified in cross section before installation. Once in place the liner is reformed or re-rounded to provide a close-fit with the existing culvert eliminating the need for grouting. Close-fit lining methods include two main types, symmetrical reduction systems and folded systems. Symmetrical reduction systems use a static die or a series of compression rollers that temporarily reduce the diameter of the liner as it is being inserted. Liners used in the folded method are generally folded into "C"-, "U"-, or "H"-shapes during manufacturing or by site-equipment before installation, inserted into the pipe, and then unfolded (Figure 6.15). Folded liners include both fold-and-form PVC liners and deformed-reformed HDPE liners. For any of the close-fit methods the host pipe must be round or semi-round with minimal deformations or discontinuities that can block insertion and limit diameter of the liner. Abrasion and corrosion resistance of the liner is very good, but no structural restoration is possible. Installation space requirements are small, and flow bypass is usually required. There is moderate safety concern for installers and the environment in the installation process, and specialized equipment and trained personnel needed.

Figure 6.15. Deformed-reformed liner installation.

e. Spray-On. Spray-on linings are typically cement mortar or epoxy, and can be applied by either mechanical means or person-entry into the culvert (Figure 6.16). Environmental and worker safety issues may be of concern with these liners, primarily during the application and set-up phases, and flow-bypassing is often necessary. Spray-on liners are commonly used for pipes 12 to 24 in. (300 to 600 mm) in diameter, but can accommodate larger sizes as well. Diameter reduction is minimal for non-structural liners and moderate for structural versions. Deformations, discontinuities and minor bends in the host pipe can be accommodated, and installation space requirements are small. Abrasion and corrosion resistance is poor for mortar linings and slight better for epoxy.

Figure 6.16. Spray-on mortar liner installation via person-entry (from CALTRANS).

f. Other Methods. The methods described above have been used long enough in the industry for their strengths and weaknesses to be understood. However, new liner products and installation methods are being developed and introduced on a regular basis. For example, one newer product type is a segmented panel system consisting of a series of molded, translucent PVC panels that are assembled inside the barrel. Similar to most hard liner rehabilitation methods structural grout is then used to fill the annular space between the panels creating the new pipe and the original pipe. What is unique about this system is the use of the smaller panels representing a portion of the pipe circumference to "build" a new pipe within the existing pipe using person-entry construction procedures. New product types and installation techniques such as this will continue to be introduced giving the designer an even wider range of possible liner rehabilitation options.

6.3.2 Person-Entry Repairs

Many rehabilitation designs for larger culverts are best accomplished by entry of workers into the structure, in particular, repairs to joints, inverts and other isolated areas. Person-entry repairs have been performed in pipes as small as 3 ft (900 mm), even in submerged conditions, with the use of confined-space technicians and divers. A one-page culvert Localized Man-Entry Repair Selection Matrix in the FLH culvert decision making procedure summarizes properties, advantages and disadvantages of some of the commonly used local repair techniques that require man-entry.

a. Joint Repairs. Damaged and/or separated and open joints in culverts with larger than a 2.5 ft (750 mm) rise can often be addressed with man-entry repairs such as joint sealing with an expansion gasket seal ring or hand-applied epoxy grout. Localized embankment voids that may have developed outside of the culvert as a result of joint damage can also be addressed with grout injection performed via the man-entry approach. Indications from various agencies nationwide are that the technique of joint sealing with an expansion gasket seal ring is a low cost repair alternative that can be applied to pipes ranging from 15 to 216 in. (375 to 5400 mm) in diameter. The technique is more applicable to RCP than flexible pipe, and can tolerate up to approximately 10 percent displacement at a joint. For joint repairs in non-circular culverts, another low-cost alternative is to use grouted repair sleeves or short Cured-in-Place Sleeves (CIPP) that are available in various shapes. Figure 6.17 shows two commonly joint repair methods, using a gasket system and a joint sealing compound.

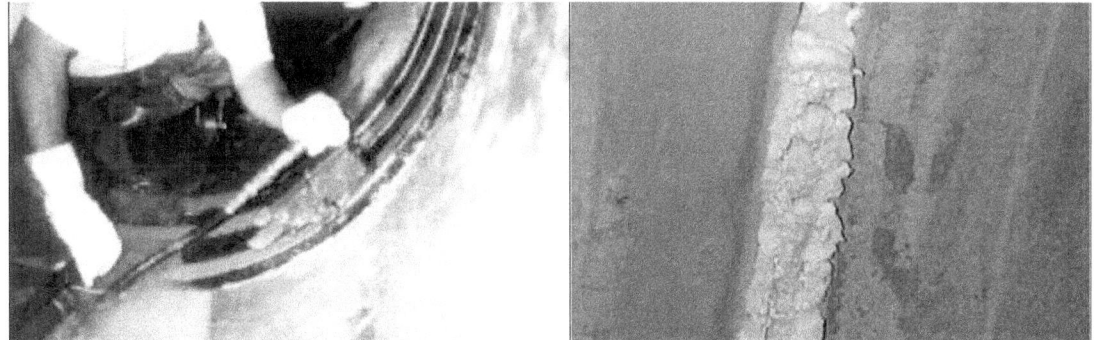

Figure 6.17. Joint repair using gaskets and joint sealing via person-entry.

b. Invert Lining. Material deterioration of culverts most often occurs in the invert, in particular with CMP and concrete. For larger culverts where deterioration is mostly confined to the invert, re-lining the bottom with steel plating, concrete and/or mortar can be a cost-effective repair (Figure 6.18). Indications from various agencies nationwide are that the invert lining technique is a medium cost repair alternative that can be applied to pipes 3 ft (900 mm) and larger in diameter. Difficulties tying into the host pipe are common and should be addressed adequately in the design. The cement used in many invert lining methods is subject to breakdown if runoff is acidic, requiring a modified high-strength concrete mix which is also desirable for abrasion resistance. Steel plating is often preferred by agencies for CMP and RCP, but corrosion is a concern for plating and anchorage hardware.

c. Spot Repairs. Isolated and limited repairs of larger culverts, such as pumping grout into voids or patching local failures, can usually be more effectively addressed by person-entry methods rather than full-length, full-circumference lining. Indications from various agencies nationwide are that the grouting voids is a medium cost repair alternative, with toxicity hazards for entry personnel during installation and difficulty in judging completeness of the repair. In addition to isolated joint repairs methods discussed above, low cost spot repairs to a culvert barrel can also commonly include epoxy or mortar injection of cracks, crack or spall patching, and epoxy coating of exposed rebar. Epoxy or mortar injection repair is not recommended for cracks greater than 0.1 in. (30 mm) wide, which can be addressed with other hand-applied techniques both above and below water. Repair material toxicity is a concern with these person-entry techniques. Crack injection and patching, and rebar coating, oftentimes may only slow deterioration.

d. Appurtenance Repairs. Many repairs have little if anything to do with the barrel and lining and do not require barrel entry. These repairs are typically to the appurtenances, and include rehabilitation or replacement of headwalls, wing walls, aprons, riprap and debris controls. Replacement of concrete headwalls and repointing masonry walls and facades are common low-cost design repairs for many culverts. Energy dissipators may also be called for in repair design, as well as modifications to improve AOP performance. Figure 6.19 shows a typical installation of concrete headwall and wingwalls on a pipe culvert.

Figure 6.18. Invert repair of CMP using steel plating.

Figure 6.19. Concrete headwall and wingwall installation on pipe culvert.

6.4 HYDRAULIC ANALYSIS OF REPAIR AND REHABILITATION ACTIONS

Culvert size and capacity plays a prominent role in the selection of rehabilitation design options for a deteriorated culvert, as well as the evaluation of changes in culvert performance due to repair designs. Understanding culvert hydraulic theory as described in this publication, particularly the concepts of inlet and outlet control, will allow the designer to correctly analyze the impact or changes to headwater and/or outlet velocity that may occur for any type of culvert repair or rehabilitation.

For example, complete culvert replacement with a different size, shape or type of culvert than was originally built can be analyzed following the normal steps and procedures for any new culvert, as detailed in Chapter 3. An important factor in this analysis is verifying that the design discharge has not changed, particularly in urbanizing areas.

Frequent overtopping often suggests that an existing culvert is undersized and of insufficient capacity with a recommendation for culvert replacement, or installation of additional barrels. However, if the culvert is in inlet control the only rehabilitation that might be needed is the addition of a bevel or tapered entrance to the existing barrel. Hydraulic analysis of any recommended repair or rehabilitation, whether it is additional barrels or the use of a tapered inlet, would again be based on Chapter 3 procedures.

The use of a liner may improve or diminish hydraulic performance. In outlet control a liner would typically reduce flow capacity due to the decreased cross-section area, but that impact may be offset by decreased barrel roughness. Alternatively, hydraulic performance of an inlet control culvert could be better with a liner if a beveled inlet condition is created between the original culvert and smaller diameter liner. In either case, the impact of area reduction from a liner is more significant in smaller culverts, particularly when using slip lining methods based on standard pipe sizes. To illustrate, the percent reduction in flow area by slip lining a 24 inch (600 mm) pipe with an 18 inch (450 mm) pipe is much more significant (44%) than slip lining a 60 inch (1500 mm) pipe with a 54 inch (1350 mm) pipe (19%). Regardless, hydraulic analysis based on Chapter 3 will allow the designer to accurately determine the consequences of installing any type of liner.

Joint repairs or invert lining typically do not have any impact on the hydraulic performance of an inlet control culvert, since the condition of the barrel is not a factor in inlet control. The exception to this is would be if the repairs increase friction loss, or decrease the flow area, enough to cause a change to outlet control. However, joint repairs typically decrease the friction loss by eliminating gaps, and invert lining often creates a smaller composite Manning's n value given the typically smoother invert lining material, with little impact on flow area. Regardless, the impact of these repairs in outlet control should be evaluated by defining a new Manning's n value and/or flow area for the rehabilitated condition and applying the energy equation according to normal procedures for outlet control analysis as described in Chapter 3.

(page intentionally left blank)

CHAPTER 7

REFERENCES

AASHTO, 1975, "Guidelines for the Hydraulic Design of Culverts," Task Force on Hydrology and Hydraulics AASHTO Highway Subcommittee on Design, American Association of State Highway and Transportation Officials, Washington, D.C.

AASHTO, 2002, "Standard Specifications for Highway Bridges," 17th edition, American Association of State Highway and Transportation Officials, Washington, D.C.

AASHTO, 2007, "Highway Drainage Guidelines. 4th edition, American Association of State Highway and Transportation Officials, Washington, D.C.

AASHTO, 2011, "Roadside Design Guide, 4th edition, American Association of State Highway and Transportation Officials, Washington, D.C.

ACPA, 1981, "Concrete Pipe Handbook," American Concrete Pipe Association (ACPA), 8320 Old Courthouse Road, Vienna, VA 22180, June.

AISI, 1983, "Handbook of Steel Drainage and Highway Construction Products," American Iron and Steel Institute (AISI), 1000 - 16th, N.W., Washington, D.C. 20036.

Apelt, C.J., 1981, "Hydraulics of Minimum Energy Culverts and Bridge Waterways," C.J. Apelt, Institute of Engineers, Australia, 11 National Circuit, Barton A.C.T. 2600 Australia 0 85825 157 4.

Barfuss, S., and Tullis, J.P., 1988, "Friction Factor Test on High Density Polyethylene Pipe," Hydraulics Report No. 208, Utah Water Research Laboratory, Utah State University, Logan, UT.

Bishop, R.R., and Jeppson, R.W., 1975, "Hydraulic Characteristics of PVC Sewer Pipe in Sanitary Sewers," Utah State University, Logan, UT, September.

Bossy, H.G., 1961 "Hydraulics of Conventional Highway Culverts," paper presented at the Tenth National Conference, Hydraulics Division, ASCE, Urbana Ill, August 16-18.

Bossy, H.G, 1963, "Hydraulic Design of Highway Culverts," Draft HDS, Unpublished notes furnished by Hydraulics Branch, Bridge Division, Office of Engineering, FHWA, Washington, D.C. 20590, April.

Chase, Donald V., 1999, "Hydraulic Characteristics of CON/SPAN Bridge Systems, University of Dayton.

Chow, V.T., 1959. Open Channel Flow, McGraw-Hill-Civil Engineering Series, New York.

Cottman, N.H., 1981, "Experiences in the Use of Minimum and Constant Energy Bridges and Culverts," Institution of Engineers, Australia, 11 National Circuit Barton A.C.T. 26000 Australia 0 85825 157 4.

Clyde, C.G., 1980, "Manning Friction Coefficient Testing of 4-, 10-, 12-, and 15-inch Corrugated Plastic Pipe," Hydraulics Report No. 36, Utah Water Research Laboratory, Utah State University, Logan, UT.

Davis, C.V., 1952, Handbook of Applied Hydraulics, McGraw-Hill Book Co., Inc., New York, N.Y., 1952.

DeCou, G and Davies, P, 2007, "Evaluation of Abrasion Resistance of Pipe and Pipe Lining Materials," Final Report, FHWA/CA/TL – CA01-0173, EA 680442, California Department of Transportation, Sacramento, CA.

Durow, B.P., 1982, "Flood Routing - Culvert Design," The South African Institution of Engineers, March.

FEMA, 2005, "Conduits Through Embankment Dams, Federal Emergency Management Agency, Publication number L-266, Washington, D.C.

FHWA, 1961, "Design Charts for Open Channel Flow, Hydraulic Design Series No. 3 Hydraulics Branch, Bridge Division, Office of Engineering, Washington, D.C. 20590.

FHWA, 1965, "Hydraulic Charts for the Selection of Highway Culverts," HEC No. 5, Hydraulics Branch, Bridge Division, Office of Engineering, FHWA, Washington, D.C. 20590 (L.A. Herr and H.G. Bossy).

FHWA, 1972a, "Hydraulic Design of Improved Inlets for Culverts," HEC No. 13, Hydraulics Branch, Bridge Division, Office of Engineering, FHWA, Washington, D.C. 20590, August (L.J. Harrison, J.L. Morris, J.M. Normann and F.L. Johnson).

FHWA, 1972b, "Computation of Uniform and Nonuniform Flow in Prismatic Channels," Office of Research (P.N. Zelensky).

FHWA, 1974, "Hydraulic Design of Large Structural Plate Corrugated Metal Culverts," Hydraulics Branch, Bridge Division, Office of Engineering, Washington, D.C., January (J.M. Normann).

FHWA, 1978, "Hydraulics of Bridge Waterways," Hydraulic Design Series No. 1, revised second edition (J.N. Bradley).

FHWA, 1979, "Design of Urban Highway Drainage - The State of the Art," FHWA-TS-79-225, Hydraulics Branch, Bridge Division, Office of Engineering, FHWA, Washington, D.C. 20590, August (S.W. Jens).

FHWA, 1980, "Hydraulic Flow Resistance Factors for Corrugated Metal Conduits," Report FHWA-TS-80-216, (J.M. Normann).

FHWA, 1981, "Design of Encroachments on Floodplains Using Risk Analysis," Report FHWA EPD-86-112 (M.L. Corry, J.S. Jones, and P.L. Thompson).

FHWA, 1983, "Structural Design Manual for Improved Inlets and Culverts," FHWA-IP-83-6, (T.J. McGrath and F.J. Heger).

FHWA, 1986a, "Bridge Waterways Analysis Model Research Report," Report FHWA RD-86-108 (J.O. Shearman, W.H. Kirby, V.R. Schneider, and H.N. Flippo).

FHWA, 1986b, "Culvert Inspection Manual," Report FHWA IP-86-2 (J.D. Arnoult).

FHWA, 2002, "Highway Hydrology," Hydraulic Design Series No. 2, FHWA-NHI-02-001 (R.H. McCuen, P.A. Johnson, and R.M. Ragan).

FHWA, 2005a, "Debris Control Structures, Evaluation and Countermeasures, Hydraulic Engineering Circular 9, Report FHWA IF-04-016 (J.B. Bradley, D.L. Richards, C.D. Bahner).

FHWA, 2005b, "Culvert Pipe Liner Guide and Specifications," Report FHWA-CFL/TD-05-003 (C.I. Thornton, M.D. Robeson, L.G. Girard, B.A. Smith).

FHWA, 2006a, "Hydraulic Design of Energy Dissipators for Culverts and Channels," Hydraulic Engineering Circular 14, Report NHI-06-086 (P.L. Thompson and R.T. Kilgore).

FHWA, 2006b, "Bridge Inspectors Reference Manual," Report FHWA NHI 03-001 October, 2002 Revised 2006 (T.W. Ryan, R.A. Hartle, J.E. Mann, L.J. Danovich).

FHWA, 2006c, "Effects of Inlet Geometry on Hydraulic Performance of Box Culverts, Report FHWA-HRT-06-138 (J.S. Jones, K.Kerenyi, and S. Stein).

FHWA, 2008a, "Introduction to Highway Hydraulics," Hydraulic Design Series No. 4, report number NHI-08-090, (Schall, J.D., E.V. Richardson, and J.L. Morris).

FHWA, 2008b, "Project Development and Design Manual," prepared by Federal Lands Highway (FLH).

FHWA, 2009a, "Urban Drainage Design Manual," Hydraulic Engineering Circular 22, Report NHI-10-09 (S.A. Brown, J.D. Schall, J.L. Morris, C.L. Doherty, S.M. Stein, J.C. Warner).

FHWA, 2009b, "Bridge Scour and Stream Instability Countermeasures - Experience, Selection, and Design Guidelines," Third Edition, Report FHWA NHI 09-111, Federal Highway Administration, Hydraulic Engineering Circular No. 23, U.S. Department of Transportation, Washington, D.C. (Lagasse, P.F., L.W. Zevenbergen, J.D. Schall, and P.E. Clopper).

FHWA, 2010a, "Culvert Design for Aquatic Organism Passage," Hydraulic Engineering Circular 26, Report HIF-11-008 (R.T. Kilgore, B.S. Bergendahl and R.H. Hotchkiss).

FHWA, 2010b, "Culvert Assessment and Decision-making Procedures Manual," Report FHWA-CFL/TD-10-005 (J.H. Hunt, S.M. Zerges, B.C. Roberts, B. Bergendahl).

FHWA, 2012a, "Stream Stability at Highway Structures," HEC-20 Fourth Edition, Report No. HIF-FHWA-12-004 (Arneson, L.A., P.F. Lagasse, L.W. Zevenbergen, W.J. Spitz).

FHWA, 2012b, "Hydraulic Design of Safe Bridges," Report FHWA-HIF-12-018 Hydraulic Design Series No. 7, First Edition, Washington, D.C. (Zevenbergen, L.W., L.A. Arneson, J.H. Hunt, A.C. Miller).

Kaiser, 1984, "Hydraulic Design Detail," Kaiser Aluminum and Chemical Corporation DP-131, Edition 1.

King, H.W., and Brater, E.F, 1976, Handbook of Hydraulics, Sixth Edition, McGraw-Hill Book Co.

Linsley, R.F., Franzini, J.B., 1979, Water-Resources Engineering, Third Edition, McGraw-Hill, Inc.

Mays, L.W., M.E. Walker, M.S. Bennet, and R.P. Arbuckle, 1983, "Hydraulic Performance of Culverts with Safety Grates," FHWA/TX-82/55T301-1F; PB83-219626, Texas University, Austin Center for Transportation Research, Austin, TX.

Malone, T. and Parr, A.D., 2008, Bend Losses in Rectangular Culverts, Final Report to the Kansas Department of Transportation, University of Kansas, Dept of Civil, Env and Arch Engrg, Report K-TRANS KU-05-5.

May, D.K., Peterson, A.W., and Rajaratnam, N, 1986, "A Study of Manning's Roughness Coefficient for Commercial Concrete and Plastic Pipes," T. Blench Hydraulics Laboratory, University of Alberta, Edmonton, Alberta, January 26.

Mavis, F.T., 1939, The Construction of Nomographic Charts, Scranton, International Textbook, 132 pages.

Metcalf & Eddy, 1972, "Wastewater Engineering," McGraw-Hill Book Co., New York, N.Y.

Morris, H.H. and Wiggert, J.M., 1972, Applied Hydraulics in Engineering, John Wiley & Sons.

NBS, 1955, "Hydraulics of Short Pipes, Hydraulic Characteristics of Commonly Used Pipe Entrances, First Progress Report," National Bureau of Standards Report No. 4444, Washington, D.C., December (John L. French).

NBS, 1956, "Hydraulics of Culverts, Second Progress Report, Pressure and Resistance Characteristics of a Model Pipe Culvert," National Bureau of Standards Report No. 4911, Washington, D.C., October (John L. French).

NBS, 1957, "Hydraulics of Culverts, Third Progress Report, Effects of Approach Channel Characteristics on Model Pipe Culvert Operation," National Bureau of Standards Report 5306, Washington, D.C., June (John L. French).

NBS, 1961, "Hydraulics of Improved Inlet Structures for Pipe Culverts, Fourth Progress Report," National Bureau of Standards Report No. 7178, Washington, D.C. (J. L. French).

NBS, 1966a "Hydraulics of Culverts, Fifth Progress Report, Non-enlarged Box Culvert Inlets," National Bureau of Standards Report 9327, Washington, D.C., June (John L. French).

NBS, 1966b, "Hydraulics of Culverts, Sixth Progress Report, Tapered Box Culvert Inlets," National Bureau of Standards Report No 9355, (J.L. French).

NBS, 1967, "Hydraulics of Culverts, Seventh Progress Report, Tapered Box Inlets with Fall Concentration in the Inlet Structure," National Bureau of Standards Report No. 9528 (J. L. French and H.G. Bossy).

NRC, 1964, "Corrosion Performance of Aluminum Culvert," National Research Council Highway Research Record No. 56, Highway Research Board, National Research Council, Washington, D.C., 1964, pp. 98-115 (T.A. Lowe and A.H. Koepf).

NCHRP, 2011, "Hydraulic Loss Coefficients for Culverts," Project 15-24 (Principal Investigator B.L. Tullis).

Neale, L.C. and Price, R.E., 1964, "Flow Characteristics of PVC Sewer Pipe," ASCE Journal of the Sanitary Engineering Division, Div. Proc 90SA3, p 109-129, 1964.

Struab, L.G., Bowers, C.E. and Pilch, M., 1960, "Resistance to Flow in Two Types of Concrete Pipe," Technical Paper No. 22, Series B, University of Minnesota, St. Anthony Falls Hydraulic Laboratory, Minneapolis, MN, December.

Thurman, D. R. and Horner-Devine, A.R., 2007. "Hydrodynamic Regimes and Structures in Sloped Weir Baffled Culverts and Their Influence on Juvenile Salmon Passage," Washington State Department of Transportation, WA-RD 687.1, November.

TRB, 1980, "Evaluation of the Durability of Metal Drainage Pipe," Transportation Research Board Transportation Research Record N762, pp. 25-32 (R.W. Kinchen).

TxDOT, 2011, "Hydraulic Design Manual," Texas Department of Transportation, Roadway Design Section, Design Division, Austin, TX.

Tullis, J.P., 1983, Friction Factor Tests on Spiral Rib Pipes," Hydraulic Program Report No. 83, Utah Water Research Laboratory, Utah State University, Logan, UT, April.

Tullis, J.P., 1986, "Friction Factor Tests on Concrete Pipe," Hydraulics Report No. 157 Utah Water Research Laboratory, Utah State University, Logan, UT, October.

Tullis, J.P., 1991a, "Friction Factor Test of 36-Inch Concrete Pipe," Hydraulics Report No 281, Utah Water Research Laboratory, Utah State University, Logan, UT, April.

Tullis, J.P., 1991b, "Friction Factor Test on 24 and 48-Inch Spiral Rib Pipe," Hydraulics Report No. 280, Utah Water Research Laboratory, Utah State University, Logan, UT, April.

Tullis, J.P., 1991c, "Friction Factor Test on 24-Inch Helical Corrugated Pipe," Hydraulics Report No. 279, Utah Water Research Laboratory, Utah State University, Logan, UT, April.

Tullis, J.P., Watkins, R.K., and Barfuss, S.L., 1990, "Innovative New Drainage Pipe," ASCE Proceedings of the International Conference Pipeline Design and Installation, March 25-27.

USBR, 1985, "Test for Friction Factors in 18- and 24-Diameter Corrugated Tubing," U.S. Bureau of Reclamation, Engineering and Research Center letter and test results, October 25.

USBR, 1987, U.S. Bureau of Reclamation, Design of Small Dams, Third Edition, U.S. Government Printing Office, Washington, D.C.

USGS, 1967, "Roughness Characteristics of Natural Channels," U.S. Geological Survey Water Supply Paper 1849 (H. H. Barnes, Jr.).

USGS, 1968, "Measurement of peak discharge at culverts by indirect methods," U.S. Geological Survey Techniques of Water-Resources Investigations, book 3, chapter A3 (Bodhaine, G.L.).

USGS, 1984, "Guide for Selecting Manning's Roughness Coefficients for Natural Channels and Flood Plains," U.S. Geological Survey Water Supply Paper 2339 (G.J. Arcement, Jr. and V.R. Schneider).

USGS, 1981, "Stability of Relocated Stream Channels," U.S. Geological Survey Report No. FHWA/RD-80/158 (J. C. Brice).

USFS, 2006, "Low-water Crossings: Geomorphic, Biological and Engineering Design Considerations," U.S. Forest Service Report 0625 1808 SDTDC (K. Clarkin, G. Keller, T. Warhol and S. Hixson).

USFS, 2008, "Stream Simulation: An Ecological Approach to Road-Stream Crossings," U.S. Forest Service Forest Service Simulation Working Group (FSSWG).

Wolman, M.G., and J.P. Miller, 1960, Magnitude and frequency of forces in geomorphic processes. *J. Geol.* 68(1):54-74.

APPENDIX A

APPENDIX A

INLET CONTROL EQUATIONS

A.1 INTRODUCTION

This appendix contains the inlet control equations used to develop the design charts of this publication. Section A.2 contains the equations for the unsubmerged and submerged inlet control equations. Section A.3 demonstrates how the Section A.2 equations are used to create dimensionless design curves for culvert shapes with coefficients (Section A.3.1) and without (Section A.3.2). Section A.4 discusses how the dimensionless design curves are used to develop the nomographs in Appendix C. Section A.5 discusses how the dimensionless design curves are used to develop the polynomial equations used in FHWA software.

A.2 INLET CONTROL EQUATIONS

The equations used to develop the inlet control nomographs in Appendix C are based on the research conducted by the National Bureau of Standards (NBS) under the sponsorship of the Bureau of Public Roads (now the Federal Highway Administration). John L. French of the NBS produced seven progress reports as a result of this research. Of these, the first (NBS 1955) and fourth (NBS 1961) through seventh reports (NBS 1966a, 1966b, and 1967) dealt with the hydraulics of pipe and box culvert entrances with and without tapered inlets. Herbert G. Bossy of the FHWA provides an excellent synthesis of the research in his paper, "Hydraulics of conventional Highway Culverts" (Bossy 1961). Additional background on the development of the equations is found in HEC-13 (FHWA 1972a) and unpublished notebooks and notes (Bossy 1963 and Normann 1974).

The two basic conditions of inlet control depend upon whether the inlet end of the culvert is or is not submerged by the upstream headwater. If the inlet is not submerged by the headwater, the inlet performs as a weir and the unsubmerged equations are used (Section A.2.1). If the inlet is submerged by the headwater, the inlet performs as an orifice and the submerged equations are used (Section A.2.2).

Between the unsubmerged and the submerged conditions, there is a transition zone for which the NBS research provided only limited information. The transition zone is defined empirically by drawing a curve between and tangent to the curves defined by the unsubmerged and submerged equations. In most cases, the transition zone is short and the curve is easily constructed.

A.2.1 Unsubmerged Inlet Control Equations

The unsubmerged equation has two forms. Form (1) is based on the specific head at critical depth, adjusted with correction factors. Form (2) is an exponential equation similar to a weir equation. Form (1) is preferable from a theoretical standpoint, but Form (2) is easier to apply and is the only documented form of equation for some of the inlet control equations. Equations (A.1) and (A.2) apply up to about $Q/AD^{0.5} = 3.5$ (1.93 SI).

$$\text{Form (1)} \quad \frac{HW_i}{D} = \frac{H_c}{D} + K \left[\frac{K_u Q}{AD^{0.5}} \right]^M + K_s S \tag{A.1}$$

Form (2) $\dfrac{HW_i}{D} = K\left[\dfrac{K_uQ}{AD^{0.5}}\right]^M$ (A.2)

Where:

HW_i	Headwater depth above inlet control section invert, ft (m)
D	Interior height of culvert barrel, ft (m)
H_c	Specific head at critical depth ($d_c + V_c^2/2g$), ft (m)
Q	Discharge, ft³/s (m³/s)
A	Full cross sectional area of culvert barrel, ft² (m²)
S	Culvert barrel slope, ft/ft (m/m)
K, M, c, Y	Constants from Tables A.1, A.2, A.3
K_u	Unit conversion 1.0 (1.811 SI)
K_s	Slope correction, -0.5 (mitered inlets +0.7)

A.2.2 Submerged Inlet Control Equations

The submerged equation (A.3) applies above about $Q/AD^{0.5} = 4.0$ (2.21 SI). The terms are defined in Sections A.2.1.

$$\dfrac{HW_i}{D} = c\left[\dfrac{K_uQ}{AD^{0.5}}\right]^2 + Y + K_sS$$ (A.3)

A.3 INLET CONTROL DIMENSIONLESS DESIGN CURVES

The equations in Section A.2 may be used to develop design curves for any conduit shape or size. Careful examination of the equation constants for a given form of equation reveals that there is very little difference between the constants for a given inlet configuration. Therefore, given the necessary conduit geometry for a new shape from the manufacturer, a similar shape is chosen and the constants are used to develop new design curves. The curves may be quasi-dimensionless, in terms of $Q/AD^{0.5}$ and HW_i/D, or dimensional, in terms of Q and HW_i for a particular conduit size. To make the curves truly dimensionless, $Q/AD^{0.5}$ must be divided by $g^{0.5}$, but this produces small decimal numbers. Note that coefficients for rectangular (box) shapes should not be used for nonrectangular (circular, arch, pipe-arch, etc.) shapes and vice-versa. A constant slope value of 2 percent (0.02) is usually selected for the development of design curves. This is because the slope effect is small and the resultant headwater is conservatively high for sites with slopes exceeding 2 percent (except for mitered inlets). The procedure is illustrated in Section A.3.1.

A.3.1 Elliptical Structural Plate Example

Develop a dimensionless design curve for elliptical structural plate corrugated metal culverts, with the long axis horizontal. Assume a thin wall projecting inlet. Use the coefficients and exponents for a corrugated metal pipe-arch, a shape similar to an ellipse.

From Table A.1, Chart 34, Scale 3:

- Unsubmerged: Equation Form (1) with K = .0340 and M = 1.5
- Submerged: c = .0496 and Y = 0.53

Unsubmerged, equation Form 1 (Equation A.1):

$$\frac{HW_i}{D} = \frac{H_c}{D} + .0340\left(\frac{Q}{AD^{0.5}}\right)^{1.5} - (0.5)(0.02)$$

Submerged (Equation A.3):

$$\frac{HW_i}{D} = 0.0496\left(\frac{Q}{AD^{0.5}}\right)^{2} + 0.53 - (.5)(0.02)$$

A direct relationship between HW_i/D and $Q/AD^{0.5}$ may be obtained for the submerged condition. For the unsubmerged condition, it is necessary to obtain the flow rate and equivalent specific head at critical depth. At critical depth, the critical velocity head is equal to one-half the hydraulic depth.

$$\frac{V_c^2}{2g} = \frac{y_h}{2} = \frac{A_p}{2T_p}$$

Therefore, specific head at critical depth divided by D is:

$$\frac{H_c}{D} = \frac{d_c}{D} + \frac{y_h}{2D} \tag{A.4}$$

Since the Froude number equals 1.0 at critical depth, V_c can be determined from the Froude number equation and set equal to V_c in the continuity equation to solve for Q_c.

$$F_r = \frac{V_c}{(gy_h)^{0.5}} = 1 \qquad V_c = Q_c/A_p = (gy_h)^{0.5} \qquad Q_c = A_p(gy_h)^{0.5}, \text{ or}$$

$$\frac{Q_c}{AD^{0.5}} = \frac{A_p}{A}\left(g \cdot \frac{y_h}{D}\right)^{0.5} \tag{A.5}$$

From geometric data supplied by the manufacturer for a horizontal ellipse (Kaiser 1984), the necessary geometry is obtained to calculate H_c/D and $Q_c/AD^{5.0}$.

d_c/D	y_h/D	(Equation A.4) H_c/D	A_p/A	(Equation A.5) $Q_c/AD^{0.5}$
0.1	0.04	0.12	0.04	0.05
0.2	0.14	0.27	0.14	0.30
0.4	0.30	0.55	0.38	1.18
0.6	0.49	0.84	0.64	2.54
0.8	0.85	1.22	0.88	4.60
0.9	1.27	1.53	0.97	6.20
1.0	--	--	1.00	--

From the unsubmerged equation and above table:

$Q_c/AD^{0.5}$	$.0340 \times (Q_c/AD^{0.5})^{1.5}$	$+H_c/D$	$-0.5S=$	HW_i/D
0.05	0.0004	0.12	0.01	0.11
0.3	0.0054	0.27	0.01	0.27
1.18	0.044	0.55	0.01	0.58
2.54	0.138	0.84	0.01	0.55
4.6	0.336	1.22	0.01	1.54
6.2	0.525	1.53	0.01	2.05
--	--			

For the submerged equation, any value of $Q/AD^{0.5}$ may be selected, since critical depth is not involved:

$Q_c/AD^{0.5}$	$.0496 \times (Qc/AD0.5)2$	$+Y$	$-0.5S=$	HW_i/D
1	0.05	0.53	0.01	*0.57
2	0.2	0.53	0.01	*0.72
4	0.79	0.53	0.01	1.31
6	1.79	0.53	0.01	2.31
8	3.17	0.53	0.01	3.69
				*Obviously Unsubmerged

Note that overlapping values of HW_i/D were calculated in order to define the transition zone between the unsubmerged and the submerged states of flow. The results of the above calculations are plotted in Figure A.1. A transition line is drawn between the unsubmerged and the submerged curves. The scales are dimensionless in Figure A.1, but the figures could be used to develop dimensional curves for any selected size of elliptical conduit by multiplying: $Q/AD^{0.5}$ by $AD^{0.5}$ and HW_i/D by D.

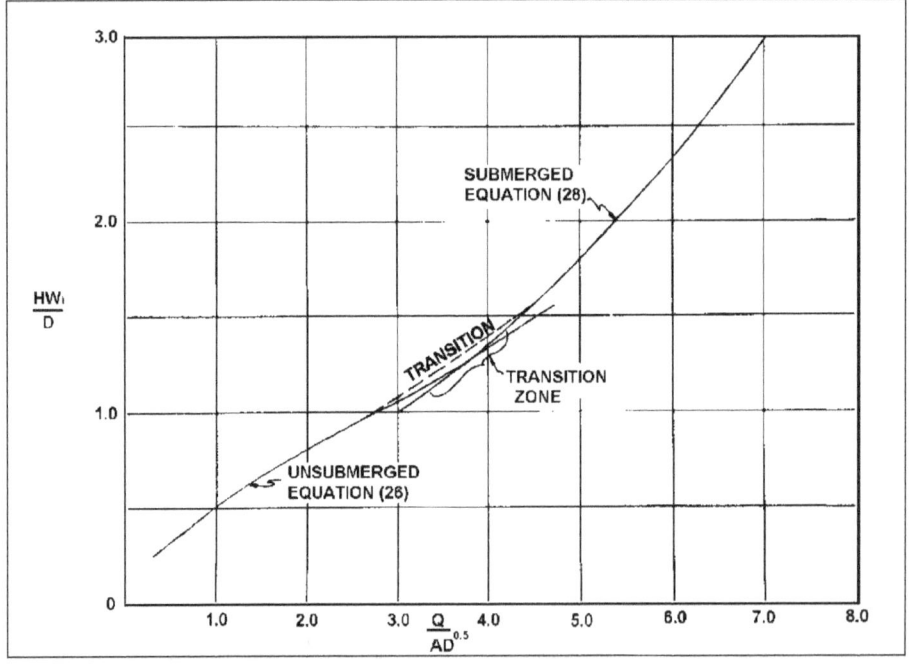

Figure A.1. Dimensionless performance curve for structural plate elliptical conduit, long axis horizontal, thin wall projecting entrance.

A.3.2 Dimensionless Design Charts for Culverts without Coefficients

The dimensionless inlet control design charts provided for long span arches (Chart 52) and for circular and elliptical pipes (Chart 51) were derived using the inlet control equations in Section A.2, selected constants from Table A.1, conduit geometry obtained from various tables and manufacturer's information (FHWA 1972b, Kaiser 1984, AISI 1983).

Some inlet configurations have no hydraulic tests. In lieu of such tests, the selected edge conditions should approximate the untested configurations and lead to a good estimate of culvert performance. In some cases, it will be necessary to evaluate the inlet edge configuration at a specific flow depth. For example, some inlets may behave as mitered inlets at low headwaters and as thin wall projecting inlets at high headwaters. The designer must apply engineering judgment in selection of the proper relationships for these major structures.

Unsubmerged Conditions. **Equation (A.1)** was used to calculate HW_i/D for selected inlet edge configurations. The following constants were taken from Table A.1, Chart 34 for pipe-arches, except for the 45 degree beveled edge inlet. These constants were taken from Chart 3, Scale A, for circular pipe. No constants were available from tests on pipe-arch models with beveled edges.

Inlet Edge	K	M	K_uS
Thin Wall Projecting	0.0340	1.5	-0.01
Mitered to Embankment	0.0300	1.0	+0.01
Square Edge in Headwall	0.0083	2.0	-0.01
Beveled Edge (45° Bevels)	0.0018	2.5	-0.01

Geometric relationships for the circular and elliptical (long axis horizontal) conduits were obtained from Tables 4 and 7 (FHWA 1972b), respectively. Geometric relationships for the high and low profile long span arches were obtained from DP-131 (Kaiser 1984) and the results were checked against tables in AISI handbook (AISI 1983).

Submerged Conditions. **Equation (A.3)** was used to calculate HW_i/D for the same inlet configurations using the following constants:

Inlet Edge	c	Y	K_uS
Thin Wall Projecting	0.0496	0.53	-0.01
Mitered to Embankment	0.0463	0.75	+0.01
Square Edge in Headwall	0.0496	0.57	-0.01
Beveled Edge (45° Bevels)	0.0300	0.74	-0.01

In terms of $Q/AD^{0.5}$, all non-rectangular shapes have practically the same dimensionless curves for submerged, inlet control flow. This is not true if $Q/BD^{1.5}$ is used as the dimensionless flow parameter.

To convert $Q/BD^{1.5}$ to $Q/AD^{0.5}$, divide by A/BD for the particular shape of interest as shown in Equation (A.6). This assumes that the shape is geometrically similar, so that A/BD is nearly constant for a range of sizes.

$$\frac{Q/BD^{1.5}}{(A/BD)} = \left(\frac{Q}{BD^{1.5}}\right)\left(\frac{BD}{A}\right) = \frac{Q}{AD^{0.5}}$$ (A.6)

Dimensionless Curves. By plotting the results of the unsubmerged and submerged calculations and connecting the resultant curves with transition lines, the dimensionless design curves shown in Charts 51 and 52 of Appendix C were developed. All high and low profile arches can be represented by a single curve for each inlet edge configuration. A similar set of curves was developed for circular and elliptical shapes. It is recommended that the high and low profile arch curves in Chart 52 be used for all true arch shapes (those with a flat bottom) and that the curves in Chart 51 be used for curved shapes including circles, ellipses, pipe-arches, and pear shapes.

A.3.3 Dimensionless Critical Depth Charts

Some of the long span culverts and special culvert shapes have no critical depth charts. These special shapes are available in numerous sizes, making it impractical to produce individual critical depth curves for each culvert size and shape. Therefore, dimensionless critical depth curves were developed for the shapes which have adequate geometric relationships in the manufacturer's literature. It should be noted that these special shapes are not truly geometrically similar, and any generalized set of geometric relationships will involve some degree of error. The amount of error is unknown since the geometric relationships were developed by the manufacturers.

The manufacturers' literature contains geometric relationships which include the hydraulic depth divided by the rise (inside height) of the conduit (y_h/D) and area of the flow prism divided by the barrel area (A_p/A) for various partial depth ratios, y/D. From Equation (A.5)

$$\frac{Q}{AD^{0.5}} = \frac{A_p}{A}\left(g \bullet \frac{y_h}{D}\right)^{0.5}$$ (A.7)

Setting y/D equal to d_c/D, it is possible to determine A_p/A and y_h/D at a given relative depth and then to calculate $Q_c/AD^{0.5}$. Dimensionless plots of d_c/D versus $Q_c/AD^{0.5}$ have been developed for the following culvert materials and shapes:

- Chart 20, corrugated metal box culverts (see Second edition HDS 5)
- Chart 44, corrugated metal arches (see Second edition HDS 5)
- Chart 53, Structural plate corrugated metal ellipses, long axis horizontal
- Chart 54, Structural plate corrugated metal arches, low and high profile

A.4 INLET CONTROL NOMOGRAPHS

The nomographs in Appendix C were developed using the equations in Section A.2 and the constants shown in Table A.1. The unsubmerged and submerged equations for a given shape, material and edge configuration were plotted using the dimensionless design curve procedures shown in Section A.3.1. A constant slope value of 2 percent (0.02) was used for the development of these design curves. This is because the slope effect is small and the resultant headwater is conservatively high for sites with slopes exceeding 2 percent (except for mitered inlets). A smooth transition was drawn by hand. This curve was the data used for constructing a nomograph. Dr. F. T. Mavis describes the process of making nomographs in

"The Construction of Nomographic Charts" (Mavis 1939). Nomographs were used extensively in engineering prior to the introduction of microcomputers in the early 1980s.

In formulating inlet and outlet control design nomographs, a certain degree of error is introduced into the design process. This error is due to the fact that the nomograph construction involves graphical fitting techniques resulting in scales which do not exactly match the equations. Checks by the authors of the first edition and others indicate that all of the nomographs from HEC-5 have precisions of $\pm$ 10 percent of the equation values in terms of headwater (inlet control) or head loss (outlet control). The nomographs for tapered inlets have errors of less than 5%, again in terms of headwater.

A.5 INLET CONTROL POLYNOMIAL EQUATIONS

The polynomial equations were developed to be used in software. The equations in Section A.2 with the constants shown in the tables of constants for a given shape, material and edge configuration were plotted using the dimensionless design curve procedures shown in Section A.3.1. The coordinates of selected points can be read from the curve and a best fit statistical analysis performed. A polynomial equation of the following form has been found to provide an adequate fit.

$$\frac{HW_i}{D} = A + B\left[\frac{Q}{BD^{1.5}}\right] + C\left[\frac{Q}{BD^{1.5}}\right]^2 + ... + X\left[\frac{Q}{BD^{1.5}}\right]^n + K_s S$$

For fitting the polynomial equations, $K_s = 0$ was used for most equations so that the slope correction could be applied by the software. The flow factor can be based on $AD^{0.5}$ rather than $BD^{1.5}$. The constants for the best fit equations are found in the HY-8 User Manual provided with HY-8. For equations that included $K_s S$, the A term is adjusted so that $K_u S = 0$. HY-8 uses the polynomial equations for all shapes that have constants determined in the laboratory or by FHWA. These include:

- Table A.1 - circles, boxes and tapered inlets (NBS, Bossy 1961)
- Table A.2 - pipe-arches, ellipses, metal boxes and arches (Bossy 1961)
- Table A.3 - South Dakota DOT RCB (FHWA 2006c)
- Table A.4 - open bottom concrete boxes (Chase 1999)
- Table A.5 - embedded circular shapes (NCHRP 2011)
- Table A.6 - embedded elliptical shapes (NCHRP 2011)

For shapes without constants, HY-8 uses Chart 52 developed using the procedures of Section A.3.2.

Table A.1. Constants for Inlet Control Equations for Charts in Appendix G.

Chart No	Shape and Material	Nomograph Scale	Inlet Configuration	Equation Form	Unsubmerged K	Unsubmerged M	Submerged c	Submerged Y	References
1	Circular Concrete	1	Square edge w/headwall	1	0.0098	2.0	0.0398	0.67	1, 2
1	Circular Concrete	2	Groove end w/headwall	1	0.0018	2.0	0.0292	0.74	1, 2
1	Circular Concrete	3	Groove end projecting	1	0.0045	2.0	0.0317	0.69	1, 2
2	Circular CM	1	Headwall	1	0.0078	2.0	0.0379	0.69	1, 2
2	Circular CM	2	Mitered to slope	1	0.0210	1.33	0.0463	0.75	1, 2
2	Circular CM	3	Projecting	1	0.0340	1.50	0.0553	0.54	1, 2
3	Circular	A	Beveled ring, 45° bevels	1	0.0018	2.50	0.0300	0.74	2
3	Circular	B	Beveled ring, 33.7° bevels*	1	0.0018	2.50	0.0243	0.83	2
8	Rect. Box Concrete	1	30° to 75° wingwall flares	1	0.026	1.0	0.0347	0.81	1, 3
8	Rect. Box Concrete	2	90° and 15° wingwall flares	1	0.061	0.75	0.0400	0.80	1, 3
8	Rect. Box Concrete	3	0° wingwall flares	1	0.061	0.75	0.0423	0.82	1, 3
9	Rect. Box Concrete	1	45° wingwall flare d = .043D	2	0.510	0.667	0.0309	0.80	3
9	Rect. Box Concrete	2	18° to 33.7° wingwall flare d = .083D	2	0.486	0.667	0.0249	0.83	3
10	Rect. Box Concrete	1	90° headwall w/3/4" chamfers	2	0.515	0.667	0.0375	0.79	3
10	Rect. Box Concrete	2	90° headwall w/45° bevels	2	0.495	0.667	0.0314	0.82	3
10	Rect. Box Concrete	3	90° headwall w/33.7° bevels	2	0.486	0.667	0.0252	0.865	3
11	Rect. Box Concrete	1	3/4" chamfers; 45° skewed headwall	2	0.545	0.667	0.04505	0.73	3
11	Rect. Box Concrete	2	3/4" chamfers; 30° skewed headwall	2	0.533	0.667	0.0425	0.705	3
11	Rect. Box Concrete	3	3/4" chamfers; 15° skewed headwall	2	0.522	0.667	0.0402	0.68	3
11	Rect. Box Concrete	4	45° bevels; 10°-45° skewed headw.	2	0.498	0.667	0.0327	0.75	3
12	Rect. Box 3/4" chamf. Conc.	1	45° non-offset wingwall flares	2	0.497	0.667	0.0339	0.803	3
12	Rect. Box 3/4" chamf. Conc.	2	18.4° non-offset wingwall flares	2	0.493	0.667	0.0361	0.806	3
12	Rect. Box 3/4" chamf. Conc.	3	18.4° non-offset wingwall flares 30° skewed barrel	2	0.495	0.667	0.0386	0.71	3
13	Rect. Box Top Bev. Conc.	1	45° wingwall flares - offset	2	0.497	0.667	0.0302	0.835	3
13	Rect. Box Top Bev. Conc.	2	33.7° wingwall flares - offset	2	0.495	0.667	0.0252	0.881	3
13	Rect. Box Top Bev. Conc.	3	18.4° wingwall flares - offset	2	0.493	0.667	0.0227	0.887	3
55	Circular	1	Smooth tapered inlet throat	2	0.534	0.555	0.0196	0.90	4
55	Circular	2	Rough tapered inlet throat	2	0.519	0.64	0.0210	0.90	4
56	Ellipital Face	1	Tapered inlet-beveled edges	2	0.536	0.622	0.0368	0.83	4
56	Ellipital Face	2	Tapered inlet-square edges	2	0.5035	0.719	0.0478	0.80	4
56	Ellipital Face	3	Tapered inlet-thin edge projecting	2	0.547	0.80	0.0598	0.75	4
57	Rectangular Concrete	1	Tapered inlet throat	2	0.475	0.667	0.0179	0.97	4
58	Rectangular Concrete	1	Side tapered-less favorable edges	2	0.56	0.667	0.0446	0.85	4
58	Rectangular Concrete	2	Side tapered-more favorable edges	2	0.56	0.667	0.0378	0.87	4
59	Rectangular Concrete	1	Slope tapered-less favorable edges	2	0.50	0.667	0.0446	0.65	4
59	Rectangular Concrete	2	Slope tapered-more favorable edges	2	0.50	0.667	0.0378	0.71	4

[1]Bossy 1963, [2]FHWA 1974, [3]NBS 5th, [4]HEC 13

Table A.2. Constants for Inlet Control Equations for Discontinued Charts (see 2005 HDS 5).

Chart No.	Shape and Material	Inlet Configuration	Nomograph Scale	Equation Form	Unsubmerged K	Unsubmerged M	Submerged c	Submerged Y	References
16-19	Boxes CM	90° headwall	2	1	0.0083	2.0	0.0379	0.69	FHWA 1974
16-19	Boxes CM	Thick wall projecting	3	1	0.0145	1.75	0.0419	0.64	FHWA 1974
16-19	Boxes CM	Thin wall projecting	5	1	0.0340	1.5	0.0496	0.57	FHWA 1974
29	Horizontal Ellipse Concrete	Square edge w/headwall	1	1	0.0100	2.0	0.0398	0.67	FHWA 1974
29	Horizontal Ellipse Concrete	Groove end w/headwall	2	1	0.0018	2.5	0.0292	0.74	FHWA 1974
29	Horizontal Ellipse Concrete	Groove end projecting	3	1	0.0045	2.0	0.0317	0.69	FHWA 1974
30	Vertical Ellipse Concrete	Square edge w/headwall	1	1	0.0100	2.0	0.0398	0.67	FHWA 1974
30	Vertical Ellipse Concrete	Groove end w/headwall	2	1	0.0018	2.5	0.0292	0.74	FHWA 1974
30	Vertical Ellipse Concrete	Groove end projecting	3	1	0.0095	2.0	0.0317	0.69	FHWA 1974
34	Pipe Arch 18" Corner radius CM	90° headwall	1	1	0.0083	2.0	0.0379	0.69	FHWA 1974
34	Pipe Arch 18" Corner radius CM	Mitered to slope	2	1	0.0300	1.0	0.0463	0.75	FHWA 1974
34	Pipe Arch 18" Corner radius CM	Projecting	3	1	0.0340	1.5	0.0496	0.57	FHWA 1974
35	Pipe Arch 18" Corner radius CM	Projecting	1	1	0.0300	1.5	0.0496	0.57	Bossy 1963
35	Pipe Arch 18" Corner radius CM	No Bevels	2	1	0.0088	2.0	0.0368	0.68	Bossy 1963
35	Pipe Arch 18" Corner radius CM	33.7° Bevels	3	1	0.0030	2.0	0.0269	0.77	Bossy 1963
36	Pipe Arch 31" Corner radius CM	Projecting	1	1	0.0300	1.5	0.0496	0.57	Bossy 1963
36	Pipe Arch 31" Corner radius CM	No Bevels	2	1	0.0088	2.0	0.0368	0.68	Bossy 1963
36	Pipe Arch 31" Corner radius CM	33.7° Bevels	3	1	0.0030	2.0	0.0269	0.77	Bossy 1963
41-43	Arch CM	90° headwall	1	1	0.0083	2.0	0.0379	0.69	FHWA 1974
41-43	Arch CM	Mitered to slope	2	1	0.0300	1.0	0.0473	0.75	FHWA 1974
41-43	Arch CM	Thin wall projecting	3	1	0.0340	1.5	0.0496	0.57	FHWA 1974

Table A.3. Constants for Inlet Control Equations for South Dakota Concrete Box (HY-8 User Manual and Table 11 of FHWA 2006).

Sketch	Wingwall Flare	Top Bevel	Top Radius	Corner Fillet	RCB Inlet Configuration	Equation Form	Unsubmerged K	Unsubmerged M	Submerged c	Submerged Y
1	30°	45°	-	-	Single barrel	2	0.44	0.74	0.040	0.48
2	30°	45°	-	6"	Multiple barrel (2, 3, and 4 cells)	2	0.47	0.68	0.04	0.62
3	30°	45°	-	-	Single barrel (2:1 to 4:1 span-to-rise ratio)	2	0.48	0.65	0.041	0.57
4	30°	45°	-	-	Multiple barrels (15° skewed headwall)	2	0.69	0.49	0.029	0.95
5	30°	45°	-	-	Multiple barrels (30° to 45° skewed headwall)	2	0.69	0.49	0.027	1.02
6	0°	none	-	-	Single barrel, top edge 90°	2	0.55	0.64	0.047	0.55
7	0°	45°	-	6"	Single barrel, (0 and 6-inch corner fillets)	2	0.56	0.62	0.045	0.55
8	0°	45°	-	6"	Multiple barrels (2, 3, and 4 cells)	2	0.55	0.59	0.038	0.69
9	0°	45°	-	-	Single barrels 2:1 to 4:1 span-to-rise ratio)	2	0.61	0.57	0.041	0.67
10	0°	-	8"	6"	Single barrel (0 and 6-inch fillets)	2	0.56	0.62	0.038	0.67
11	0°	-	8"	12"	Single barrel (12-inch corner fillets)	2	0.56	0.62	0.038	0.67
12	0°	-	8"	12"	Multiple barrels (2, 3, and 4 cells)	2	0.55	0.6	0.023	0.96
13	0°	-	8"	12"	Single barrel (2:1 to 4:1 span-to-rise ratio)	2	0.61	0.57	0.033	0.79

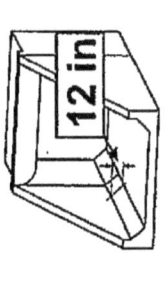

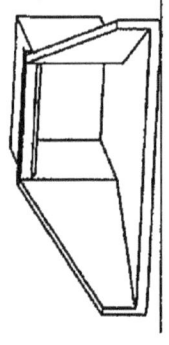

Sketches are shown in the HY-8 documentation and research report. Since sketches 2 and 8 show fillets, a 6-inch fillet is assumed.

Sketches 1 through 5 have this configuration.   Sketches 7 through 13 have this configuration.

Table A.4. Constants for Inlet Control Equations for Concrete Open-Bottom Arch (Chase 1999).

Span to Rise[1]	Wingwall Flare	Top Edge	Inlet Configuration	Equation Form	Unsub-merged K	Unsub-merged M	Sub-merged c	Sub-merged Y
2:1	0°	90°	Mitered to conform to slope	2	0.44	0.74	0.040	0.48
2:1	45°	90°	Headwall with wingwalls	2	0.47	0.68	0.04	0.62
2:1	90°	90°	Headwall	2	0.48	0.65	0.041	0.57
4:1	0°	90°	Mitered to conform to slope	2	0.69	0.49	0.029	0.95
4:1	45°	90°	Headwall with wingwalls	2	0.69	0.49	0.027	1.02
4:1	90°	90°	Headwall	2	0.56	0.62	0.045	0.55

[1]The 2:1 constants are used for ratios less than or equal to 3:1 and the 4:1 constants for ratios greater than 3:1.

Table A.5. Constants for Inlet Control Equations for Embedded Circular Shapes (NCHRP 15-24).

Embedded	Top Edge	Inlet Configuration	Unsub-merged K Form 1	Unsub-merged M Form 1	Unsub-merged K Form 2	Unsub-merged M Form 2	Sub-merged c	Sub-merged Y
0.2D	thin	Projecting End, Ponded	0.0860	0.58	0.4293	0.64	0.0303	0.58
0.2D	thin	Projecting End, Channelized	0.0737	0.45	0.4175	0.62	0.0250	0.63
0.2D	--	Mitered End 1.5H:1V	0.0431	0.58	0.4002	0.63	0.0235	0.61
0.2D	90°	Square Headwall	0.0566	0.44	0.4001	0.63	0.0198	0.69
0.2D	45°	Beveled End	0.0292	0.57	0.3869	0.63	0.0161	0.73
0.4D	thin	Projecting End, Ponded	0.0840	0.76	0.4706	0.69	0.0453	0.69
0.4D	thin	Projecting End, Channelized	0.0927	0.59	0.4789	0.66	0.0441	0.52
0.4D	--	Mitered End 1.5H:1V	0.0317	0.77	0.4185	0.68	0.0363	0.65
0.4D	90°	Square Headwall	0.0490	0.71	0.4354	0.68	0.0332	0.67
0.4D	45°	Beveled End	0.0358	0.62	0.4223	0.67	0.0245	0.75
0.5D	thin	Projecting End, Ponded	0.1057	0.69	0.4955	0.71	0.0606	0.54
0.5D	thin	Projecting End, Channelized	0.1055	0.59	0.4955	0.69	0.0570	0.48
0.5D	--	Mitered End 1.5H:1V	0.0351	0.59	0.4419	0.68	0.0504	0.44
0.5D	90°	Square Headwall	0.0595	0.59	0.0595	0.59	0.0402	0.65
0.5D	45°	Beveled End	0.0464	0.46	0.4364	0.69	0.0324	0.67

Table A.6. Constants for Inlet Control Equations for Embedded Elliptical Shape (NCHRP 15-24).

Embedded	Top Edge	Inlet Configuration	Unsub-merged K Form1	Unsub-merged M Form 1	Unsub-merged K Form 2	Unsub-merged M Form 2	Sub-merged c	Sub-merged Y
0.5D	thin	Projecting End, Ponded	0.1231	0.51	0.5261	0.65	0.0643	0.50
0.5D	thin	Projecting End, Channelized	0.0928	0.54	0.4937	0.67	0.0649	0.12
0.5D	--	Mitered End 1.5H:1V	0.0599	0.60	0.4820	0.67	0.0541	0.50
0.5D	90°	Square Headwall	0.0819	0.45	0.4867	0.66	0.0431	0.61
0.5D	45°	Beveled End	0.0551	0.52	0.4663	0.63	0.0318	0.68

(page intentionally left blank)

APPENDIX B

APPENDIX B

HYDRAULIC RESISTANCE OF CULVERT BARRELS

B.1 OVERVIEW

In outlet control, the hydraulic resistance of the culvert barrel must be calculated using a friction loss equation. Numerous equations, both theoretical and empirical, are available, including the Darcy equation and the Manning equation. The Darcy equation, shown in Equation (B.1), is theoretically correct, and is described in most hydraulic texts.

$$h_f = f \left(\frac{L}{D} \right) \left(\frac{V^2}{2g} \right) \tag{B.1}$$

h_f is the friction head loss, ft (m)
f is the Darcy resistance factor
L is the conduit length, ft (m)
D is the conduit diameter, ft (m)
V is the mean velocity, ft/s (m/s)
g is the acceleration due to gravity, 32.2 ft/s^2 (9.8 m/s^2)

The Darcy friction factor, f, is selected from a chart commonly referred to as the Moody diagram, which relates f to Reynolds number (flow velocity, conduit size, and fluid viscosity) and relative roughness (ratio of roughness element size to conduit size). To develop resistance coefficients for new and untested wall roughness configurations, the Darcy f value can be derived theoretically and then converted to a Manning's n value through use of the relationship shown in Equation (B.2).

$$n = 0.0926 \; R^{1/6} \; f^{1/2}$$
(B.2)

R is the hydraulic radius, ft

A comprehensive discussion of the Darcy f, its derivation, and its relationship to other resistance coefficients is given in reference (Morris and Wiggert 1972).

The Manning equation, an empirical relationship, is commonly used to calculate the barrel friction losses in culvert design. The usual form of the Manning equation is as follows:

$$V = \frac{1.486}{n} \; R^{2/3} \; S^{1/2}$$
(B.3)

V   is the mean velocity of flow, ft/s (m/s)
R   is the hydraulic radius, ft (m)
S   is the slope of the conduit, ft/ft, equal to the slope of the water surface in uniform flow

Substituting H_f/L for S and rearranging Equation (B.3) results in Equation (3.4b).

The Manning's n value in Equation (B.3) is based on either hydraulic test results or resistance values calculated using a theoretical equation such as the Darcy equation and then converting to the Manning's n. As is seen from Equation (B.2), the Manning's n varies with the conduit size (hydraulic radius) to the 1/6 power and has dimensions of ft$^{1/6}$.

Therefore, for very large or very small conduits, the Manning's n should be adjusted for conduit size. Most hydraulic tests for Manning's n values have been conducted on moderate size conduits, with pipes in the range of 2 to 5 ft (600 to 1500 mm) in diameter or on open channels with hydraulic radii in the range of 1 to 4 ft (0.3 to 1.2 m). For large natural channels, backwater calculations are used to match observed water surface profiles by varying the Manning's n. The resultant Manning's n accounts for channel size and roughness.

Using a constant value of Manning's n regardless of conduit size or flow rate assumes that the Manning's n is a function of only the absolute size of the wall roughness elements and is independent of conduit size and Reynolds number. This assumption is best for rough conduits where Reynolds number has little influence and the inherent variation with conduit size to the 1/6 power holds true. Thus, the Manning equation has found wide acceptance for use in natural channels and conduits with rough surfaces. For smooth pipes, other empirical resistance equations, such as the Hazen-Williams equation, are more often used.

Table B.1 summarizes the Manning's n values for materials commonly used in culvert construction. For the corrugated metal culverts, the specified range of n values is related to the size of the conduit (Section B.4). For other conduits, the range shown relates to the quality of the conduit construction (Sections B.2 and B.3). In all cases, judgment is necessary in selecting the proper Manning's n value, and the designer is directed to other references for additional guidance in special situations. Extensive tables of Manning's n values are provided in HDS 3 (FHWA 1961) and Open Channel Hydraulics (Chow 1959). For natural channels, the designer is referred to Table C.1 in Appendix C as well as to USGS 1967 and USGS 1984. Manning's n values for commonly used culvert materials are discussed in the following sections.

B.2 CONCRETE PIPE CULVERTS

Concrete pipes are manufactured (precast) using various methods, including centrifugally spun, dry cast, packerhead, tamp, and wet cast (ACPA 1981). The interior finish (wall roughness) varies with the method of manufacture. For instance, the tamped process generally results in a rougher interior finish than the wet cast process. The quality of the joints and aging (abrasion and corrosion) also affect the hydraulic resistance of concrete pipe. Laboratory tests on tamped pipe (24 to 36 in. (600 to 900 mm) average to good joints) resulted in Manning's n values of about 0.009 (Straub et al. 1960). These values are increased to 0.011 to 0.013 based on field installation and aging. Laboratory determined values of Manning's n for concrete pipes are shown in Table B.1.

B.3 CONCRETE BOX CULVERT

The hydraulic resistance of concrete box culverts is based on the method of manufacture, installation practices and aging. Concrete box culverts are either precast or cast-in-place. For precast boxes, the smoothness of the walls, the quality of the joints, and aging affect the Manning's n values. For cast-in-place boxes, the quality of the formwork, construction practices, and aging are factors. Manning's n values for cast-in-place boxes range from 0.012 to 0.022 see concrete lined open channels from HDS 3 Table 1 (FHWA 1961). For precast sections, the smoother laboratory values shown for concrete pipes can be assumed.

B.4 CORRUGATED METAL CULVERTS

The hydraulic resistance coefficients for corrugated metal conduits are based on the size and shape of the corrugations, spacing of the corrugations, type of joints, bolt or rivet roughness, method of manufacture, size of conduit, flow velocity, and aging. A complete description of the hydraulic resistance of corrugated metal conduits is presented in the publication, "Hydraulic Flow Resistance Factors for Corrugated Metal Conduits" (FHWA 1980). Information from that report has been condensed and included herein. The resistance values provided in this Appendix are based on specific criteria, including the use of a typical culvert flow rate ($Q/D^{2.5} = 4.0$). Bolt and joint effects, where appropriate, are included.

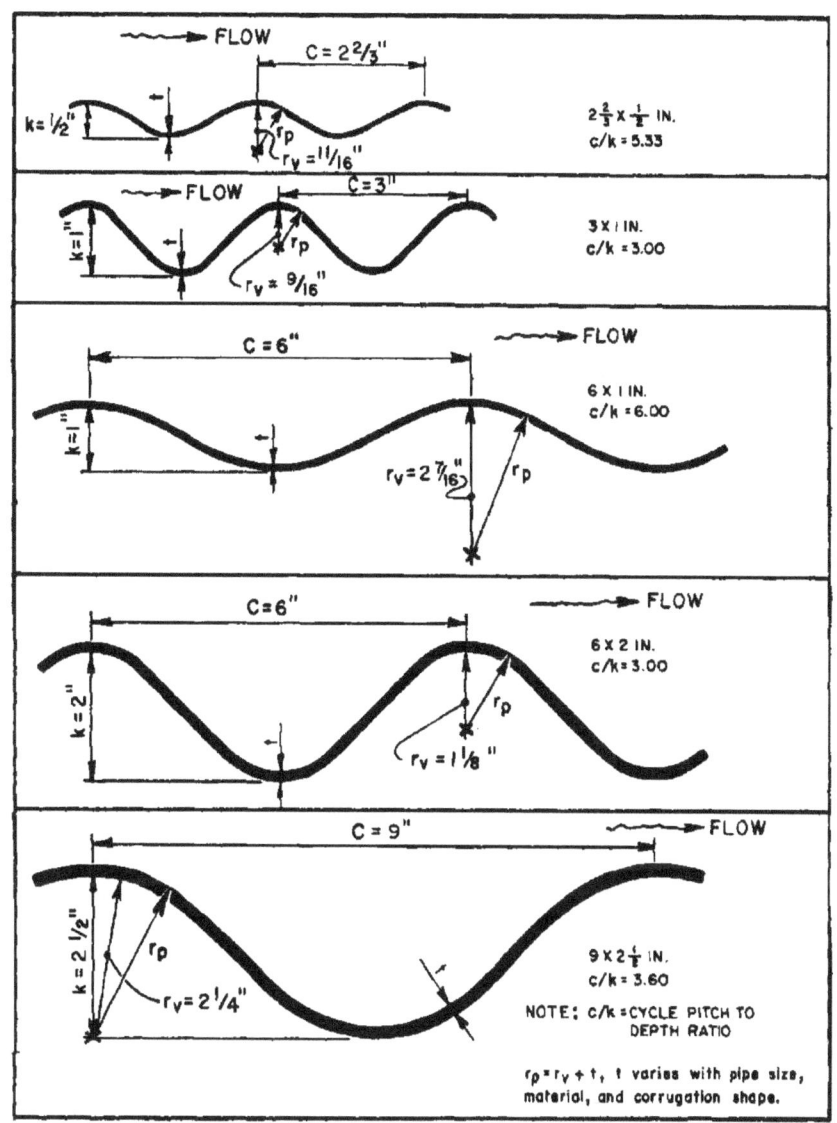

Figure B.1. Shapes of annular corrugations.

B.4.1 Annular Corrugations

Resistance factors for the annular corrugation shapes shown in Figure B-1 (FHWA 1980). Methods are also presented for estimating the hydraulic resistance of new or untested corrugation types. Those methods have been used to estimate the resistance of 5-by 1-inch corrugations, shown in Figure B-2, for which no test results are yet available (AISI 1983).

A series of charts were developed depicting the Manning's n resistance value for various corrugation shapes over a range of conduit sizes (FHWA 1980). The charts show the variation of Manning's n value with diameter, flow rate, and depth. The curves for structural plate conduits have discontinuities due to changes in the number of plates used to fabricate the conduits. Curves are presented for two flow rates, $Q/D^{2.5} = 2.0$ and $Q/D^{2.5} = 4.0$. Under design conditions, culvert flow rates approximate the $Q/D^{2.5} = 4.0$ curves.

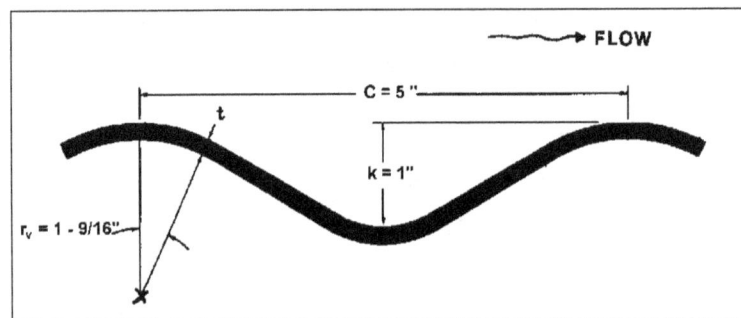

Figure B.2. Shape of 5 in by 1 in corrugation.

B.4.2 Helical Corrugations

In pipes less than about 6 ft (1800 mm) in diameter, helical corrugations may provide lower resistance values. This is due to the spiral flow which develops when such conduits flow full. As the pipe size increases, the helix angle approaches 90 degrees, and the Manning's n value is the same as for pipes with annular corrugations.

For partial flow in circular metal pipes with 2-2/3 by 1/2 in. (68 mm by 13 mm) in helical corrugations, Manning's n should be 11% higher than that for the full flow. In the case of full flow in corrugated metal pipe-arches with 2-2/3 by 1/2 in. (68 mm by 13 mm), Manning's n is the same as an equivalent diameter pipe.

B.4.3 Corrugated Metal n Value versus Diameter

- Based on the charts for annular and helical corrugations (FHWA 1980), Figure B-3 has been developed to show the variation in n value with diameter for corrugated metal conduits. The figure is based on the following assumptions which reduce the complexity of the relationships: The curves are based on $Q/D^{2.5} = 4.0$, which is typical of culvert design flow rates.
- Discontinuities inherent in the structural plate curves have been ignored in favor of a smooth curve.
- The only helically corrugated metal conduit curve shown is for 2-2/3 by 1/2 in. (68 x 13 mm) corrugations, with a 24 in. (600 mm) plate width.

To use Figure B-3, enter the horizontal scale with the circular conduit diameter and read the Manning's n from the curve for the appropriate corrugation.

B.5 SPIRAL RIB PIPE

Spiral rib pipe is smooth walled metal pipe fabricated using helical seams. Roughness elements include the joints and a helical, recessed rib running spirally around the pipe. Based on tests of 2 and 3 ft (600 to 900 mm) diameter spiral rib pipe (Tullis 1983), the pipe has essentially the same hydraulic resistance as smooth steel pipe, plus joint and aging effects. The laboratory test results indicate Manning's n values of from 0.010 to 0.011. Allowing for aging and higher joint resistance, Manning's n values in the range of 0.012 to 0.013 are recommended for design use. In using these low resistance values, the designer should ascertain that no large roughness elements such as projecting interior ribs or poor joints are present.

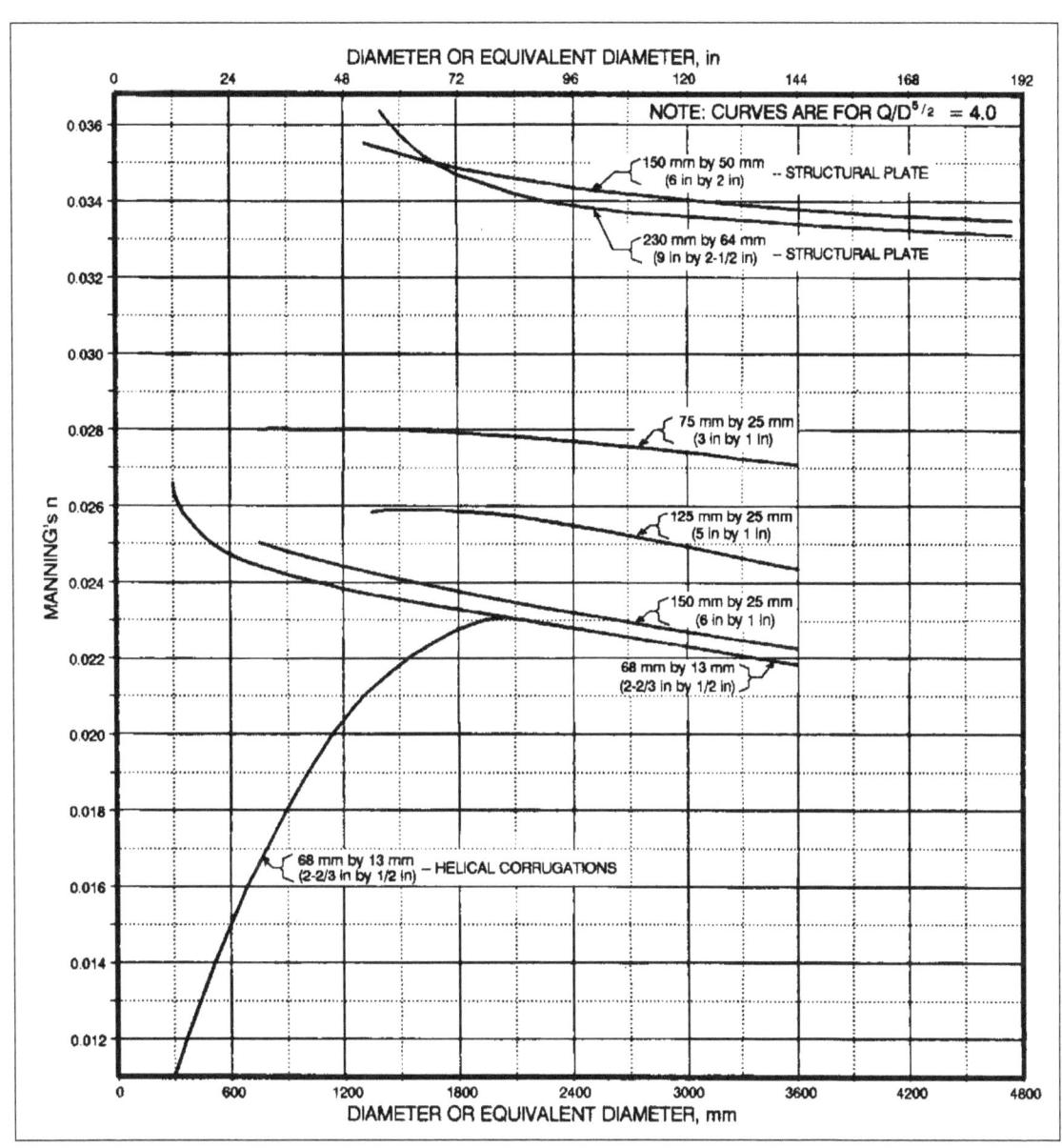

Figure B.3. Manning's n versus diameter for corrugated metal conduits.

B.1. Manning's n Values for Culverts.[1]			
Type of Culvert	Roughness or Corrugation	Manning's n	Reference
Concrete Pipe	Smooth	0.010-0.011	Straub et al. 1960 May et al. 1986 Tullis 1986 & 1991a
Concrete Boxes	Smooth	0.012-0.015	FHWA 1961
Spiral Rib Metal Pipe	Smooth	0.012-0.013	Tullis 1983 & 1991b
Corrugated Metal Pipe[2] (Helical Corrugations)	2-2/3 by 1/2 in 68 by 13 mm	0.011-0.023	FHWA 1980 Tullis 1991c
Corrugated Metal Pipe[2] (Helical Corrugations)	6 by 1 in 150 by 25 mm	0.022-0.025	FHWA 1980
Corrugated Metal Pipe[2], Pipe-Arch and Box (Annular Corrugations)	2-2/3 by 1/2 in 68 by 13 mm	0.022-0.027	FHWA 1980
Corrugated Metal Pipe[2], Pipe-Arch and Box (Annular Corrugations)	5 by 1 in 125 by 25 mm	0.025-0.026	FHWA 1980
Corrugated Metal Pipe[2], Pipe-Arch and Box (Annular Corrugations)	3 by 1 in 75 by 25 mm	0.027-0.028	FHWA 1980
Corrugated Metal Structural Plate[2] (Annular Corrugations)	6 by 2 in 150 by 50 mm	0.033-0.035	FHWA 1980
Corrugated Metal Structural Plate[2] (Annular Corrugations)	9 by 2-1/2 in 230 by 64 mm	0.033-0.037	FHWA 1980
Corrugated Polyethylene	Smooth	0.009-0.015	Barfuss & Tullis 1988 Tullis et al. 1990
Corrugated Polyethylene	Corrugated	0.018-0.025	Clyde 1980 USBR 1985
Polyvinyl chloride (PVC)	Smooth	0.009-0.011	Neale and Price 1964 Bishop and Jeppson 1975

[1]The Manning's n values indicated in this table were obtained in the laboratory and are supported by the provided reference. Actual field values for culverts may vary depending on the effect of abrasion, corrosion, deflection, and joint conditions.

[2]See Figure B.3, Manning's n varies with barrel size.

APPENDIX C

APPENDIX C

DESIGN CHARTS, TABLES, AND FORMS

Note: **Design Charts are given in SI and English Units. For example, Chart 1A is in SI Units and Chart 1B is in English Units.** English design charts have a small symbol in the upper outside corner representing the shape involved. SI charts have the chart number in a box.

(page intentionally left blank)

Table C.1. Manning's n for Small Natural Stream Channels. Surface width at flood stage less than 30 m (100 ft)

1. Fairly regular section:

   a. Some grass and weeds, little or no brush... 0.030--0.035

   b. Dense growth of weeds, depth of flow materially greater
      than weed height...0.035-0.05

   c. Some weeds, light brush on banks 0.035--0.05

   d. Some weeds, heavy brush on banks 0.05--0.07

   e. Some weeds, dense willows on banks 0.06--0.08

   f. For trees within channel, with branches submerged at high stage,
      increase all above values by 0.01--0.02

2. Irregular sections, with pools, slight channel meander; increase values
   given above about.. 0.01--0.02

3. Mountain streams, no vegetation in channel, banks usually steep, trees
   and brush along banks submerged at high stage:

   a. Bottom of gravel, cobbles, and few boulders 0.04--0.05

   b. Bottom of cobbles, with large bounders.. 0.05--0.07

Table C.2. Entrance Loss Coefficients.

Outlet Control, Full or Partly Full Entrance Head Loss

$$H_e = K_e \left[\frac{V^2}{2g} \right]$$

Type of Structure and Design of Entrance	Coefficient K_e
• Pipe, Concrete	
Projecting from fill, socket end (groove-end)	0.2
Projecting from fill, sq. cut end	0.5
Headwall or headwall and wingwalls	
Socket end of pipe (groove-end	0.2
Square-edge	0.5
Rounded (radius = D/12	0.2
Mitered to conform to fill slope	0.7
*End-Section conforming to fill slope	0.5
Beveled edges, 33.7^0 or 45^0 bevels	0.2
Side- or slope-tapered inlet	0.2
• Pipe. or Pipe-Arch. Corrugated Metal	
Projecting from fill (no headwall)	0.9
Headwall or headwall and wingwalls square-edge	0.5
Mitered to conform to fill slope, paved or unpaved slope	0.7
*End-Section conforming to fill slope	0.5
Beveled edges, 33.7^0 or 45^0 bevels	0.2
Side- or slope-tapered inlet	0.2
• Box, Reinforced Concrete	
Headwall parallel to embankment (no wingwalls)	
Square-edged on 3 edges	0.5
Rounded on 3 edges to radius of D/12 or B/12	
or beveled edges on 3 sides	0.2
Wingwalls at 30^0 to 75^0 to barrel	
Square-edged at crown	0.4
Crown edge rounded to radius of D/12 or beveled top edge	0.2
Wingwall at 10^0 to 25^0 to barrel	
Square-edged at crown	0.5
Wingwalls parallel (extension of sides)	
Square-edged at crown	0.7
Side- or slope-tapered inlet	0.2

*Note: "End Sections conforming to fill slope," made of either metal or concrete, are the sections commonly available from manufacturers. From limited hydraulic tests they are equivalent in operation to a headwall in both inlet and outlet control. Some end sections, incorporating a closed taper in their design have a superior hydraulic performance. These latter sections can be designed using the information given for the beveled inlet.

(page intentionally left blank)

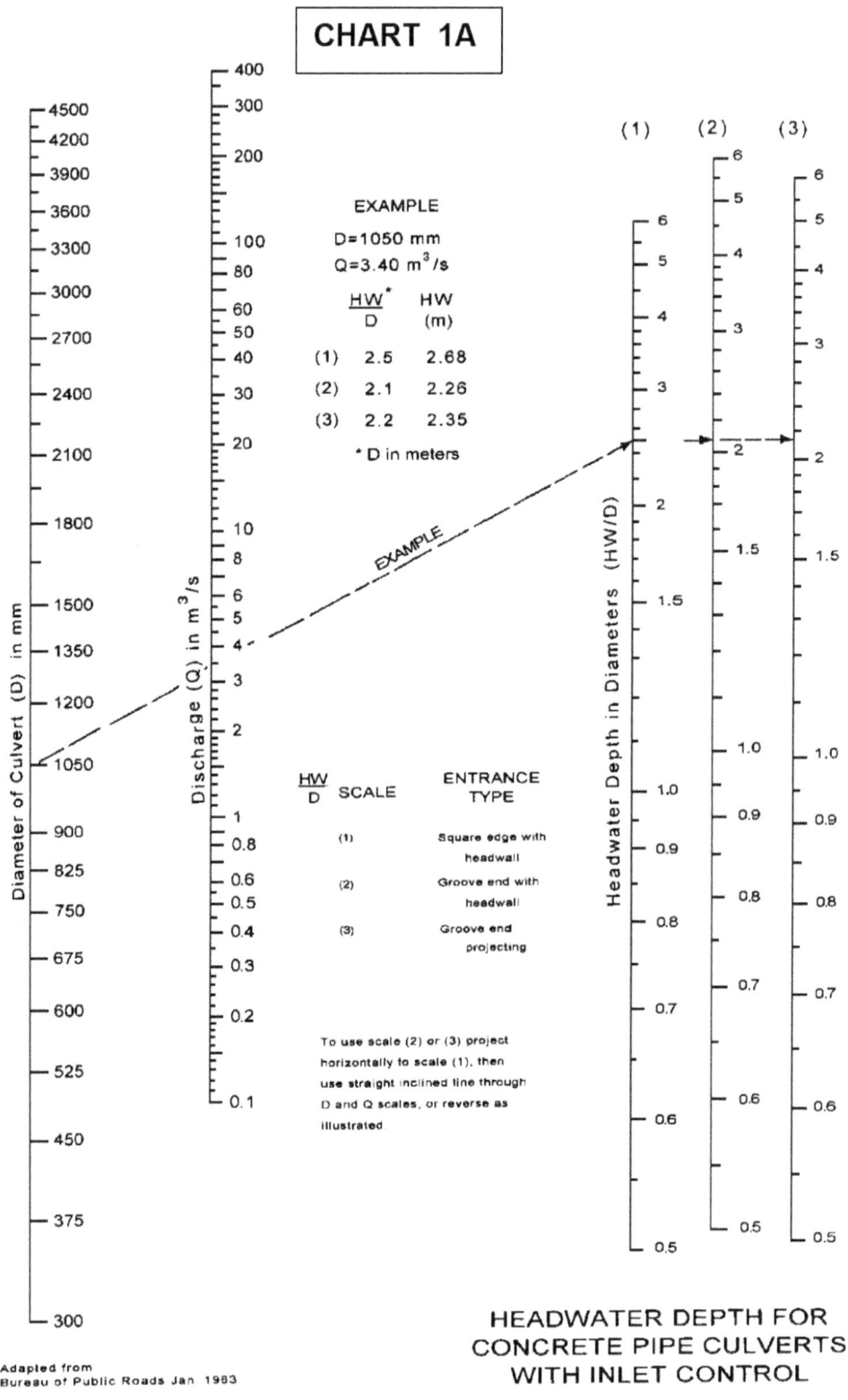

CHART 1A

EXAMPLE

D=1050 mm

Q=3.40 m³/s

	$\frac{HW^*}{D}$	HW (m)
(1)	2.5	2.68
(2)	2.1	2.26
(3)	2.2	2.35

* D in meters

Diameter of Culvert (D) in mm

Discharge (Q) in m³/s

Headwater Depth in Diameters (HW/D)

$\frac{HW}{D}$ SCALE	ENTRANCE TYPE
(1)	Square edge with headwall
(2)	Groove end with headwall
(3)	Groove end projecting

To use scale (2) or (3) project
horizontally to scale (1), then
use straight inclined line through
D and Q scales, or reverse as
illustrated

EXAMPLE

**HEADWATER DEPTH FOR
CONCRETE PIPE CULVERTS
WITH INLET CONTROL**

Adapted from
Bureau of Public Roads Jan 1983

CHART 1B

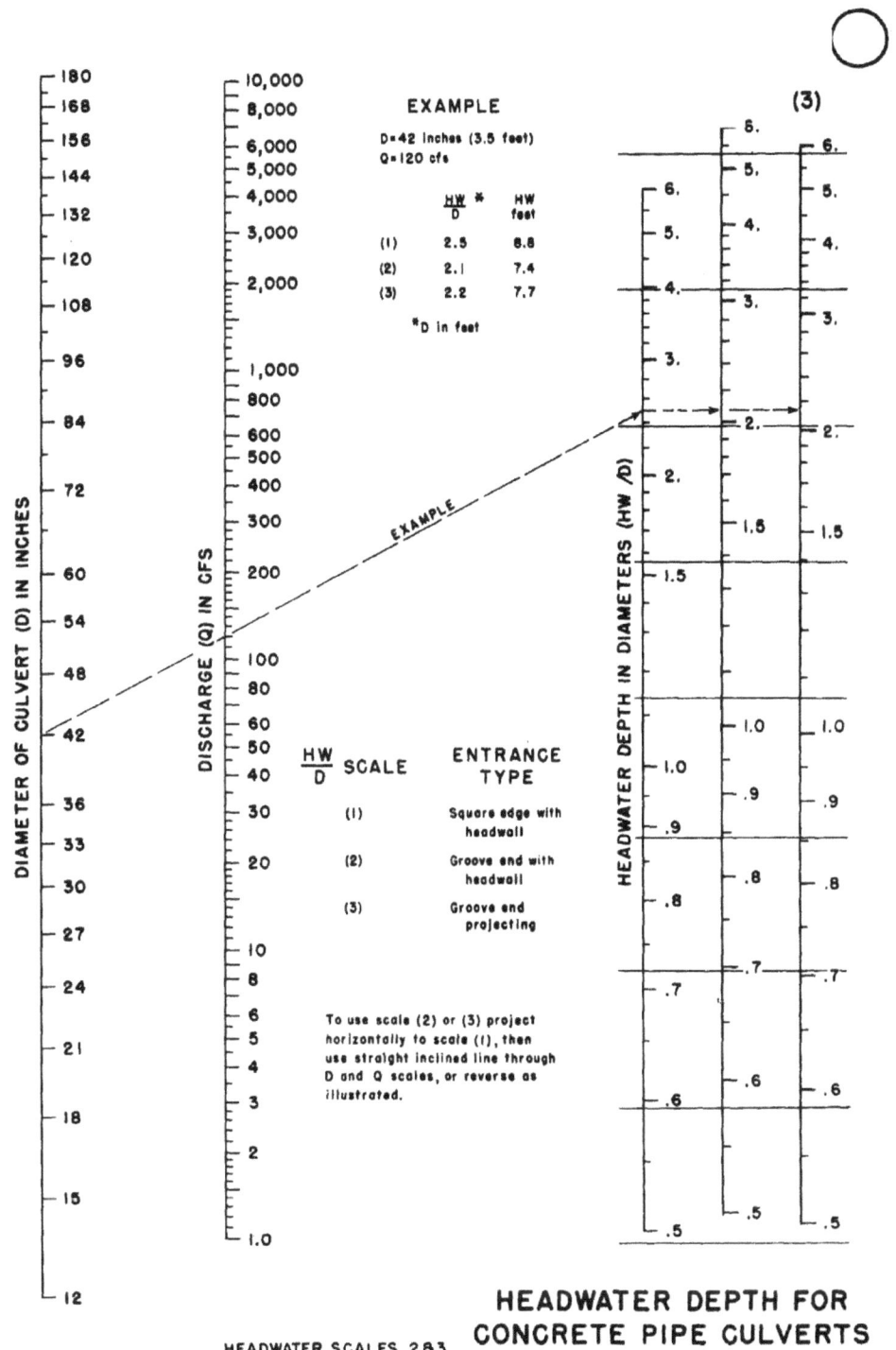

EXAMPLE

D=42 inches (3.5 feet)
Q=120 cfs

	$\frac{HW}{D}$ *	HW feet
(1)	2.5	8.8
(2)	2.1	7.4
(3)	2.2	7.7

*D in feet

DIAMETER OF CULVERT (D) IN INCHES

DISCHARGE (Q) IN CFS

HEADWATER DEPTH IN DIAMETERS (HW/D)

$\frac{HW}{D}$ SCALE	ENTRANCE TYPE
(1)	Square edge with headwall
(2)	Groove end with headwall
(3)	Groove end projecting

To use scale (2) or (3) project
horizontally to scale (1), then
use straight inclined line through
D and Q scales, or reverse as
illustrated.

EXAMPLE

(3)

HEADWATER DEPTH FOR
CONCRETE PIPE CULVERTS
WITH INLET CONTROL

HEADWATER SCALES 2 & 3
REVISED MAY 1964

BUREAU OF PUBLIC ROADS JAN. 1963

CHART 2A

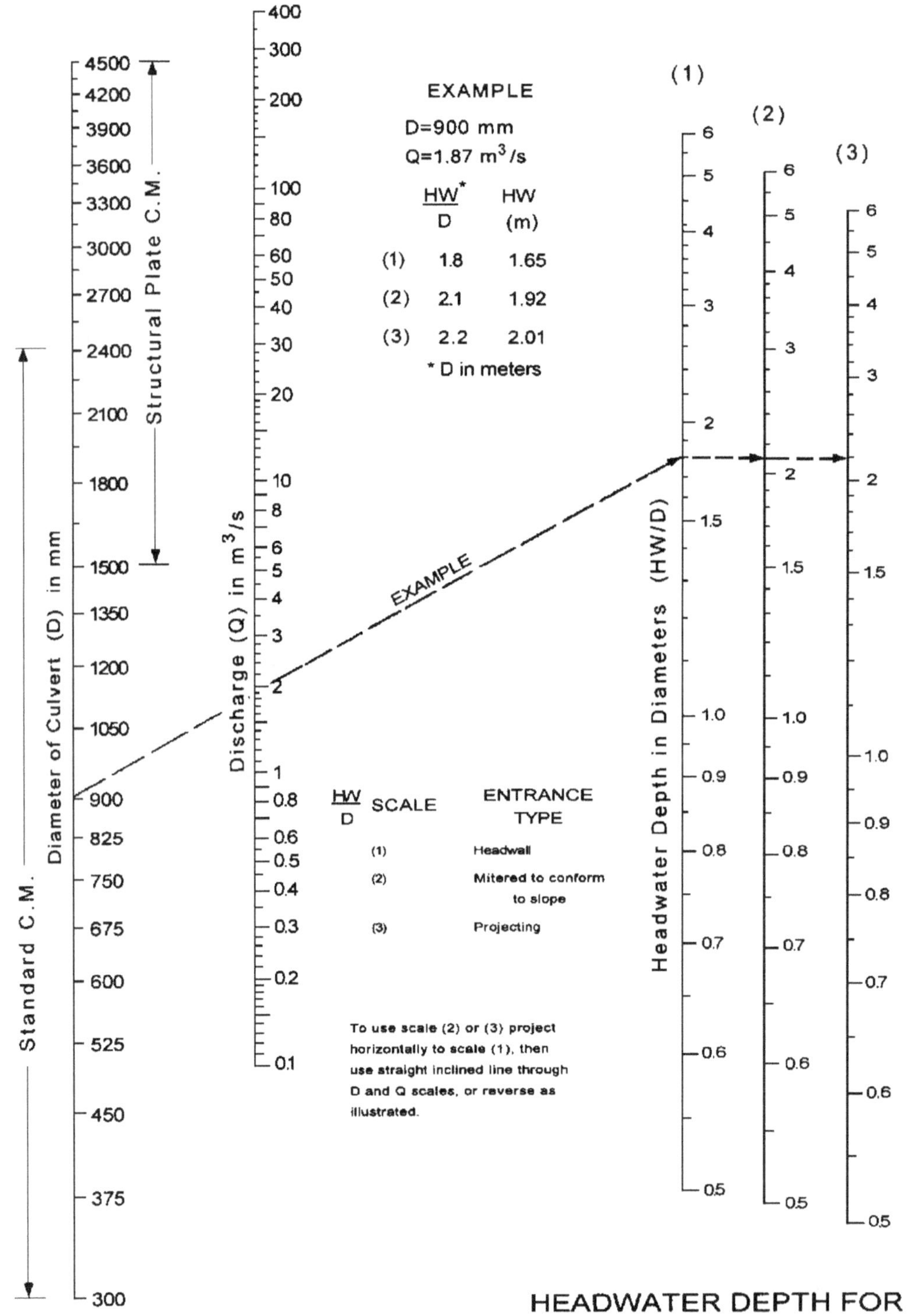

EXAMPLE

D=900 mm

Q=1.87 m³/s

	$\frac{HW^*}{D}$	HW (m)
(1)	1.8	1.65
(2)	2.1	1.92
(3)	2.2	2.01

* D in meters

(1)

(2)

(3)

400
300
200

4500
4200
3900
3600
100
3300
80
3000
60
50
2700
40
2400
30
2100
20
1800

6
5
4

3

1500
10
1350
8
6
5
4
1200
3
1050
2
900

6
5
4

3

2

825

2

2

EXAMPLE

1.5

750
1
0.8
675
0.6
0.5
0.4

1.5

1.5

$\frac{HW}{D}$ SCALE

**ENTRANCE
TYPE**

(1) Headwall

600
0.3

(2) Mitered to conform
to slope

(3) Projecting

1.0
0.9

1.0
0.9

1.0

525
0.2

0.8

0.8

0.9

450

To use scale (2) or (3) project
horizontally to scale (1), then
use straight inclined line through
D and Q scales, or reverse as
illustrated.

0.7

0.7

0.8

375

0.1

0.6

0.6

0.7

300

0.5

0.5

0.6

0.5

Diameter of Culvert (D) in mm

Structural Plate C.M.

Standard C.M.

Discharge (Q) in m³/s

Headwater Depth in Diameters (HW/D)

Adapted from
Bureau of Public Roads Jan. 1963

HEADWATER DEPTH FOR
C.M. PIPE CULVERTS
WITH INLET CONTROL

CHART 2B

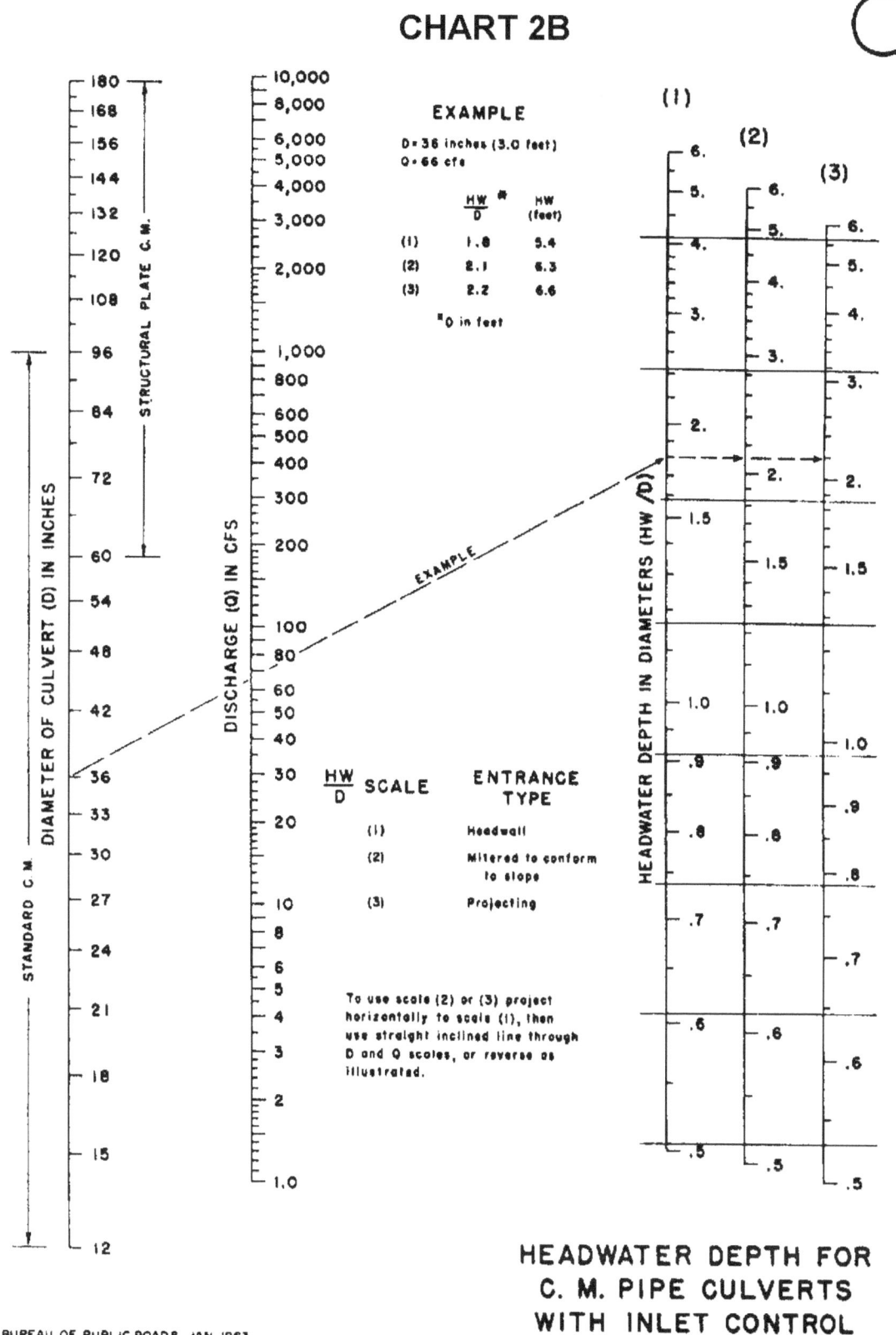

EXAMPLE

D = 36 inches (3.0 feet)
Q = 66 cfs

	HW/D #	HW (feet)
(1)	1.8	5.4
(2)	2.1	6.3
(3)	2.2	6.6

D in feet

DIAMETER OF CULVERT (D) IN INCHES

STANDARD C.M.   STRUCTURAL PLATE C.M.

DISCHARGE (Q) IN CFS

EXAMPLE

HEADWATER DEPTH IN DIAMETERS (HW/D)

(1) (2) (3)

HW/D SCALE   ENTRANCE TYPE

(1)   Headwall

(2)   Mitered to conform to slope

(3)   Projecting

To use scale (2) or (3) project
horizontally to scale (1), then
use straight inclined line through
D and Q scales, or reverse as
illustrated.

HEADWATER DEPTH FOR
C. M. PIPE CULVERTS
WITH INLET CONTROL

CHART 3A

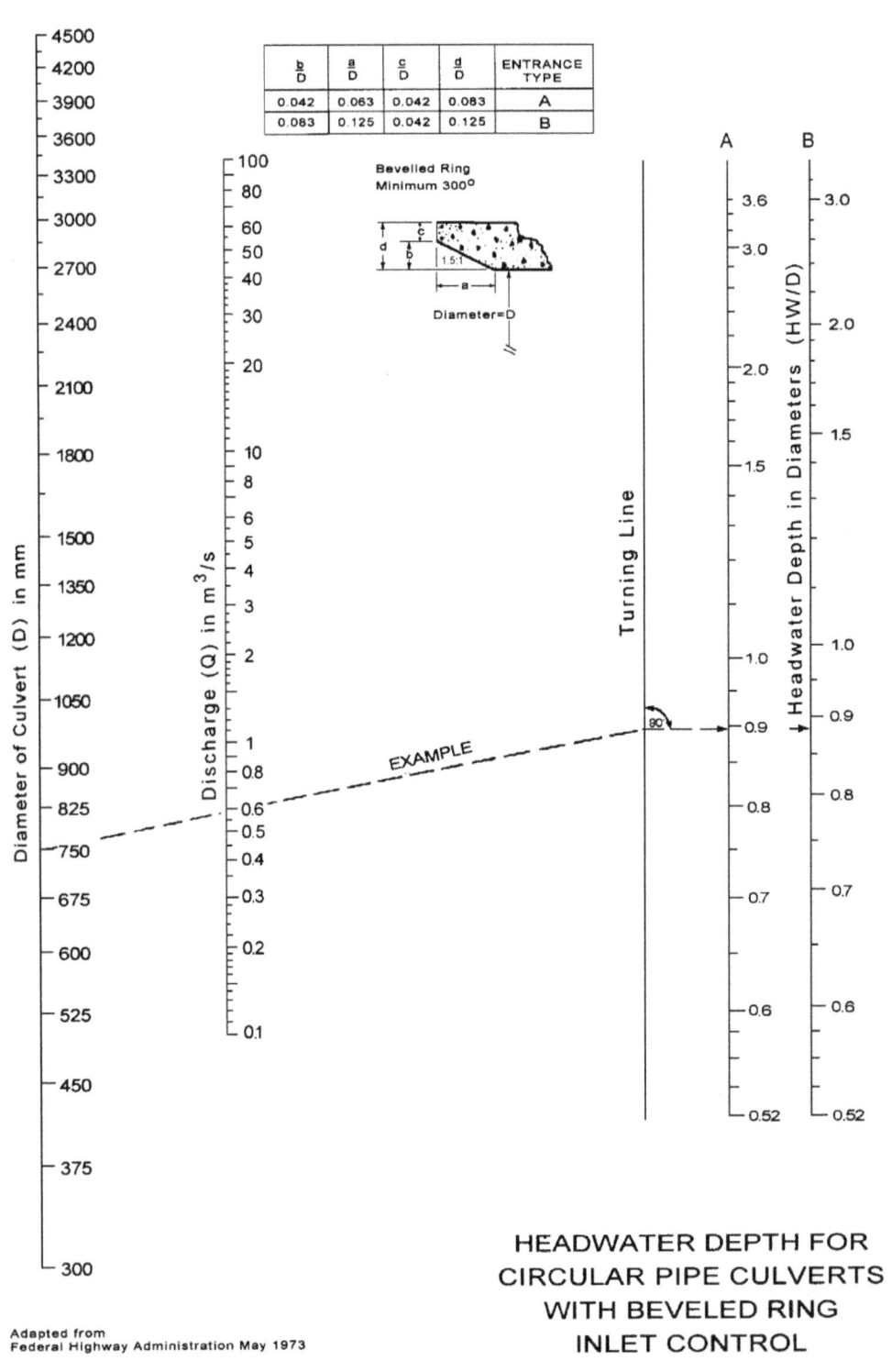

b/D	a/D	c/D	d/D	ENTRANCE TYPE
0.042	0.063	0.042	0.083	A
0.083	0.125	0.042	0.125	B

Bevelled Ring
Minimum 300°

Diameter=D

Diameter of Culvert (D) in mm

Discharge (Q) in m³/s

Turning Line

EXAMPLE

90

Headwater Depth in Diameters (HW/D)

A

B

Adapted from
Federal Highway Administration May 1973

HEADWATER DEPTH FOR
CIRCULAR PIPE CULVERTS
WITH BEVELED RING
INLET CONTROL

CHART 3B

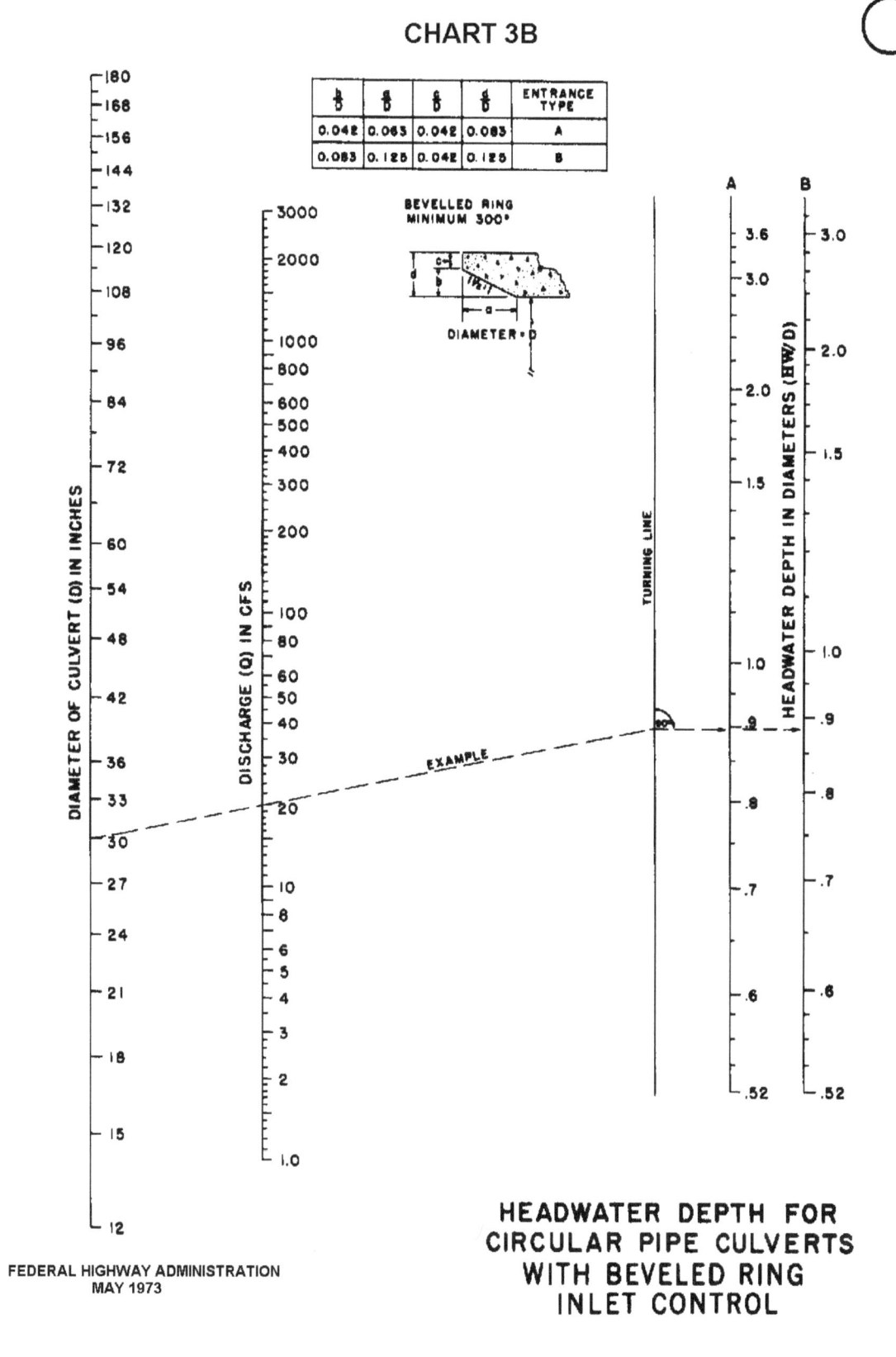

$\frac{t}{D}$	$\frac{d}{D}$	$\frac{c}{D}$	$\frac{a}{D}$	ENTRANCE TYPE
0.042	0.063	0.042	0.083	A
0.083	0.125	0.042	0.125	B

BEVELLED RING
MINIMUM 300°

DIAMETER = D

DIAMETER OF CULVERT (D) IN INCHES

DISCHARGE (Q) IN CFS

TURNING LINE

HEADWATER DEPTH IN DIAMETERS (HW/D)

EXAMPLE

FEDERAL HIGHWAY ADMINISTRATION
MAY 1973

HEADWATER DEPTH FOR CIRCULAR PIPE CULVERTS WITH BEVELED RING INLET CONTROL

CHART 4A

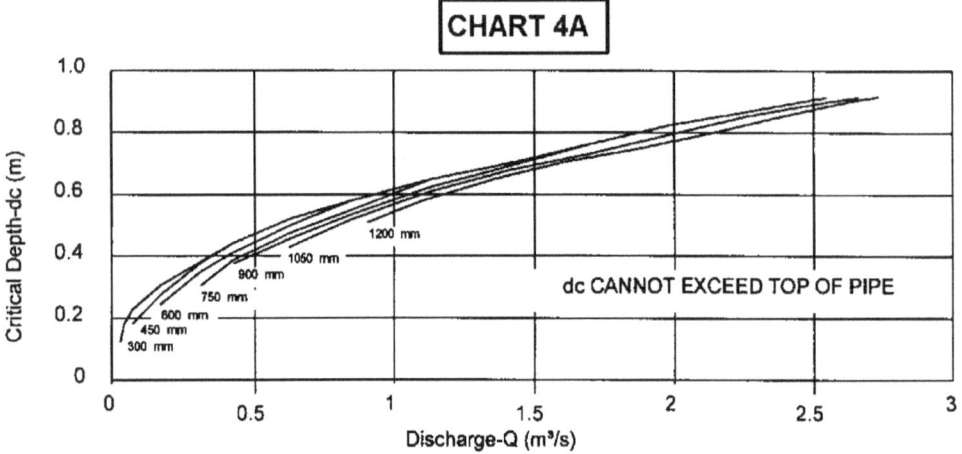

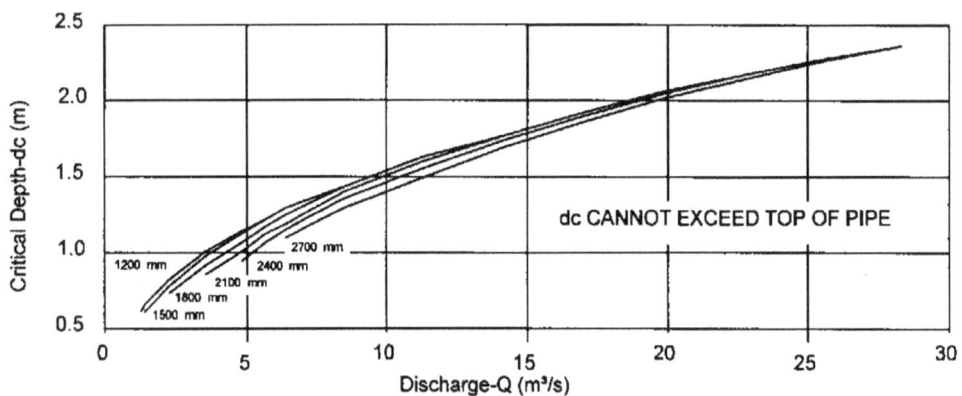

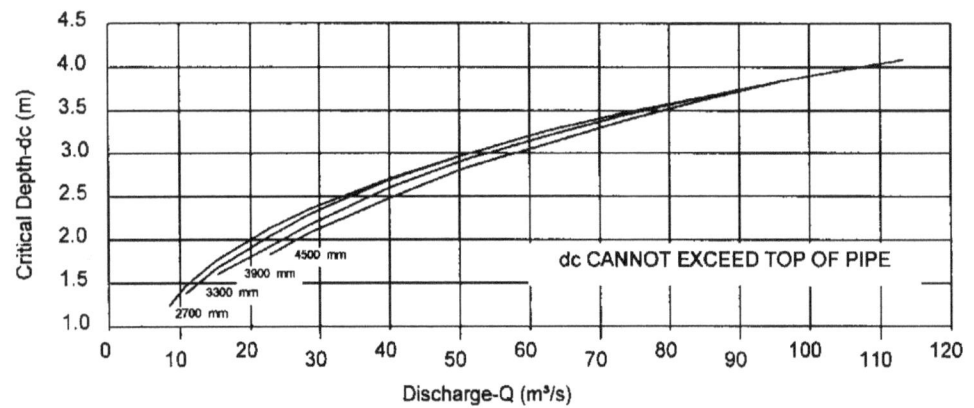

Adapted from Bureau of Public Roads

Critical Depth-Circular Pipe

CHART 4B

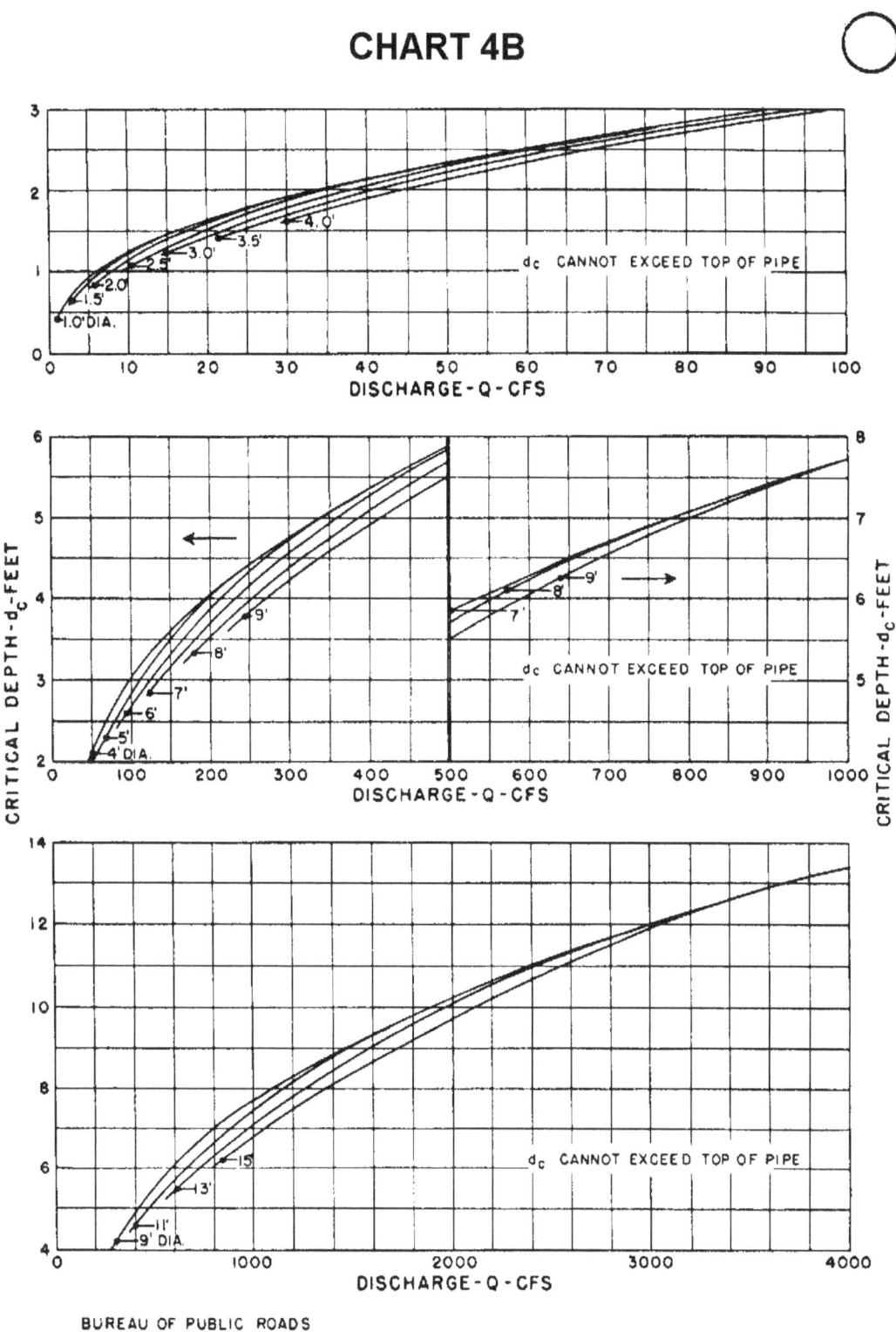

BUREAU OF PUBLIC ROADS

JAN. 1964

CRITICAL DEPTH
CIRCULAR PIPE

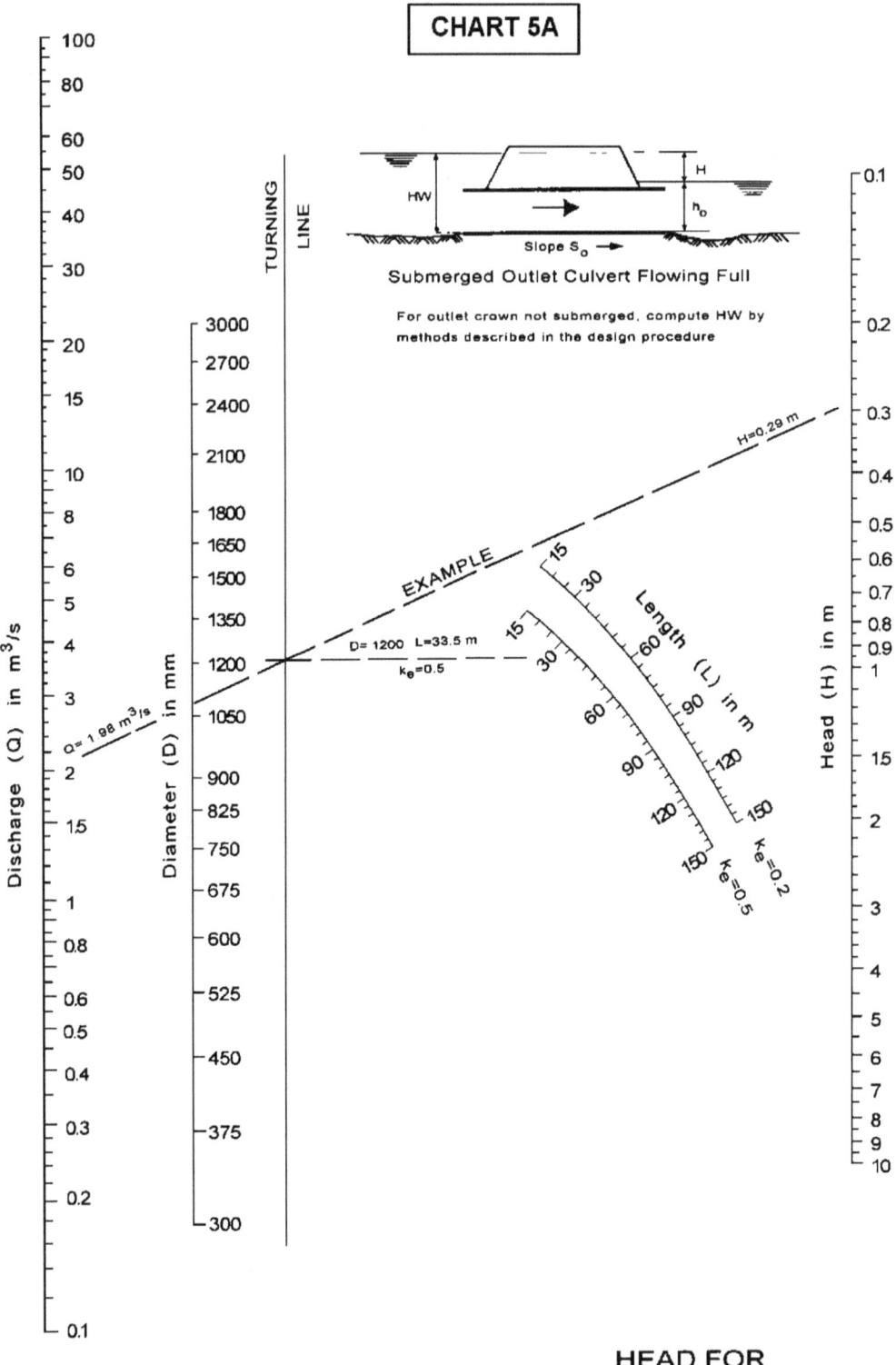

CHART 5A

Submerged Outlet Culvert Flowing Full

For outlet crown not submerged, compute HW by
methods described in the design procedure

Discharge (Q) in m³/s

100
80
60
50
40
30
20
15
10
8
6
5
4
3
2
1.5
1
0.8
0.6
0.5
0.4
0.3
0.2
0.1

TURNING LINE

Diameter (D) in mm

3000
2700
2400
2100
1800
1650
1500
1350
1200
1050
900
825
750
675
600
525
450
375
300

EXAMPLE

$Q = 1.98 \ m^3/s$

D= 1200   L=33.5 m
$k_e = 0.5$

Length (L) in m

15
30
60
90
120
150

$k_e = 0.2$

15
30
60
90
120
150

$k_e = 0.5$

H=0.29 m

Head (H) in m

0.1
0.2
0.3
0.4
0.5
0.6
0.7
0.8
0.9
1
1.5
2
3
4
5
6
7
8
9
10

**HEAD FOR
CONCRETE PIPE CULVERTS
FLOWING FULL**
n=0.012

Adapted from
Bureau of Public Roads Jan. 1963

CHART 5B

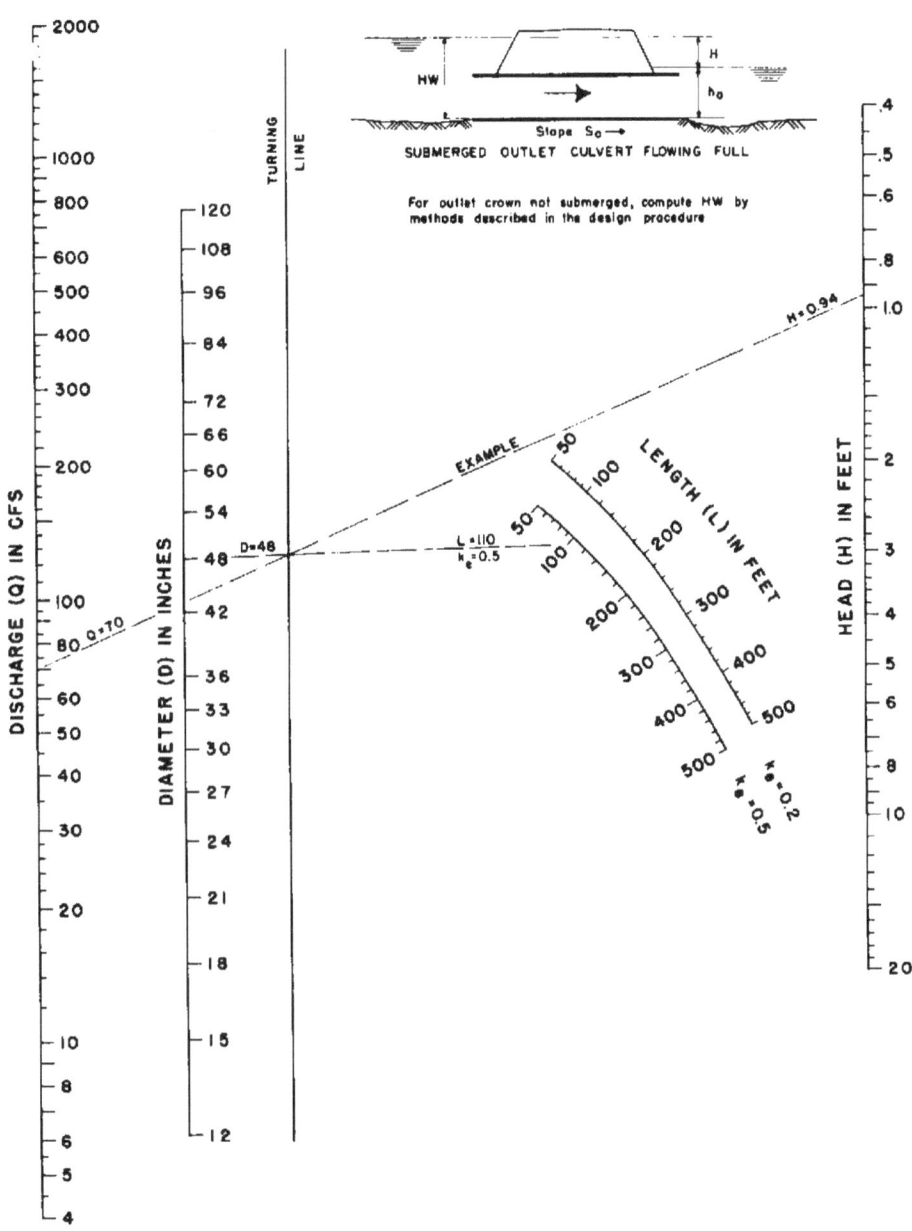

**HEAD FOR
CONCRETE PIPE CULVERTS
FLOWING FULL
n = 0.012**

BUREAU OF PUBLIC ROADS JAN. 1963

CHART 6A

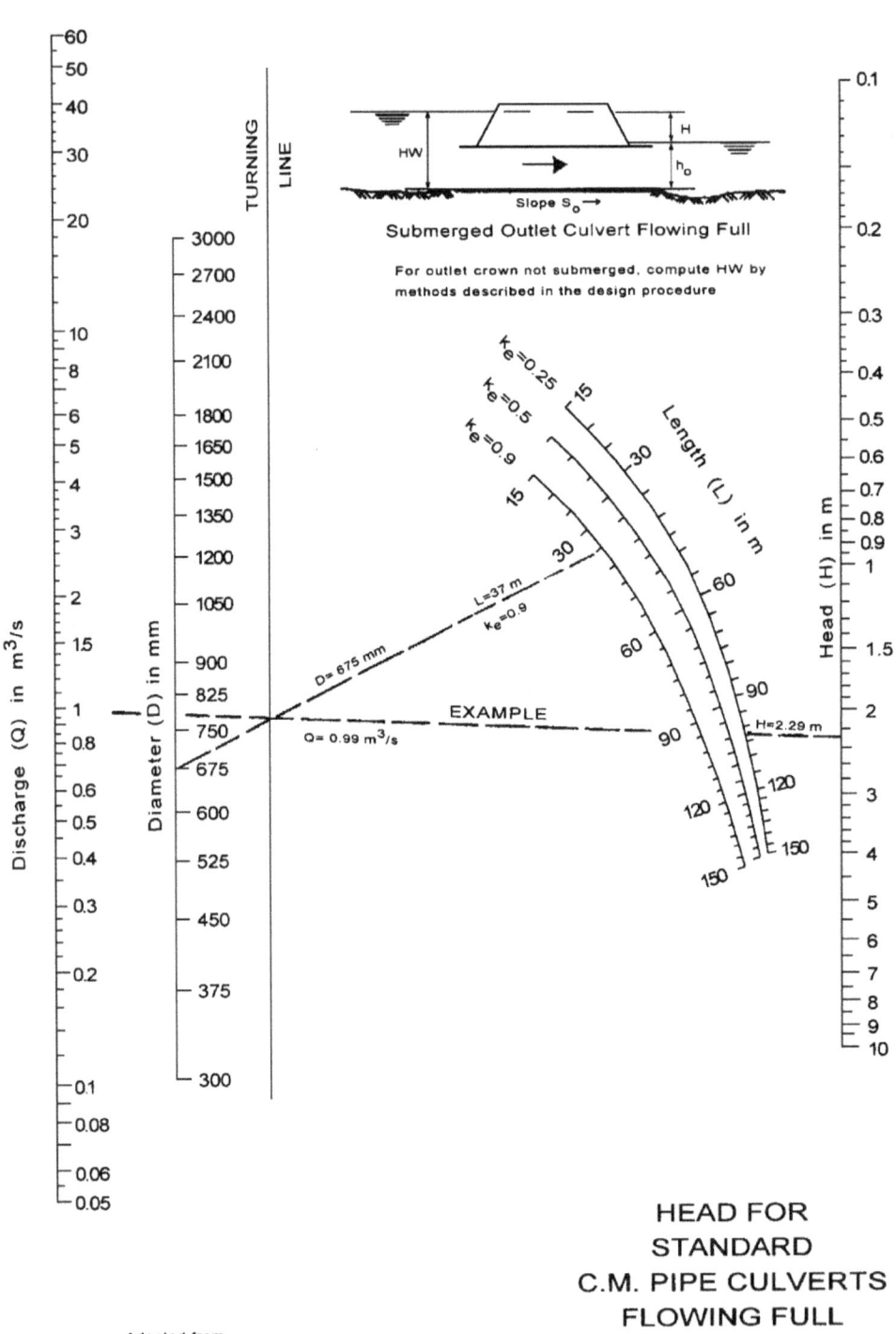

Submerged Outlet Culvert Flowing Full

For outlet crown not submerged, compute HW by methods described in the design procedure

HEAD FOR
STANDARD
C.M. PIPE CULVERTS
FLOWING FULL
n=0.024

CHART 6B

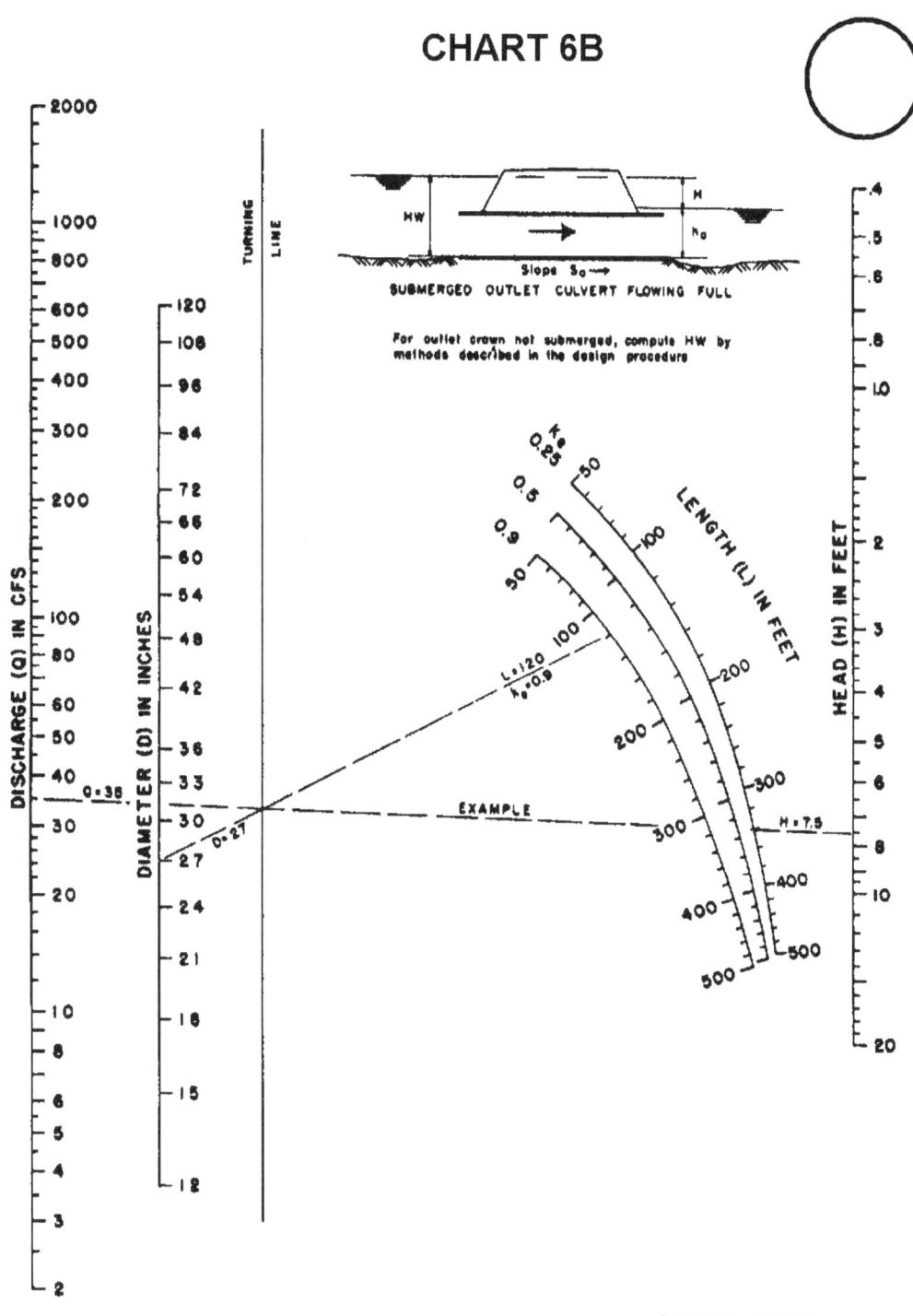

SUBMERGED OUTLET CULVERT FLOWING FULL

For outlet crown not submerged, compute HW by methods described in the design procedure

EXAMPLE

L=120 100
Ke=0.9

0.38
0.27

H = 7.5

DISCHARGE (Q) IN CFS

DIAMETER (D) IN INCHES

TURNING LINE

LENGTH (L) IN FEET

HEAD (H) IN FEET

Ke
0.25
0.5
0.9

HEAD FOR
STANDARD
C. M. PIPE CULVERTS
FLOWING FULL
n = 0.024

CHART 7A

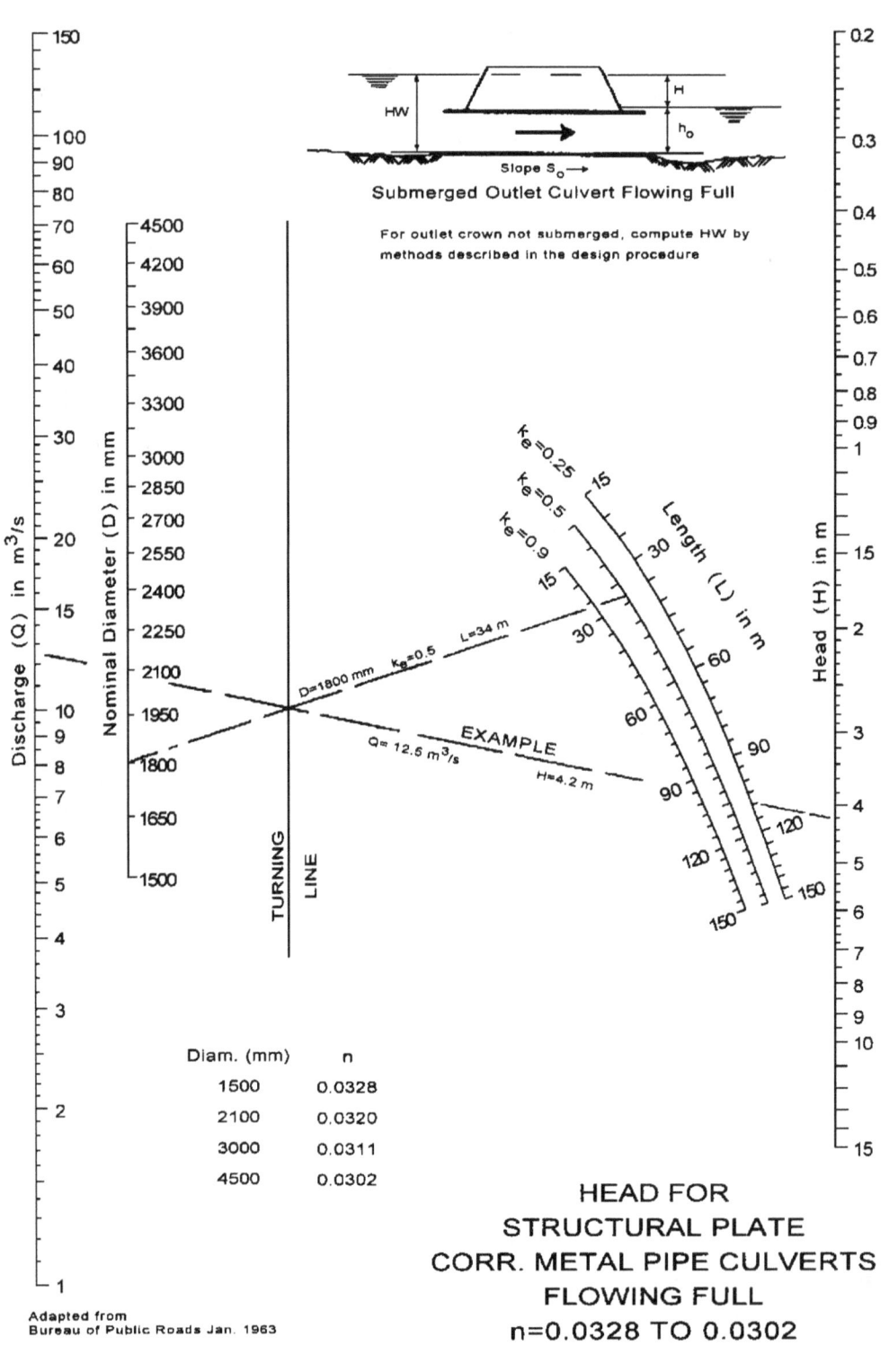

Submerged Outlet Culvert Flowing Full

For outlet crown not submerged, compute HW by methods described in the design procedure

Discharge (Q) in m³/s

Nominal Diameter (D) in mm

TURNING LINE

k_e=0.25

k_e=0.5

k_e=0.9

Length (L) in m

Head (H) in m

D=1800 mm k_e=0.5   L=34 m

EXAMPLE

Q= 12.5 m³/s   H=4.2 m

Diam. (mm)	n
1500	0.0328
2100	0.0320
3000	0.0311
4500	0.0302

Adapted from
Bureau of Public Roads Jan. 1963

HEAD FOR
STRUCTURAL PLATE
CORR. METAL PIPE CULVERTS
FLOWING FULL
n=0.0328 TO 0.0302

CHART 7B

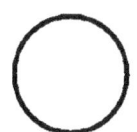

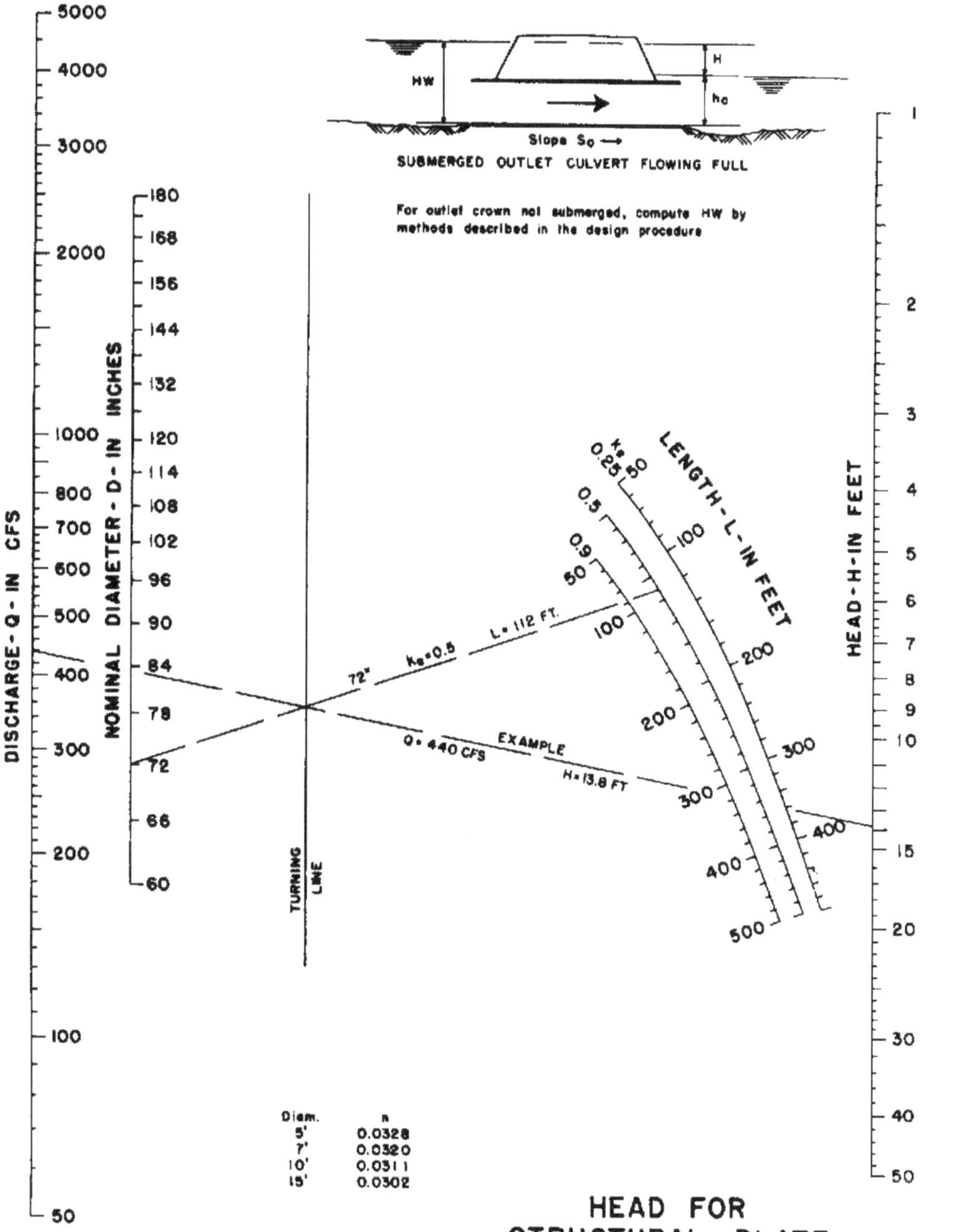
HW
H
h_o
Slope S_o →
SUBMERGED OUTLET CULVERT FLOWING FULL
For outlet crown not submerged, compute HW by methods described in the design procedure
DISCHARGE - Q - IN CFS
5000
4000
3000
2000
1000
800
700
600
500
400
300
200
100
50
NOMINAL DIAMETER - D - IN INCHES
180
168
156
144
132
120
114
108
102
96
90
84
78
72
66
60
TURNING LINE
72"
k_e=0.5
L= 112 FT.
Q= 440 CFS
EXAMPLE
H=13.8 FT
k_e
0.25
0.5
0.9
LENGTH - L - IN FEET
50
100
200
300
400
500
HEAD - H - IN FEET
1
2
3
4
5
6
7
8
9
10
15
20
30
40
50
Diam. n
5' 0.0328
7' 0.0320
10' 0.0311
15' 0.0302

HEAD FOR
STRUCTURAL PLATE
CORR. METAL PIPE CULVERTS
FLOWING FULL
n = 0.0328 TO 0.0302

BUREAU OF PUBLIC ROADS JAN. 1963

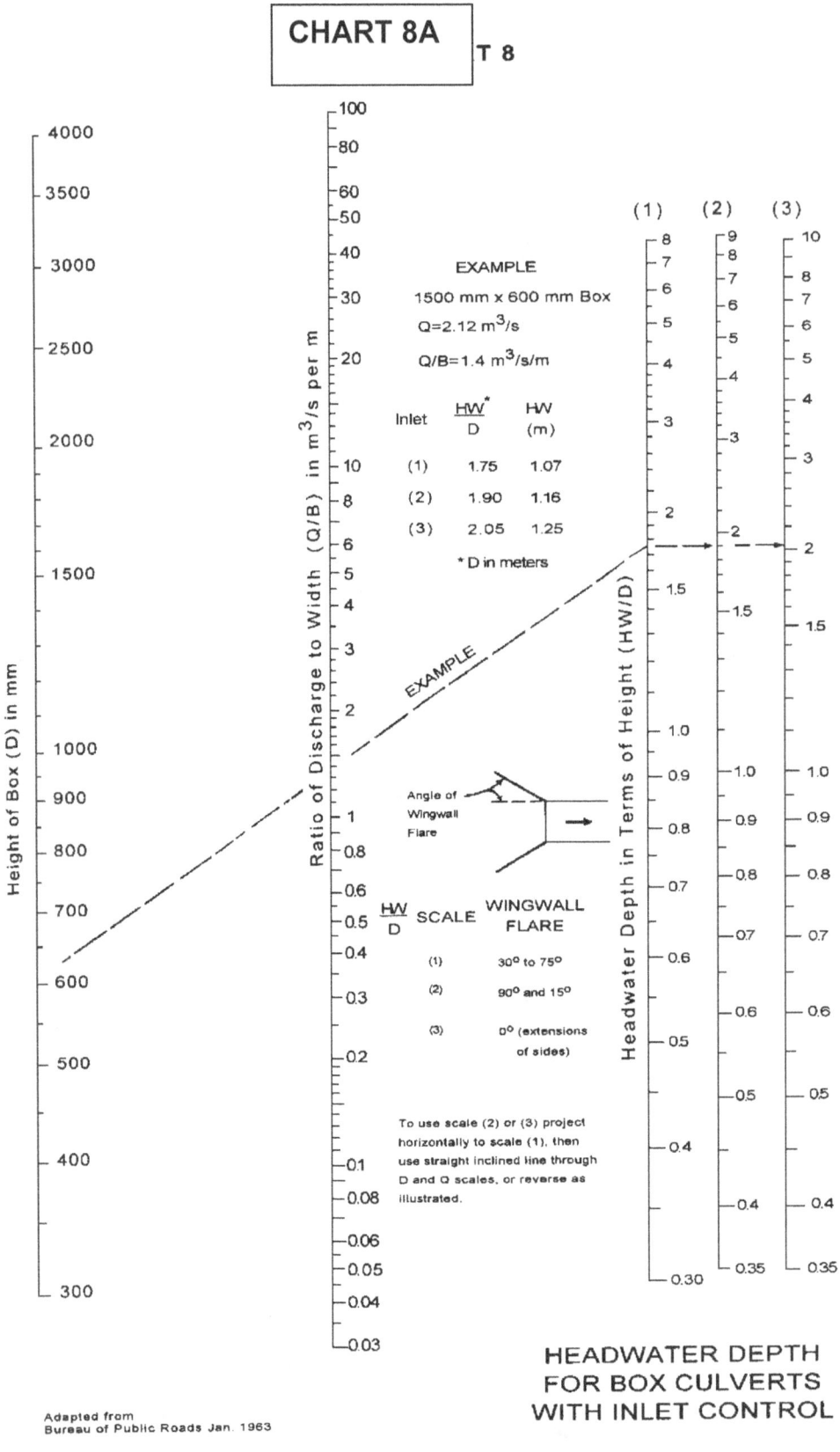

HEADWATER DEPTH
FOR BOX CULVERTS
WITH INLET CONTROL

Adapted from
Bureau of Public Roads Jan. 1963

CHART 8B

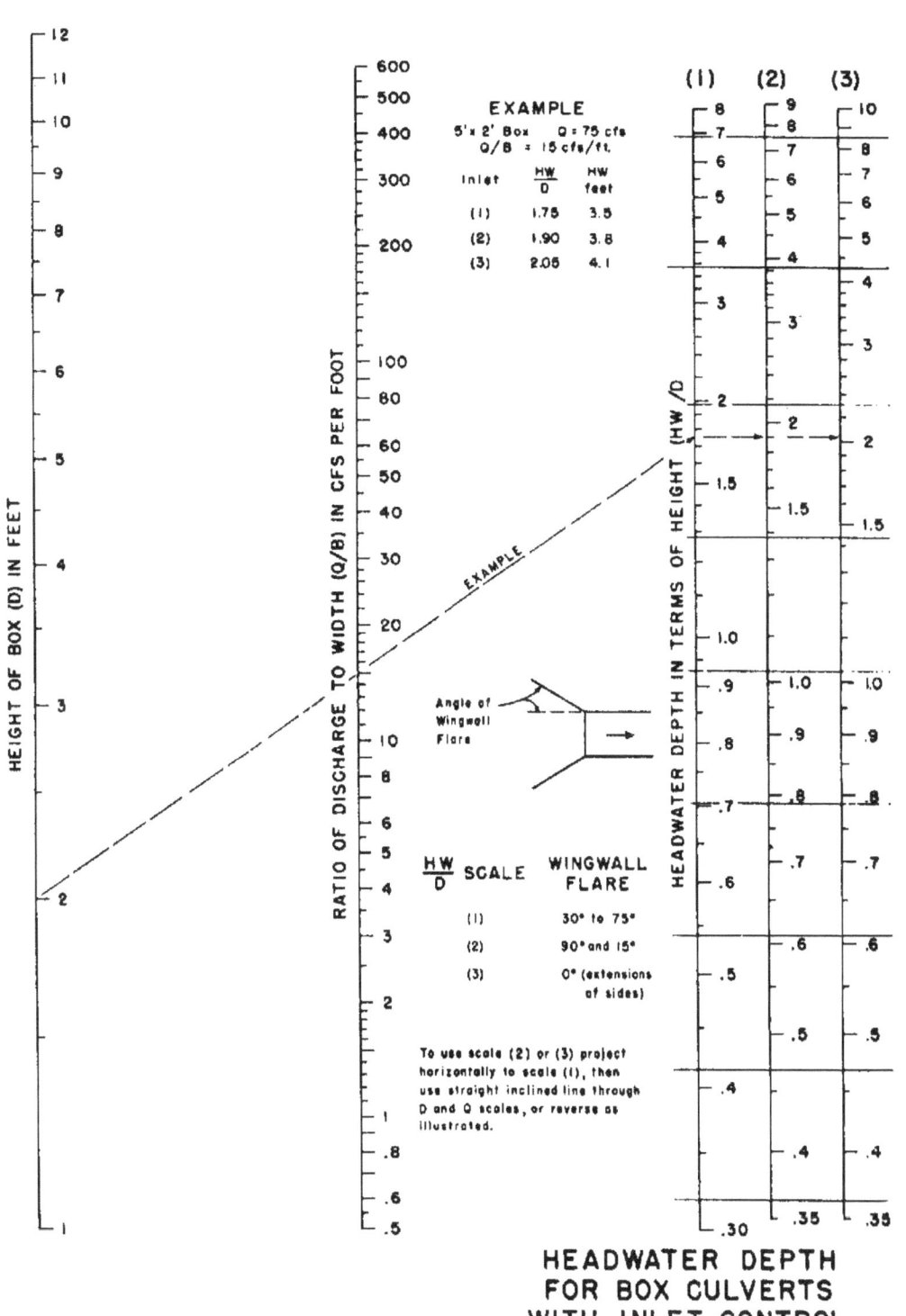

HEIGHT OF BOX (D) IN FEET

RATIO OF DISCHARGE TO WIDTH (Q/B) IN CFS PER FOOT

HEADWATER DEPTH IN TERMS OF HEIGHT (HW/D)

(1) (2) (3)

EXAMPLE

5'x 2' Box Q = 75 cfs
Q/B = 15 cfs/ft.

Inlet	HW/D	HW feet
(1)	1.75	3.5
(2)	1.90	3.8
(3)	2.05	4.1

EXAMPLE

Angle of Wingwall Flare

HW/D SCALE	WINGWALL FLARE
(1)	30° to 75°
(2)	90° and 15°
(3)	0° (extensions of sides)

To use scale (2) or (3) project horizontally to scale (1), then use straight inclined line through D and Q scales, or reverse as illustrated.

HEADWATER DEPTH FOR BOX CULVERTS WITH INLET CONTROL

BUREAU OF PUBLIC ROADS JAN. 1963

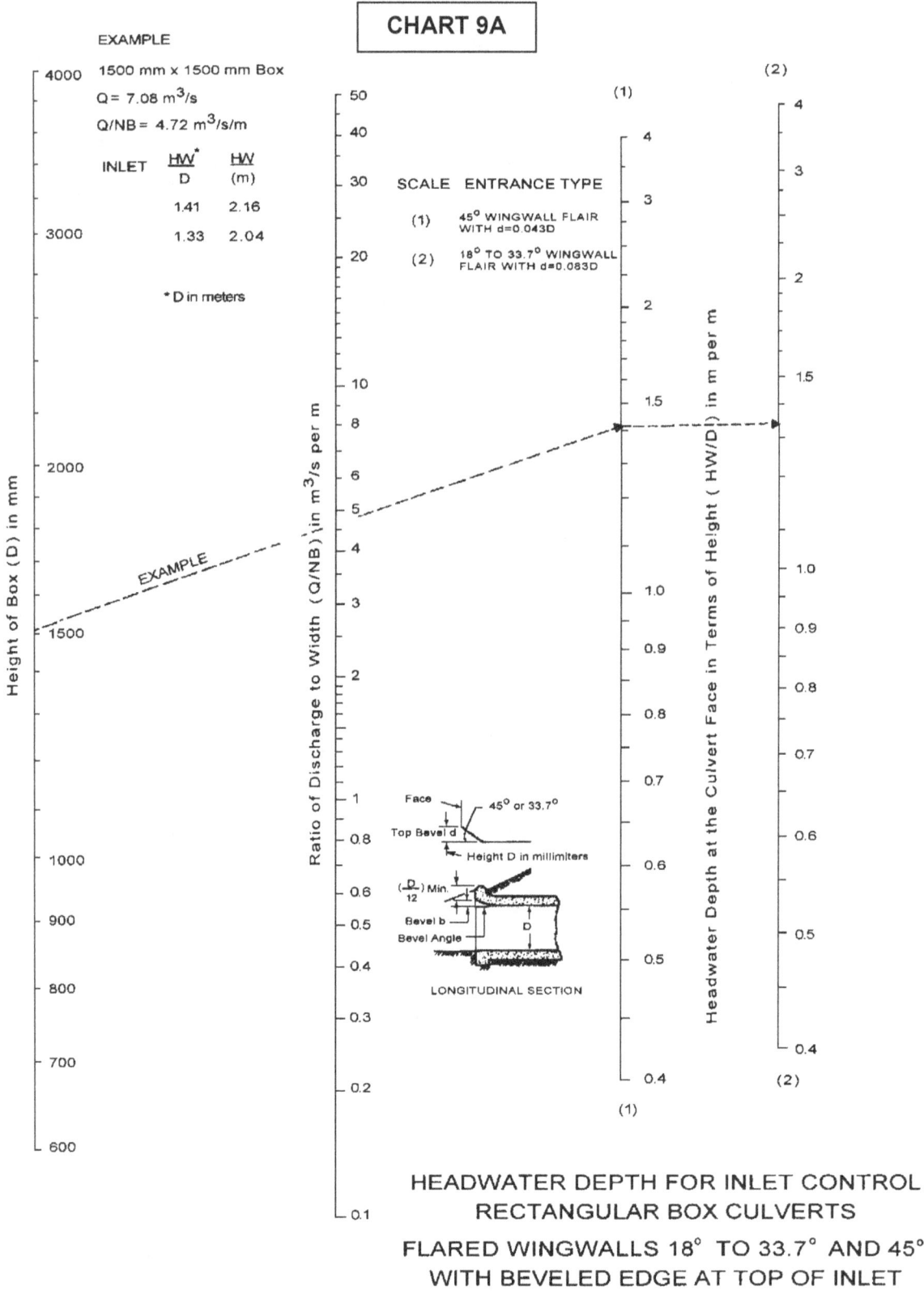

CHART 9A

(1)   (2)

EXAMPLE

1500 mm x 1500 mm Box

Q = 7.08 m³/s

Q/NB = 4.72 m³/s/m

INLET	$\frac{HW^*}{D}$	$\frac{HW}{(m)}$
	1.41	2.16
	1.33	2.04

* D in meters

SCALE   ENTRANCE TYPE

(1)   45° WINGWALL FLAIR
        WITH d=0.043D

(2)   18° TO 33.7° WINGWALL
        FLAIR WITH d=0.083D

Height of Box (D) in mm

Ratio of Discharge to Width (Q/NB)\in m³/s per m

Headwater Depth at the Culvert Face in Terms of Height (HW/D) in m per m

EXAMPLE

Face

45° or 33.7°

Top Bevel d

Height D in millimiters

$(\frac{D}{12})$ Min.

Bevel b

Bevel Angle

D

LONGITUDINAL SECTION

**HEADWATER DEPTH FOR INLET CONTROL
RECTANGULAR BOX CULVERTS
FLARED WINGWALLS 18° TO 33.7° AND 45°
WITH BEVELED EDGE AT TOP OF INLET**

CHART 9B

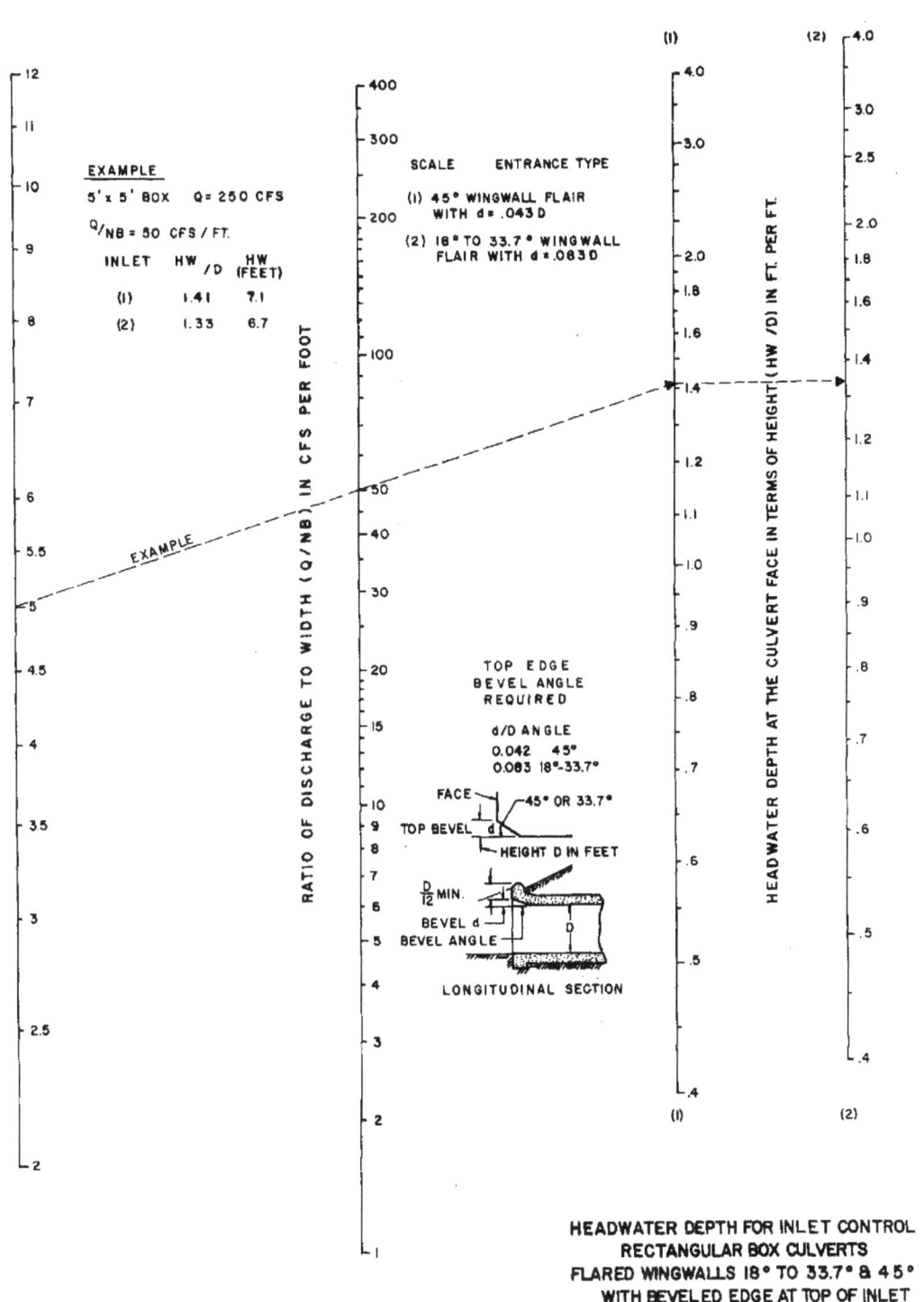

EXAMPLE

5' x 5' BOX   Q = 250 CFS

Q/NB = 50 CFS / FT.

INLET	HW/D	HW (FEET)
(1)	1.41	7.1
(2)	1.33	6.7

SCALE   ENTRANCE TYPE

(1) 45° WINGWALL FLAIR
    WITH d = .043 D

(2) 18° TO 33.7° WINGWALL
    FLAIR WITH d = .083 D

RATIO OF DISCHARGE TO WIDTH (Q/NB) IN CFS PER FOOT

HEADWATER DEPTH AT THE CULVERT FACE IN TERMS OF HEIGHT (HW/D) IN FT. PER FT.

TOP EDGE
BEVEL ANGLE
REQUIRED

d/D	ANGLE
0.042	45°
0.083	18°-33.7°

FACE   45° OR 33.7°
TOP BEVEL   d
            HEIGHT D IN FEET
D/12 MIN.
BEVEL d
BEVEL ANGLE   D

LONGITUDINAL SECTION

HEADWATER DEPTH FOR INLET CONTROL
RECTANGULAR BOX CULVERTS
FLARED WINGWALLS 18° TO 33.7° & 45°
WITH BEVELED EDGE AT TOP OF INLET

CHART 10A

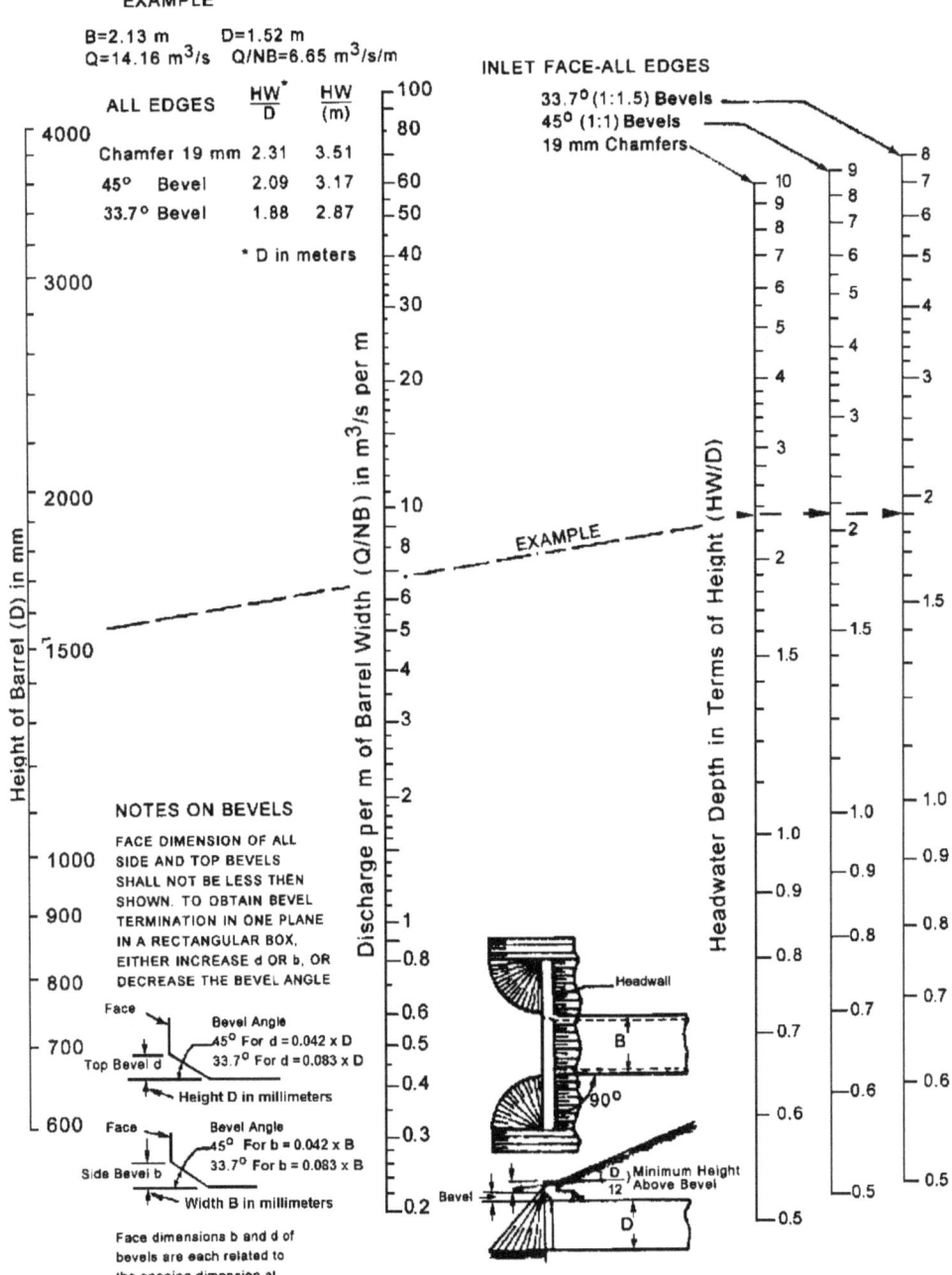

EXAMPLE

B=2.13 m   D=1.52 m
Q=14.16 m³/s   Q/NB=6.65 m³/s/m

ALL EDGES	HW*/D	HW (m)
Chamfer 19 mm	2.31	3.51
45° Bevel	2.09	3.17
33.7° Bevel	1.88	2.87

* D in meters

INLET FACE-ALL EDGES

33.7° (1:1.5) Bevels
45° (1:1) Bevels
19 mm Chamfers

Height of Barrel (D) in mm

Discharge per m of Barrel Width (Q/NB) in m³/s per m

Headwater Depth in Terms of Height (HW/D)

EXAMPLE

NOTES ON BEVELS

FACE DIMENSION OF ALL
SIDE AND TOP BEVELS
SHALL NOT BE LESS THEN
SHOWN. TO OBTAIN BEVEL
TERMINATION IN ONE PLANE
IN A RECTANGULAR BOX,
EITHER INCREASE d OR b, OR
DECREASE THE BEVEL ANGLE

Face
Top Bevel d
Bevel Angle
45° For d = 0.042 x D
33.7° For d = 0.083 x D
Height D in millimeters

Face
Side Bevel b
Bevel Angle
45° For b = 0.042 x B
33.7° For b = 0.083 x B
Width B in millimeters

Face dimensions b and d of
bevels are each related to
the opening dimension at
right angles to the edge

Headwall

B

90°

Bevel

D/12 Minimum Height
Above Bevel

D

Adapted from
Federal Highway Administration
May 1973

**HEADWATER DEPTH FOR INLET CONTROL
RECTANGULAR BOX CULVERTS
90° HEADWALL
CHAMFERED OR BEVELED INLET EDGES**

CHART 10B

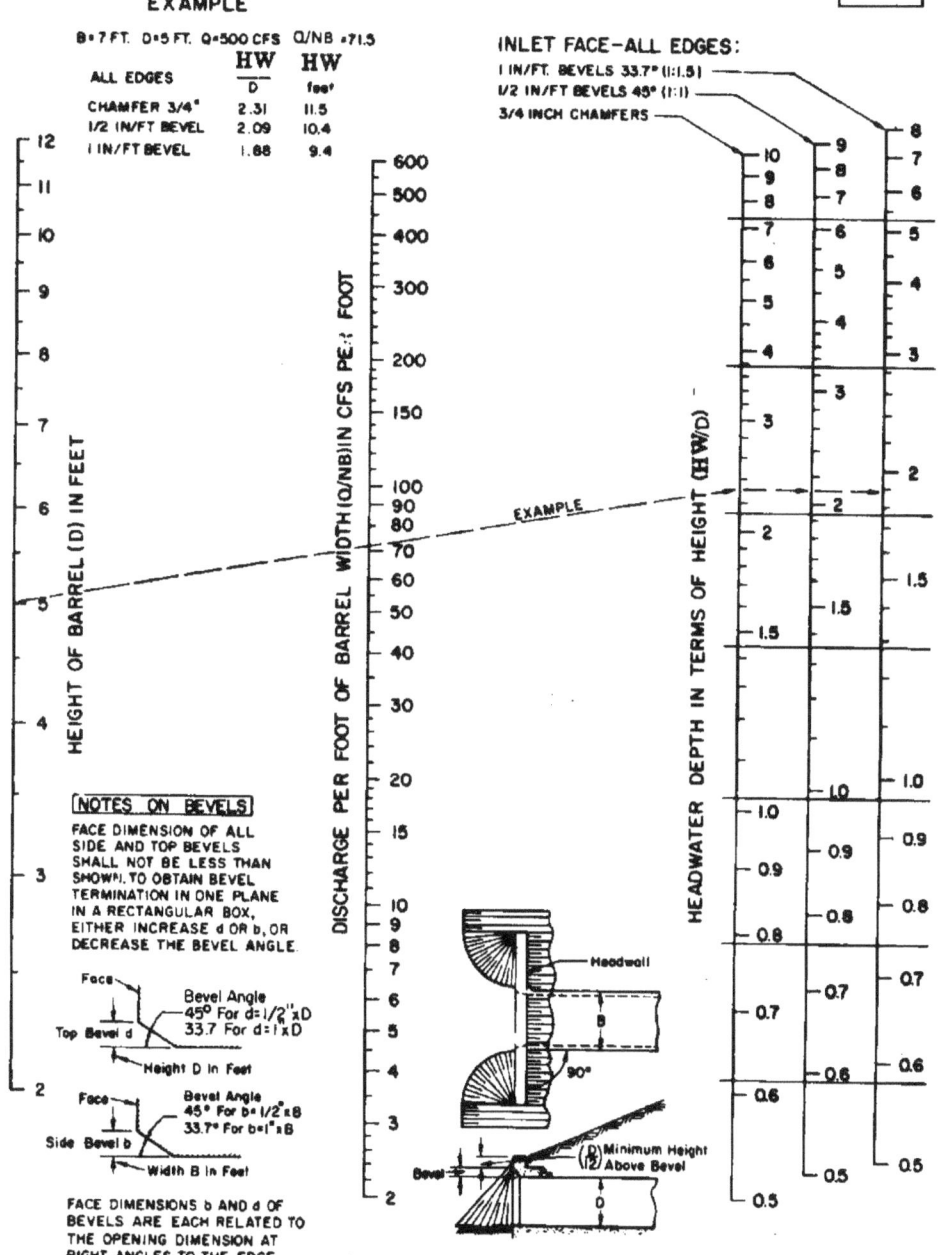

EXAMPLE

B = 7 FT. D = 5 FT. Q = 500 CFS Q/NB = 71.5

	HW/D	HW feet
ALL EDGES		
CHAMFER 3/4"	2.31	11.5
1/2 IN/FT BEVEL	2.09	10.4
1 IN/FT BEVEL	1.88	9.4

HEIGHT OF BARREL (D) IN FEET

DISCHARGE PER FOOT OF BARREL WIDTH (Q/NB) IN CFS PER FOOT

INLET FACE—ALL EDGES:
1 IN/FT. BEVELS 33.7° (1:1.5)
1/2 IN/FT BEVELS 45° (1:1)
3/4 INCH CHAMFERS

HEADWATER DEPTH IN TERMS OF HEIGHT (HW/D)

EXAMPLE

NOTES ON BEVELS

FACE DIMENSION OF ALL SIDE AND TOP BEVELS SHALL NOT BE LESS THAN SHOWN. TO OBTAIN BEVEL TERMINATION IN ONE PLANE IN A RECTANGULAR BOX, EITHER INCREASE d OR b, OR DECREASE THE BEVEL ANGLE.

Face
Top Bevel d
Bevel Angle 45° For d = 1/2" x D
33.7 For d = 1 x D
Height D In Feet

Face
Side Bevel b
Bevel Angle 45° For b = 1/2" x B
33.7° For b = 1" x B
Width B In Feet

FACE DIMENSIONS b AND d OF BEVELS ARE EACH RELATED TO THE OPENING DIMENSION AT RIGHT ANGLES TO THE EDGE

Headwall
90°
Bevel
(D) Minimum Height (1/2) Above Bevel

HEADWATER DEPTH FOR INLET CONTROL
RECTANGULAR BOX CULVERTS
90° HEADWALL
CHAMFERED OR BEVELED INLET EDGES

CHART 11A

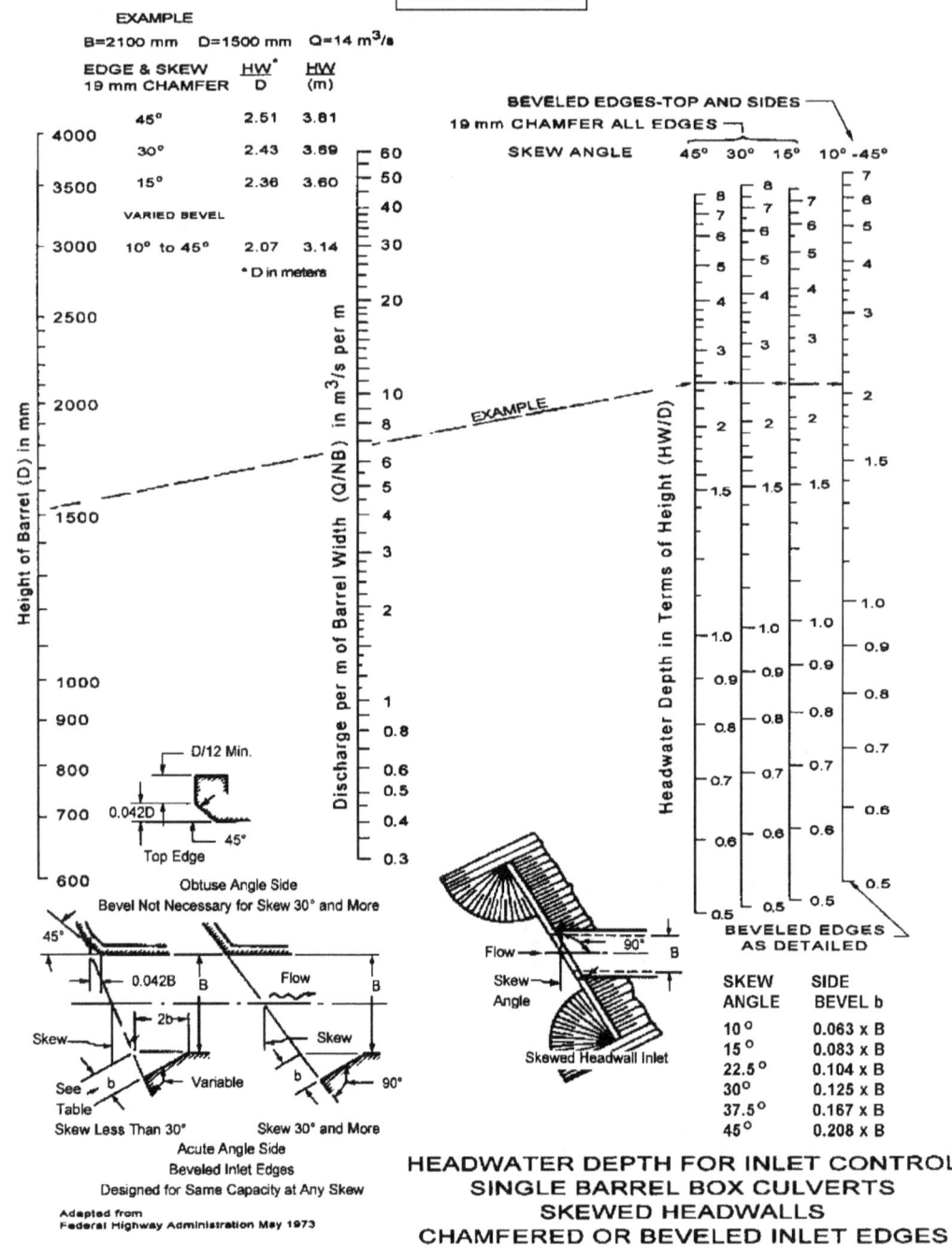

EXAMPLE

B=2100 mm   D=1500 mm   Q=14 m³/s

EDGE & SKEW 19 mm CHAMFER	$\frac{HW^*}{D}$	HW (m)
45°	2.51	3.81
30°	2.43	3.69
15°	2.36	3.60
VARIED BEVEL		
10° to 45°	2.07	3.14

* D in meters

Height of Barrel (D) in mm

Discharge per m of Barrel Width (Q/NB) in m³/s per m

BEVELED EDGES-TOP AND SIDES
19 mm CHAMFER ALL EDGES

SKEW ANGLE   45°   30°   15°   10° -45°

Headwater Depth in Terms of Height (HW/D)

BEVELED EDGES
AS DETAILED

D/12 Min.

0.042D

45°

Top Edge

Obtuse Angle Side
Bevel Not Necessary for Skew 30° and More

45°

0.042B

2b

Skew

See
Table

Flow

Skew

Variable   b   90°

Skew Less Than 30°   Skew 30° and More

Acute Angle Side
Beveled Inlet Edges
Designed for Same Capacity at Any Skew

Flow

Skew
Angle

90°   B

Skewed Headwall Inlet

SKEW ANGLE	SIDE BEVEL b
10°	0.063 x B
15°	0.083 x B
22.5°	0.104 x B
30°	0.125 x B
37.5°	0.167 x B
45°	0.208 x B

HEADWATER DEPTH FOR INLET CONTROL
SINGLE BARREL BOX CULVERTS
SKEWED HEADWALLS
CHAMFERED OR BEVELED INLET EDGES

CHART 11B

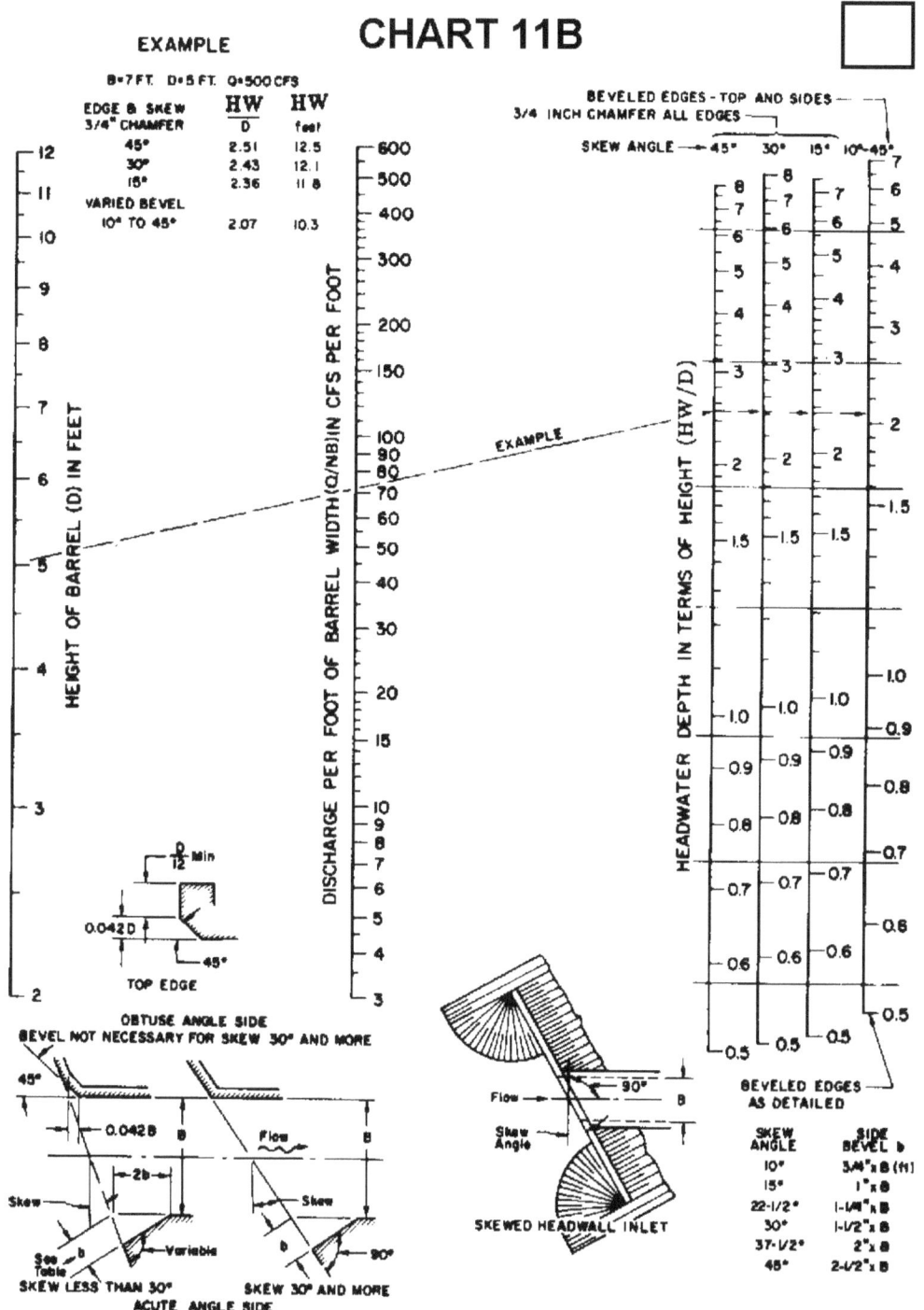

EXAMPLE

B=7FT. D=5FT. Q=500 CFS

EDGE & SKEW 3/4" CHAMFER	HW/D	HW feet
45°	2.51	12.5
30°	2.43	12.1
15°	2.36	11.8
VARIED BEVEL 10° TO 45°	2.07	10.3

HEIGHT OF BARREL (D) IN FEET

DISCHARGE PER FOOT OF BARREL WIDTH (Q/NB) IN CFS PER FOOT

EXAMPLE

BEVELED EDGES - TOP AND SIDES

3/4 INCH CHAMFER ALL EDGES

SKEW ANGLE → 45° 30° 15° 10°-45°

HEADWATER DEPTH IN TERMS OF HEIGHT (HW/D)

D/12 Min

0.042 D

45°

TOP EDGE

OBTUSE ANGLE SIDE

BEVEL NOT NECESSARY FOR SKEW 30° AND MORE

45°

0.042 B

2b

Skew

See Table

Variable

Flow

Skew

b

90°

SKEW LESS THAN 30°

SKEW 30° AND MORE

ACUTE ANGLE SIDE

BEVELED INLET EDGES

DESIGNED FOR SAME CAPACITY AT ANY SKEW

Flow

90°

B

Skew Angle

SKEWED HEADWALL INLET

BEVELED EDGES AS DETAILED

SKEW ANGLE	SIDE BEVEL b
10°	3/4"x B (ft)
15°	1"x B
22-1/2°	1-1/4"x B
30°	1-1/2"x B
37-1/2°	2"x B
45°	2-1/2"x B

FEDERAL HIGHWAY ADMINISTRATION
MAY 1973

HEADWATER DEPTH FOR INLET CONTROL
SINGLE BARREL BOX CULVERTS
SKEWED HEADWALLS
CHAMFERED OR BEVELED INLET EDGES

CHART 12A

INLET & WW NORMAL	HW*/D	HW (m)
45° WW	2.18	3.32
18.4° WW	2.27	3.47
SKEWED 15°-45° 18.4 OR MORE WW	2.20	3.35

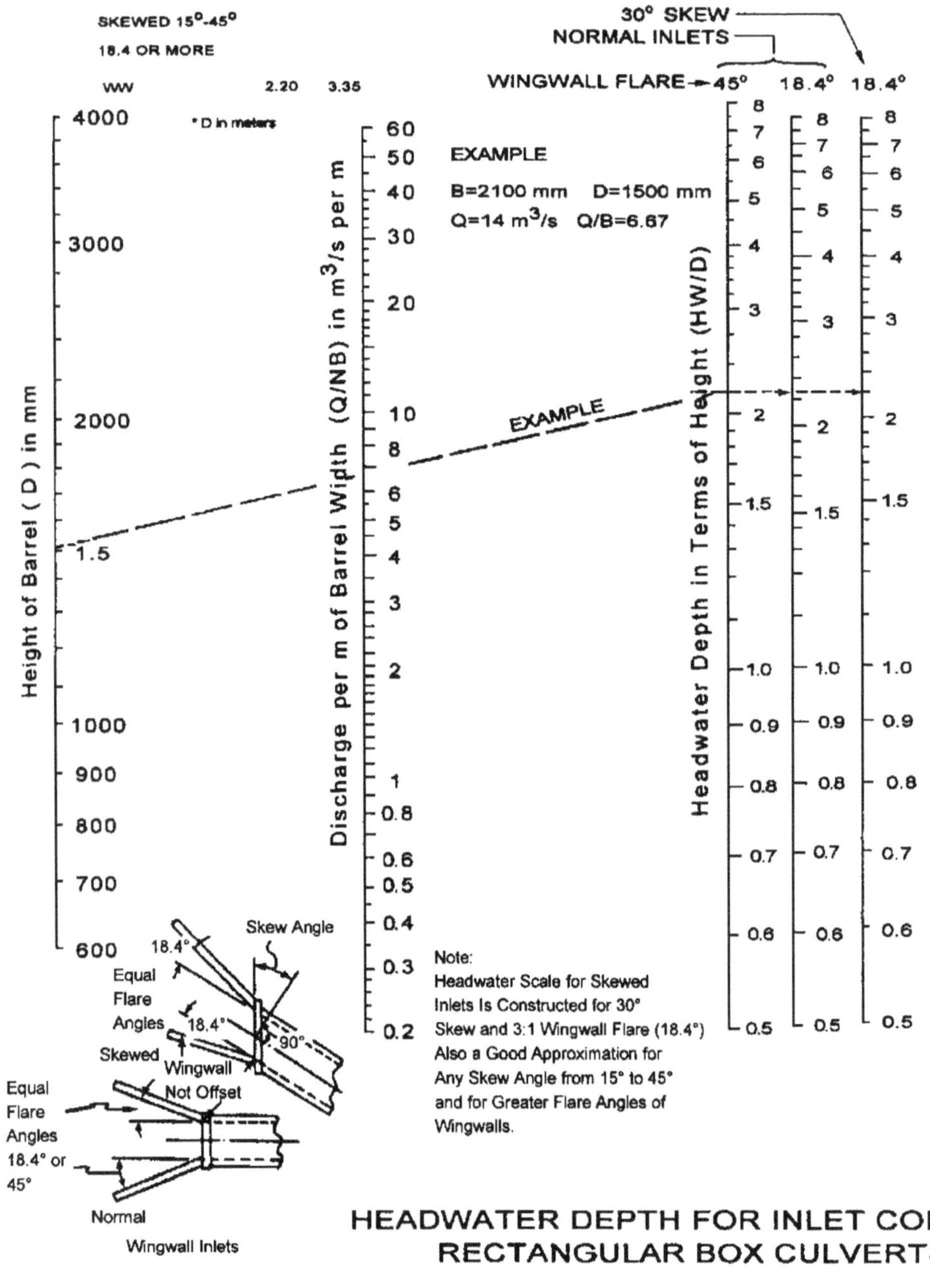

30° SKEW
NORMAL INLETS
WINGWALL FLARE 45° 18.4° 18.4°
* D in meters
EXAMPLE
B=2100 mm D=1500 mm
Q=14 m³/s Q/B=6.67
Height of Barrel (D) in mm
4000
3000
2000
1.5
1000
900
800
700
600
Discharge per m of Barrel Width (Q/NB) in m³/s per m
60
50
40
30
20
10
8
6
5
4
3
2
1
0.8
0.6
0.5
0.4
0.3
0.2
EXAMPLE
Headwater Depth in Terms of Height (HW/D)
8
7
6
5
4
3
2
1.5
1.0
0.9
0.8
0.7
0.6
0.5
Skew Angle
18.4°
Equal Flare Angles
18.4°
90°
Skewed
Wingwall Not Offset
Equal Flare Angles 18.4° or 45°
Normal
Wingwall Inlets
Note: Headwater Scale for Skewed Inlets Is Constructed for 30° Skew and 3:1 Wingwall Flare (18.4°) Also a Good Approximation for Any Skew Angle from 15° to 45° and for Greater Flare Angles of Wingwalls.

HEADWATER DEPTH FOR INLET CONTROL RECTANGULAR BOX CULVERTS FLARED WINGWALLS NORMAL AND SKEWED INLETS 19 mm CHAMFERED AT TOP OF OPENING

Adapted from
Bureau of Public Roads Office of R & D
August 1968

CHART 12B

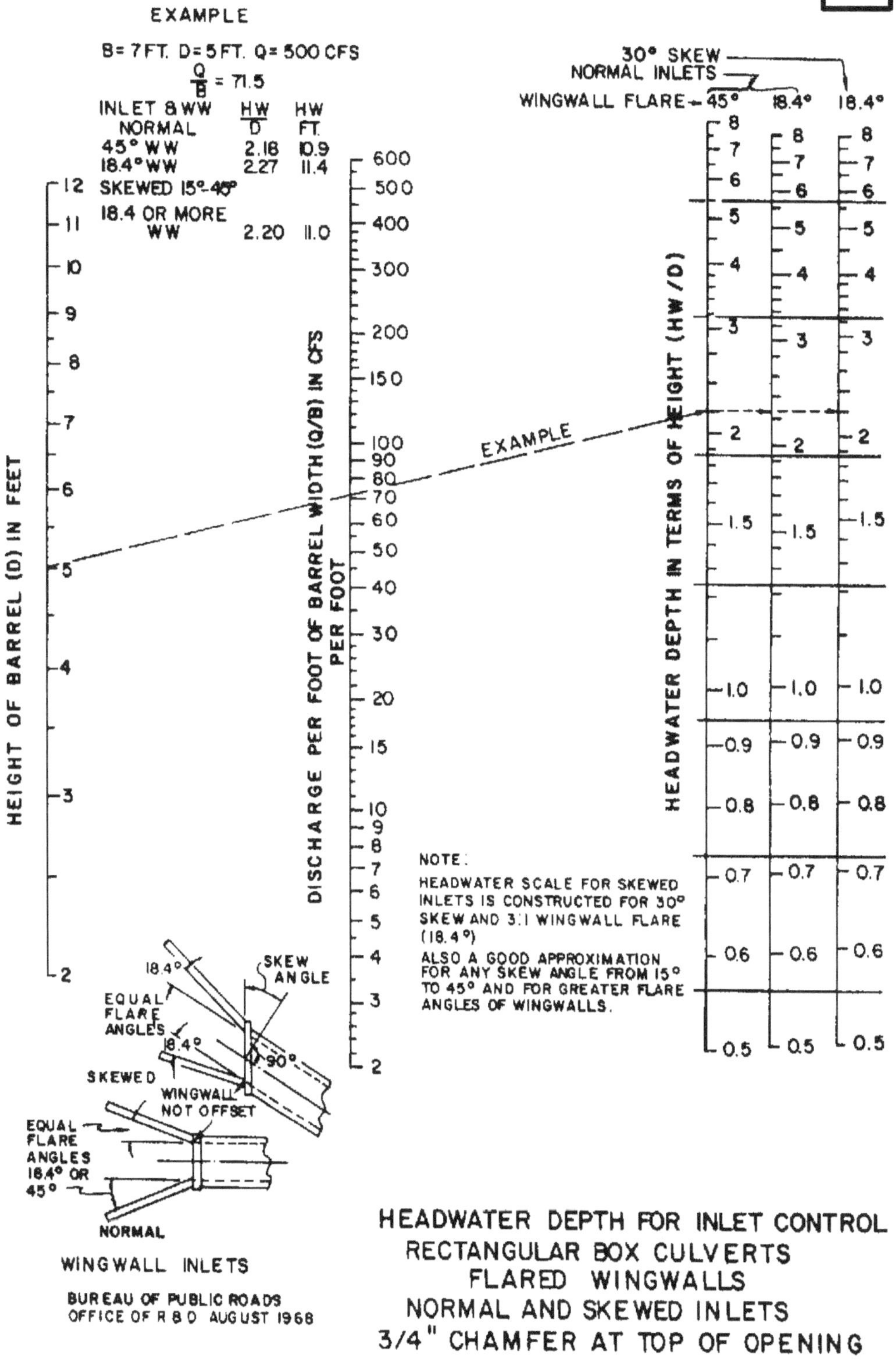

EXAMPLE

B= 7 FT. D= 5 FT. Q= 500 CFS

$$\frac{Q}{B} = 71.5$$

INLET & WW NORMAL	HW/D	HW FT.
45° WW	2.18	10.9
18.4° WW	2.27	11.4
SKEWED 15°-45° 18.4 OR MORE WW	2.20	11.0

HEIGHT OF BARREL (D) IN FEET

12
11
10
9
8
7
6
5
4
3
2

DISCHARGE PER FOOT OF BARREL WIDTH (Q/B) IN CFS PER FOOT

600
500
400
300
200
150
100
90
80
70
60
50
40
30
20
15
10
9
8
7
6
5
4
3
2

EXAMPLE

30° SKEW
NORMAL INLETS
WINGWALL FLARE— 45°   18.4°   18.4°

HEADWATER DEPTH IN TERMS OF HEIGHT (HW/D)

45°	18.4°	18.4°
8	8	8
7	7	7
6	6	6
5	5	5
4	4	4
3	3	3
2	2	2
1.5	1.5	1.5
1.0	1.0	1.0
0.9	0.9	0.9
0.8	0.8	0.8
0.7	0.7	0.7
0.6	0.6	0.6
0.5	0.5	0.5

NOTE:
HEADWATER SCALE FOR SKEWED INLETS IS CONSTRUCTED FOR 30° SKEW AND 3:1 WINGWALL FLARE (18.4°)

ALSO A GOOD APPROXIMATION FOR ANY SKEW ANGLE FROM 15° TO 45° AND FOR GREATER FLARE ANGLES OF WINGWALLS.

SKEW ANGLE

18.4°

EQUAL FLARE ANGLES

18.4°   90°

SKEWED

WINGWALL NOT OFFSET

EQUAL FLARE ANGLES 18.4° OR 45°

NORMAL

WINGWALL INLETS

BUREAU OF PUBLIC ROADS
OFFICE OF R & D AUGUST 1968

HEADWATER DEPTH FOR INLET CONTROL
RECTANGULAR BOX CULVERTS
FLARED WINGWALLS
NORMAL AND SKEWED INLETS
3/4" CHAMFER AT TOP OF OPENING

CHART 13A

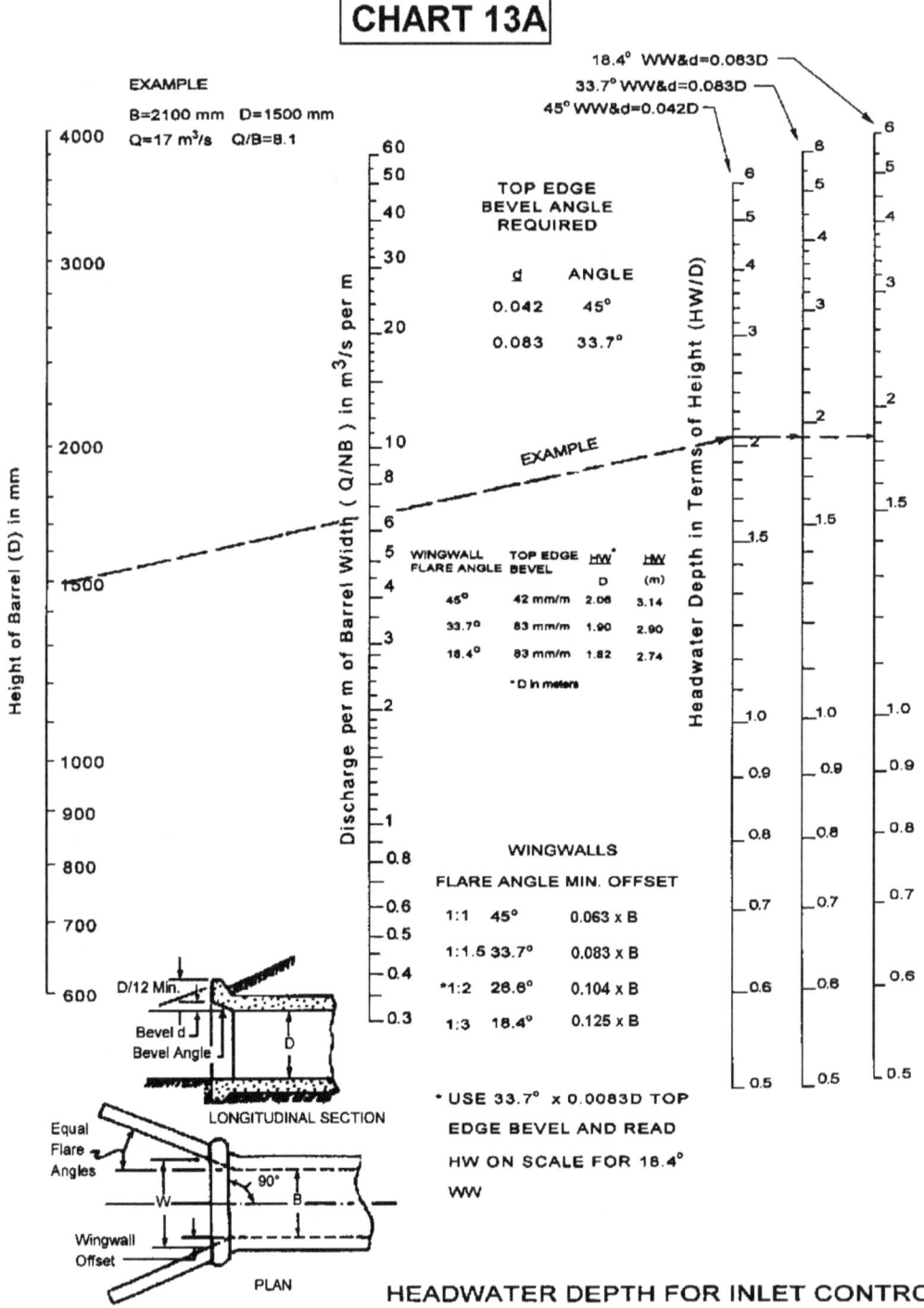

18.4° WW&d=0.083D

33.7° WW&d=0.083D

45° WW&d=0.042D

EXAMPLE

B=2100 mm   D=1500 mm

Q=17 m³/s   Q/B=8.1

Height of Barrel (D) in mm

4000

3000

2000

1500

1000
900
800

700

600

D/12 Min.

Bevel d
Bevel Angle

LONGITUDINAL SECTION

Equal
Flare
Angles

90°

W   B

Wingwall
Offset

PLAN

Discharge per m of Barrel Width (Q/NB) in m³/s per m

60
50
40

30

20

10
8

6

5
4

3

2

1

0.8

0.6
0.5

0.4

0.3

TOP EDGE BEVEL ANGLE REQUIRED

d	ANGLE
0.042	45°
0.083	33.7°

EXAMPLE

WINGWALL FLARE ANGLE	TOP EDGE BEVEL	HW* / D	HW (m)
45°	42 mm/m	2.06	3.14
33.7°	83 mm/m	1.90	2.90
18.4°	83 mm/m	1.82	2.74

* D in meters

WINGWALLS

FLARE ANGLE		MIN. OFFSET
1:1	45°	0.063 x B
1:1.5	33.7°	0.083 x B
*1:2	26.6°	0.104 x B
1:3	18.4°	0.125 x B

* USE 33.7° x 0.0083D TOP
EDGE BEVEL AND READ
HW ON SCALE FOR 18.4°
WW

Headwater Depth in Terms of Height (HW/D)

6

5

4

3

2

1.5

1.0
0.9

0.8

0.7

0.6

0.5

6

5

4

3

2

1.5

1.0
0.9

0.8

0.7

0.6

0.5

6

5

4

3

2

1.5

1.0
0.9

0.8

0.7

0.6

0.5

HEADWATER DEPTH FOR INLET CONTROL RECTANGULAR BOX CULVERTS OFFSET FLARED WINGWALLS AND BEVELED EDGE AT TOP OF INLET

Adapted from
Bureau of Public Roads Office of R & D
August 1968

CHART 13B

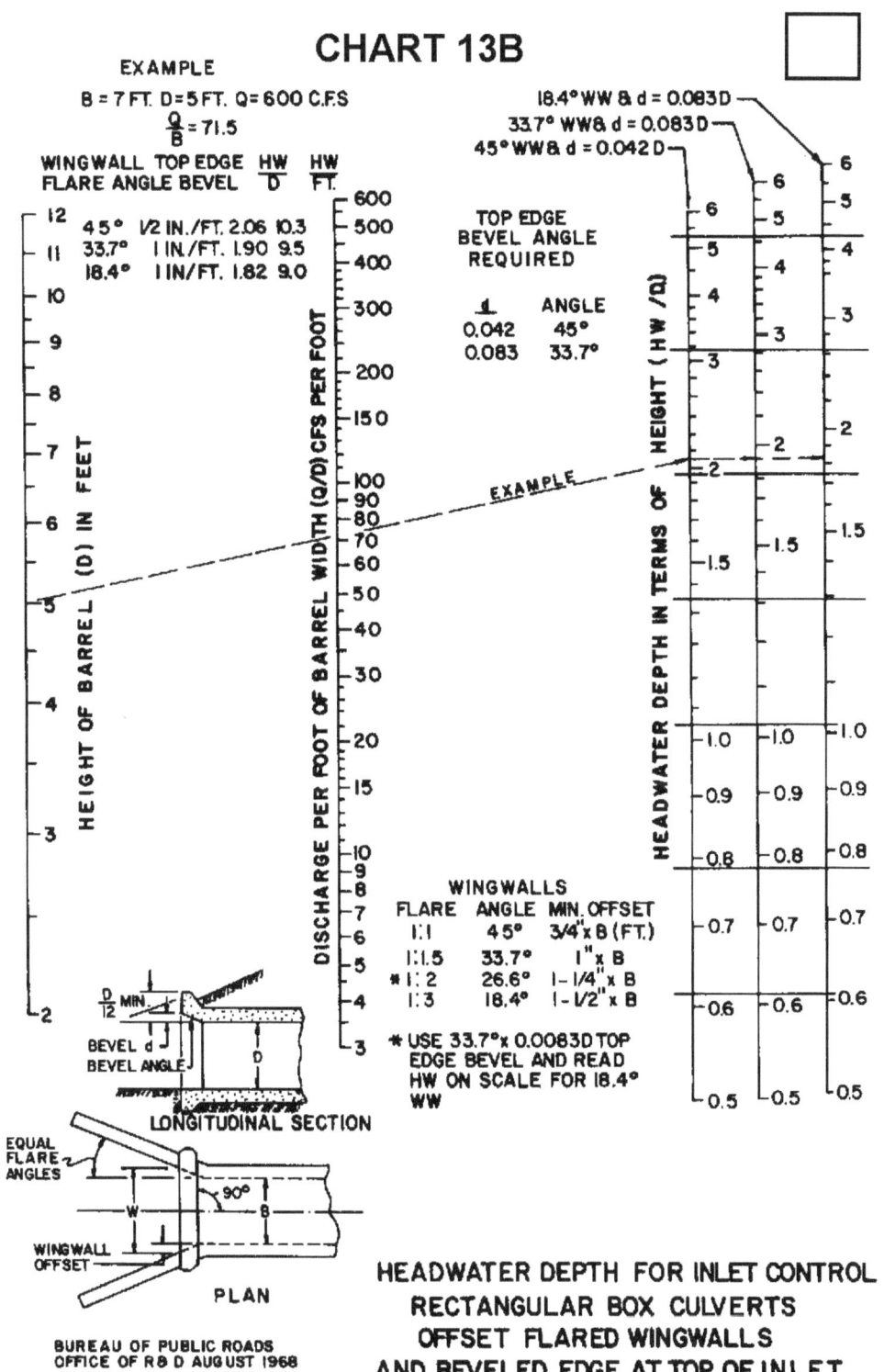

HEADWATER DEPTH FOR INLET CONTROL
RECTANGULAR BOX CULVERTS
OFFSET FLARED WINGWALLS
AND BEVELED EDGE AT TOP OF INLET

CHART 14 A

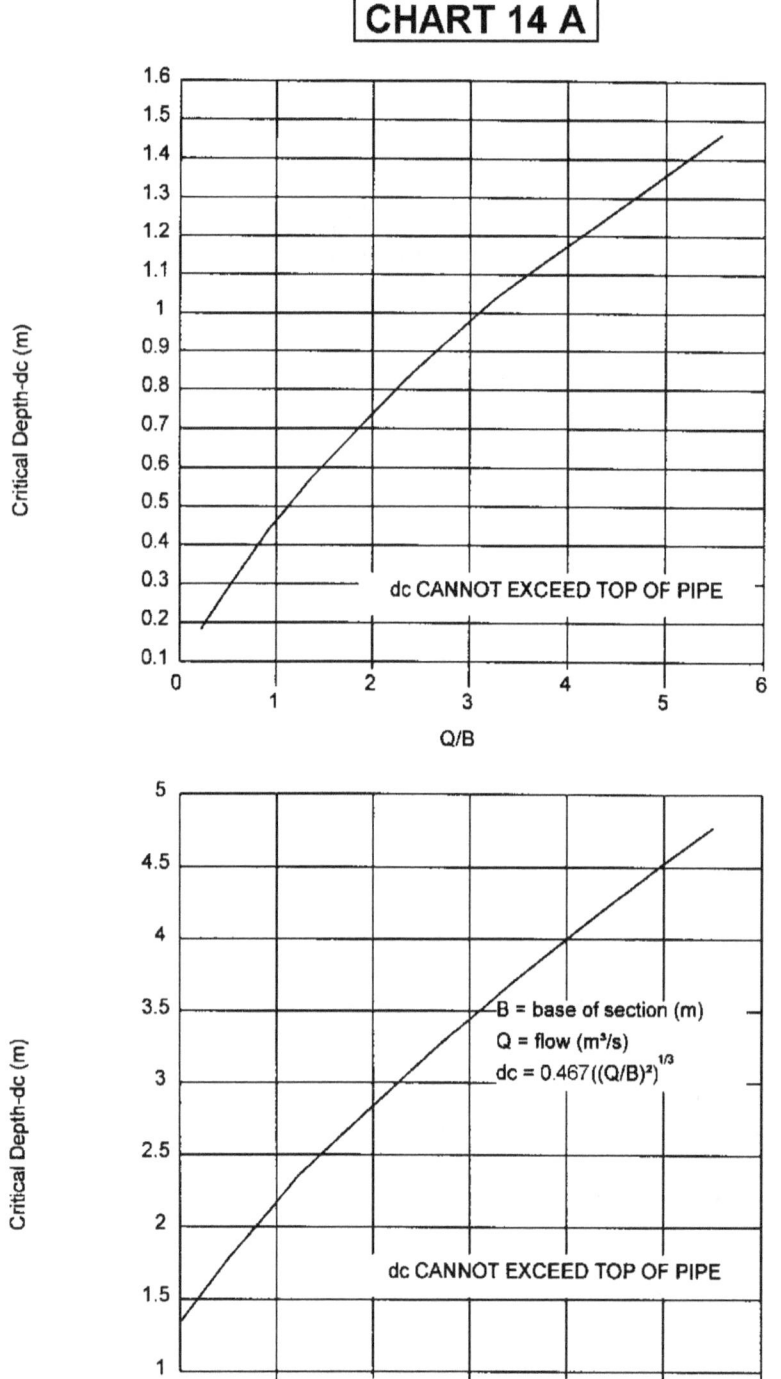

Critical Depth-dc (m)

dc CANNOT EXCEED TOP OF PIPE

Q/B

Critical Depth-dc (m)

B = base of section (m)
Q = flow (m³/s)
$dc = 0.467((Q/B)^2)^{1/3}$

dc CANNOT EXCEED TOP OF PIPE

Q/B

Adapted from Bureau of Public Roads

Critical Depth-Rectangular Section

CHART 14B

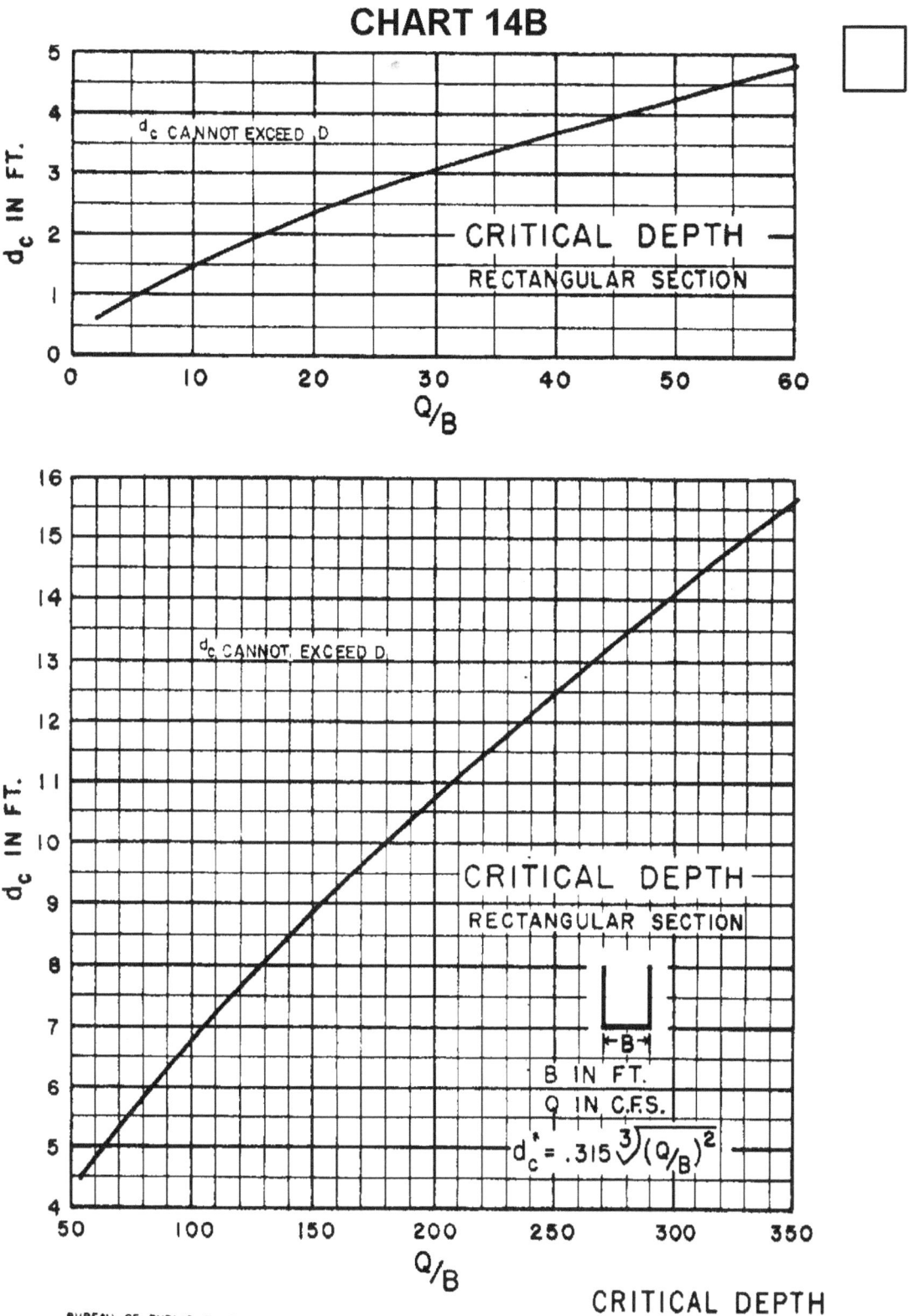

CRITICAL DEPTH
RECTANGULAR SECTION

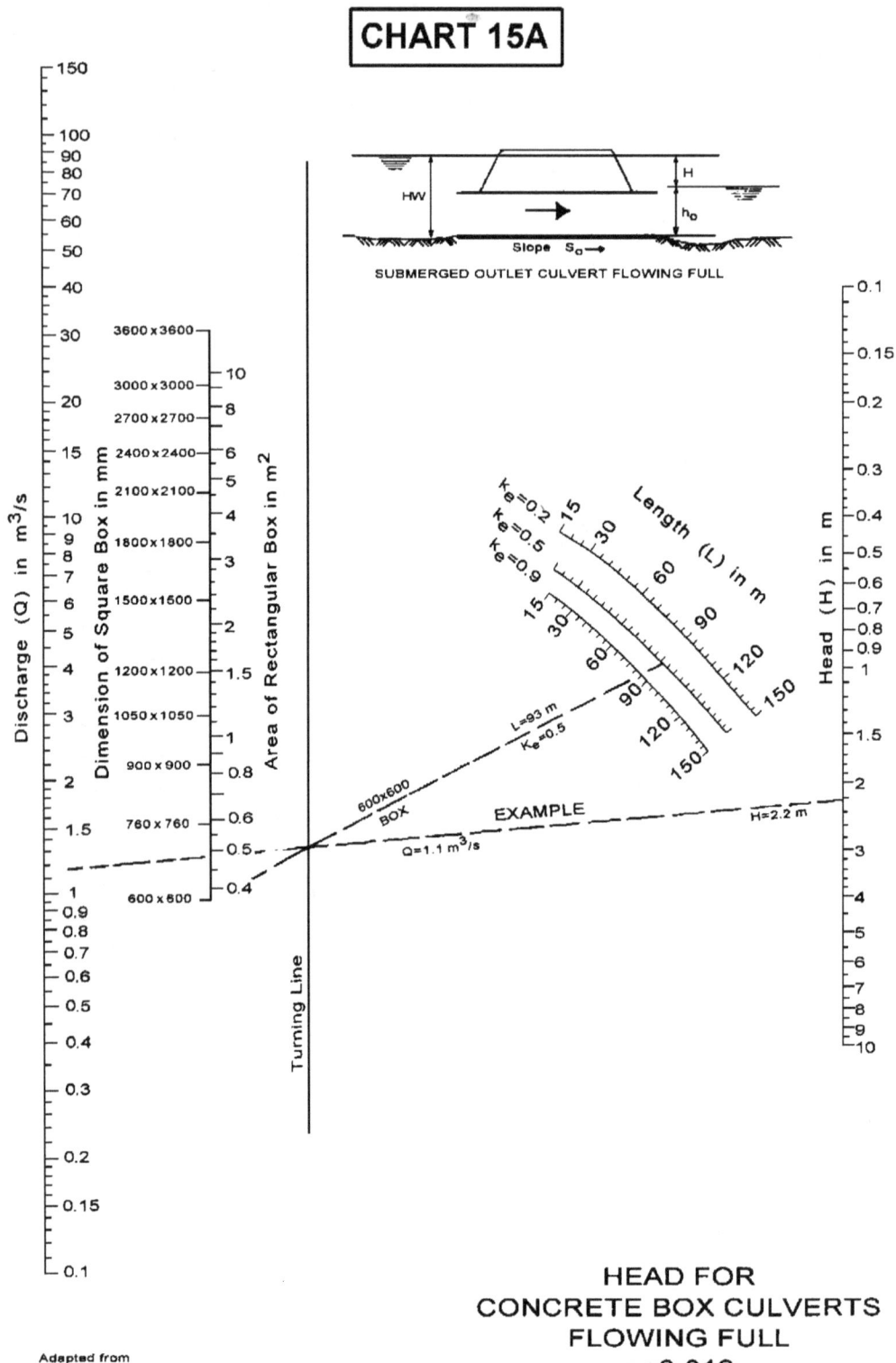

CHART 15A

SUBMERGED OUTLET CULVERT FLOWING FULL

Discharge (Q) in m³/s

Dimension of Square Box in mm

Area of Rectangular Box in m²

Turning Line

Length (L) in m

Head (H) in m

Ke=0.2
Ke=0.5
Ke=0.9

L=93 m
Ke=0.5

600x600
BOX

EXAMPLE

Q=1.1 m³/s

H=2.2 m

HEAD FOR
CONCRETE BOX CULVERTS
FLOWING FULL
n=0.012

Adapted from
Burea of Public Roads Jan. 1963

CHART 15B

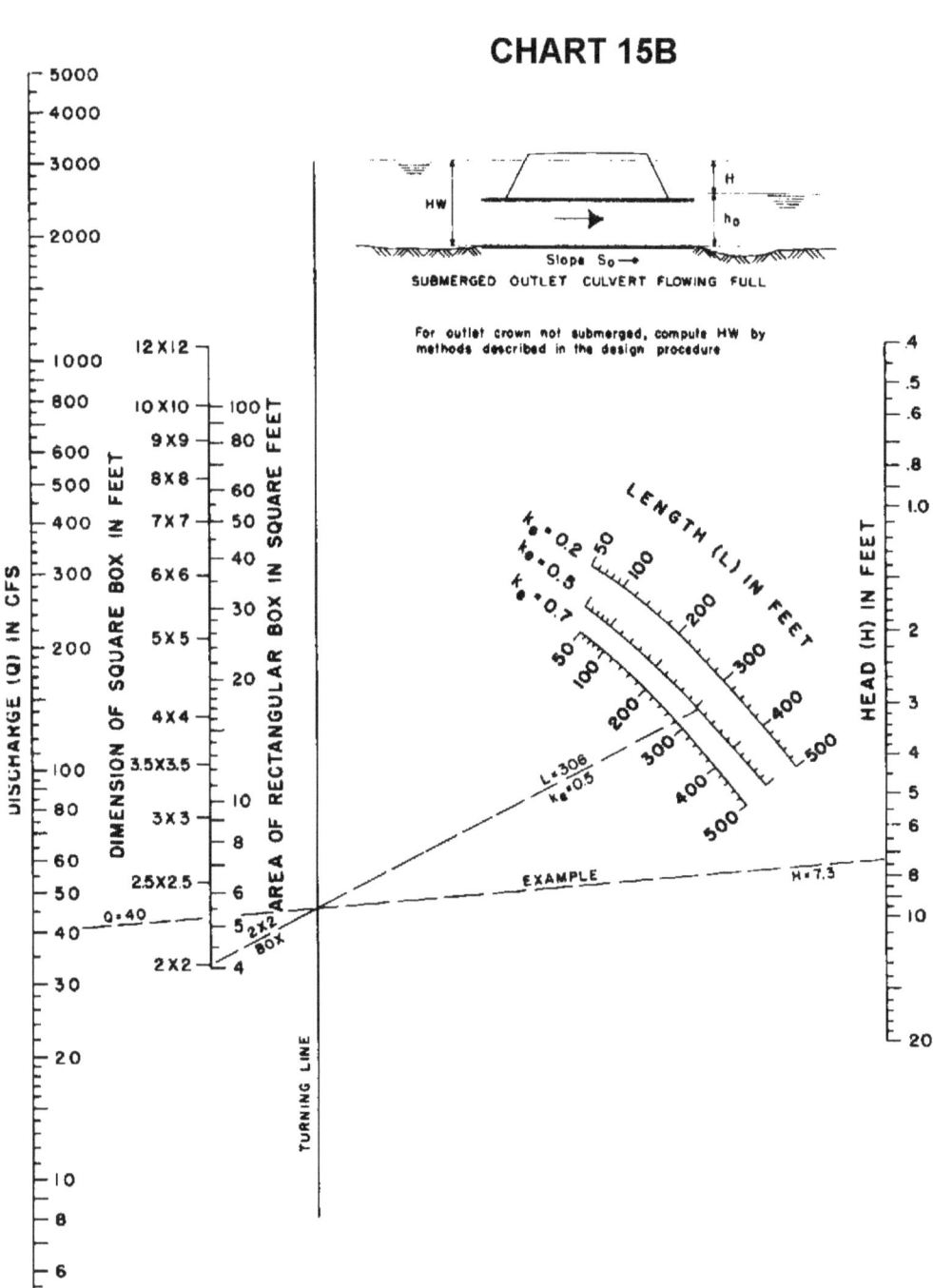

SUBMERGED OUTLET CULVERT FLOWING FULL

For outlet crown not submerged, compute HW by methods described in the design procedure

DISCHARGE (Q) IN CFS

DIMENSION OF SQUARE BOX IN FEET

AREA OF RECTANGULAR BOX IN SQUARE FEET

TURNING LINE

$k_e = 0.2$
$k_e = 0.5$
$k_e = 0.7$

LENGTH (L) IN FEET

HEAD (H) IN FEET

L = 306
$k_e = 0.5$

EXAMPLE

H = 7.3

Q = 40

2X2 BOX

HEAD FOR
CONCRETE BOX CULVERTS
FLOWING FULL
n = 0.012

CHART 51A

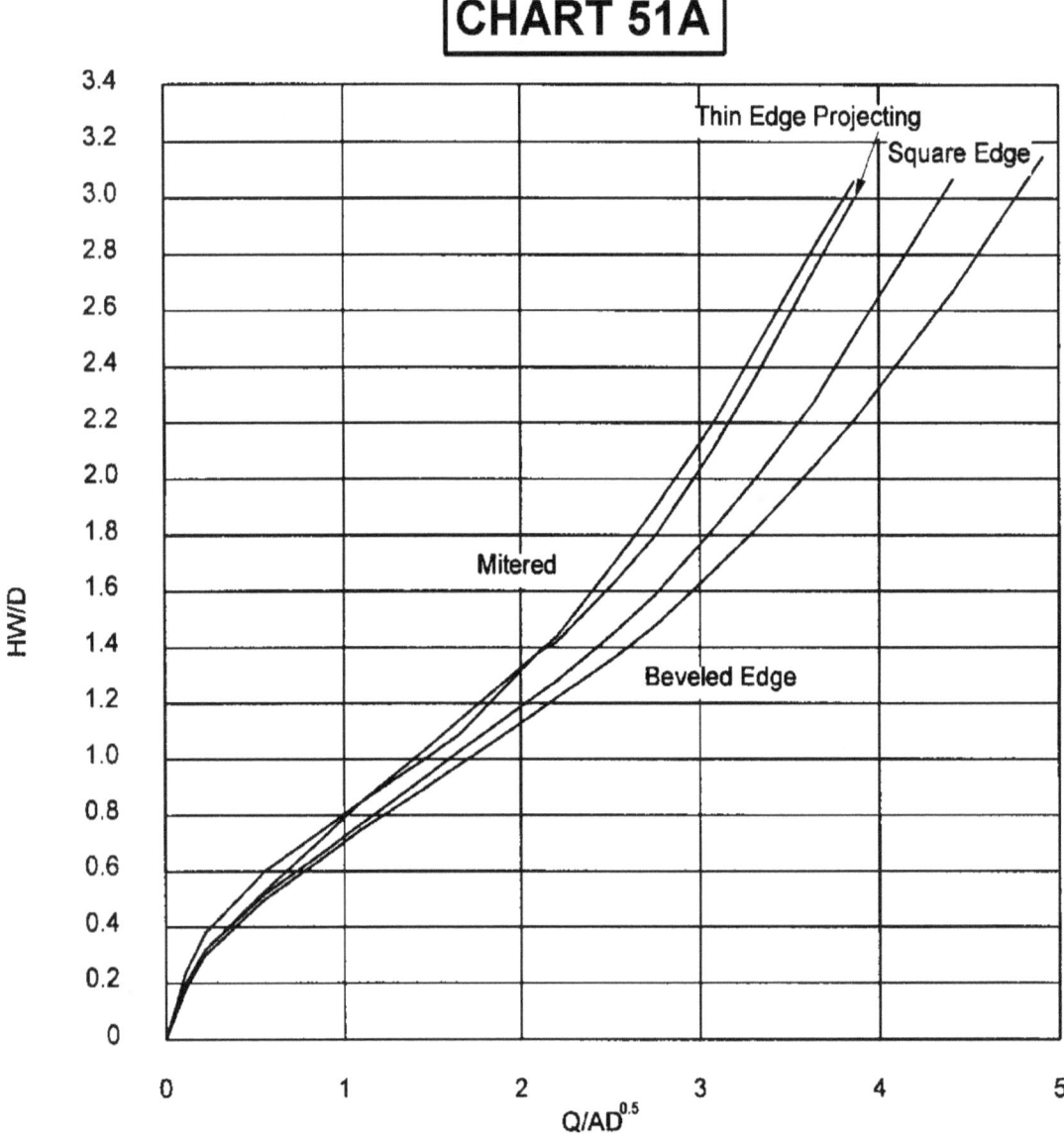

Inlet Control-Headwater Depth for Circular or Elliptical Structural Plate Corrugated Metal Conduits

CHART 51B
(English Units)

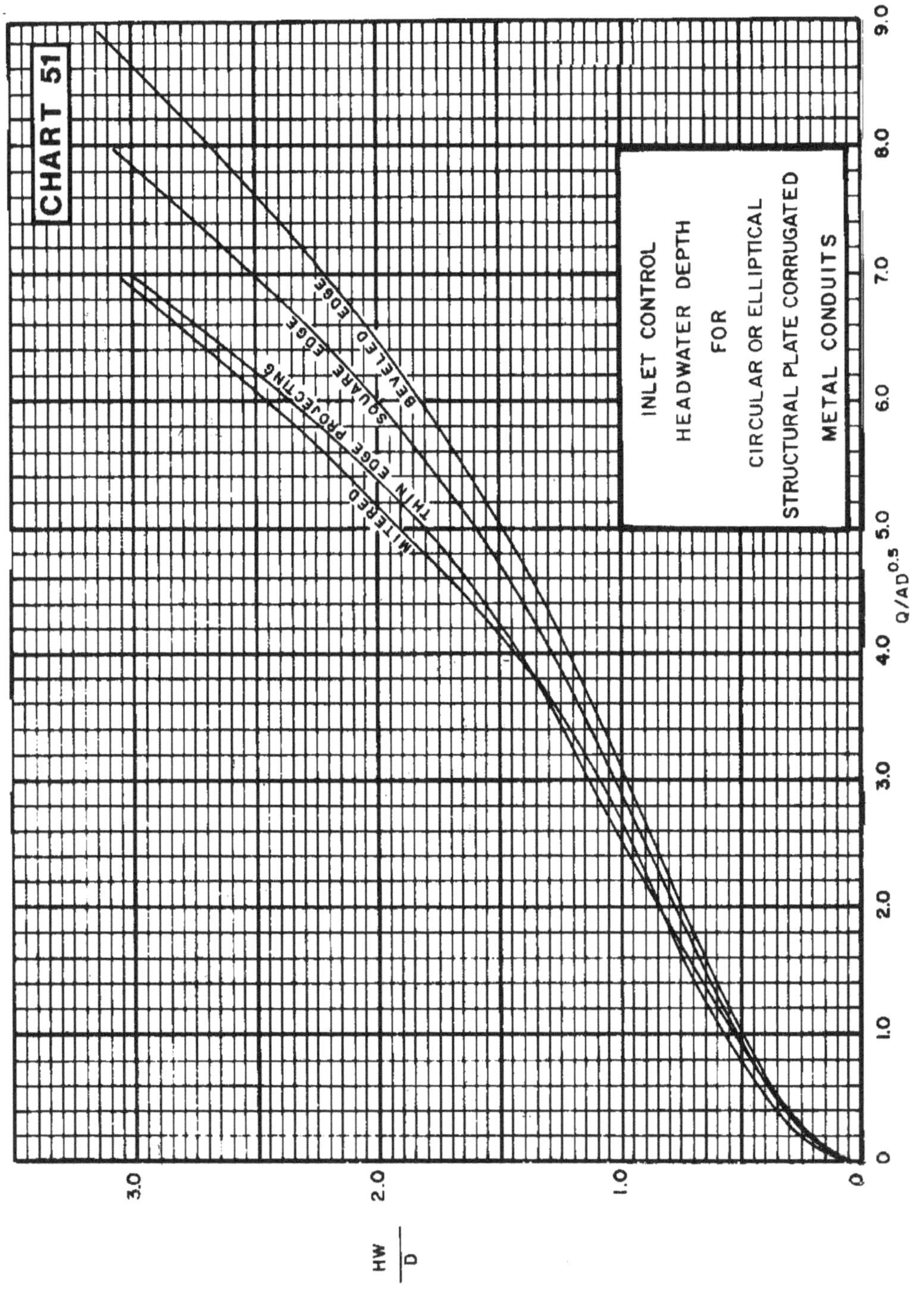

CHART 51

INLET CONTROL
HEADWATER DEPTH
FOR
CIRCULAR OR ELLIPTICAL
STRUCTURAL PLATE CORRUGATED
METAL CONDUITS

MITERED EDGE PROJECTING

THIN EDGE PROJECTING

SQUARE EDGE

BEVELED EDGE

$Q/AD^{0.5}$

HW / D

3.0

2.0

1.0

0   1.0   2.0   3.0   4.0   5.0   6.0   7.0   8.0   9.0

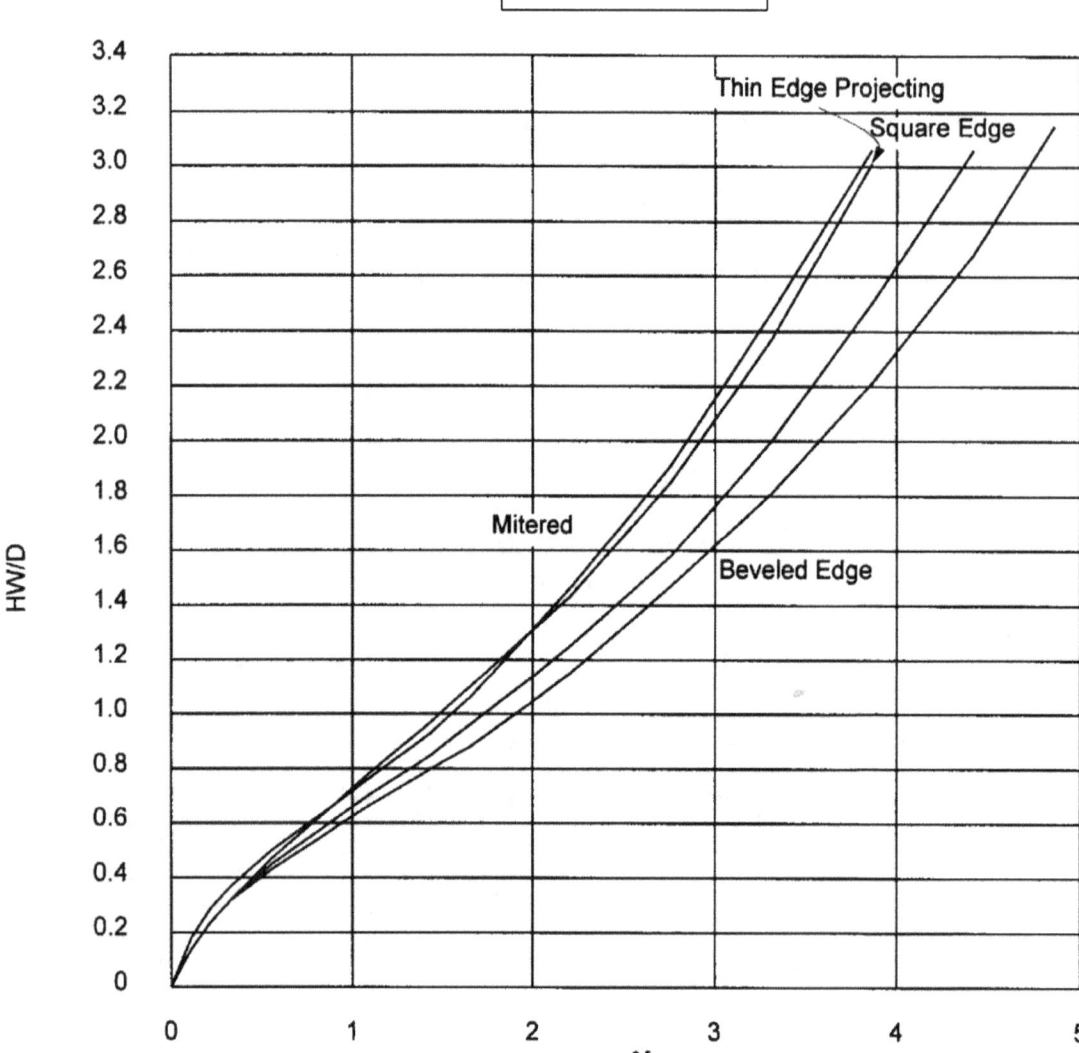

CHART 52A

Inlet Control-Headwater Depth for High and Low Profile Structural Plate Corrugated Metal Arch

CHART 52B
(English Units)

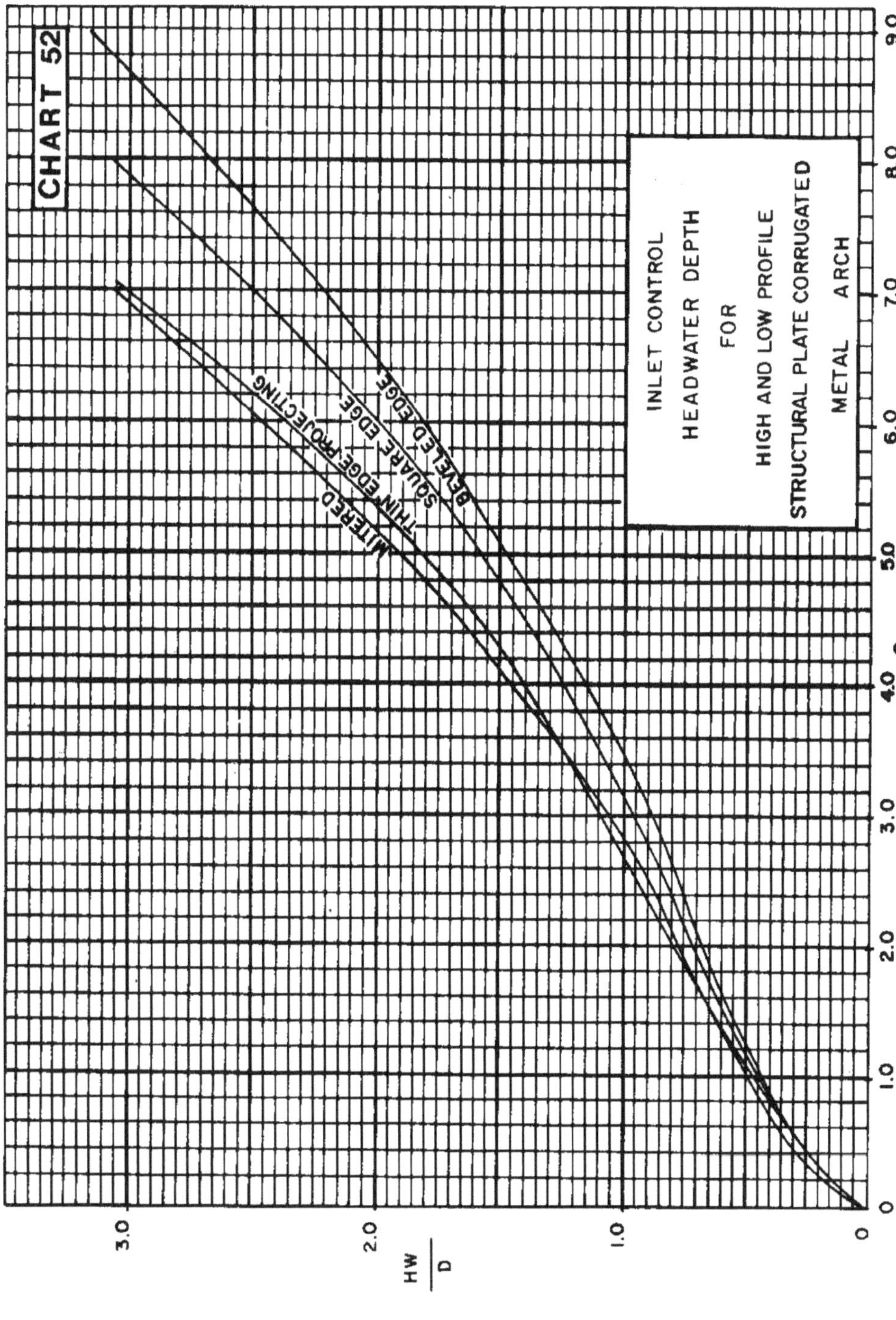

CHART 52

INLET CONTROL
HEADWATER DEPTH
FOR
HIGH AND LOW PROFILE
STRUCTURAL PLATE CORRUGATED
METAL ARCH

MITERED EDGE PROJECTING
THIN EDGE PROJECTING
SQUARE EDGE
BEVELED EDGE

$\frac{HW}{D}$

$Q/_{AD^{0.5}}$

0   1.0   2.0   3.0   4.0   5.0   6.0   7.0   8.0   9.0

0   1.0   2.0   3.0

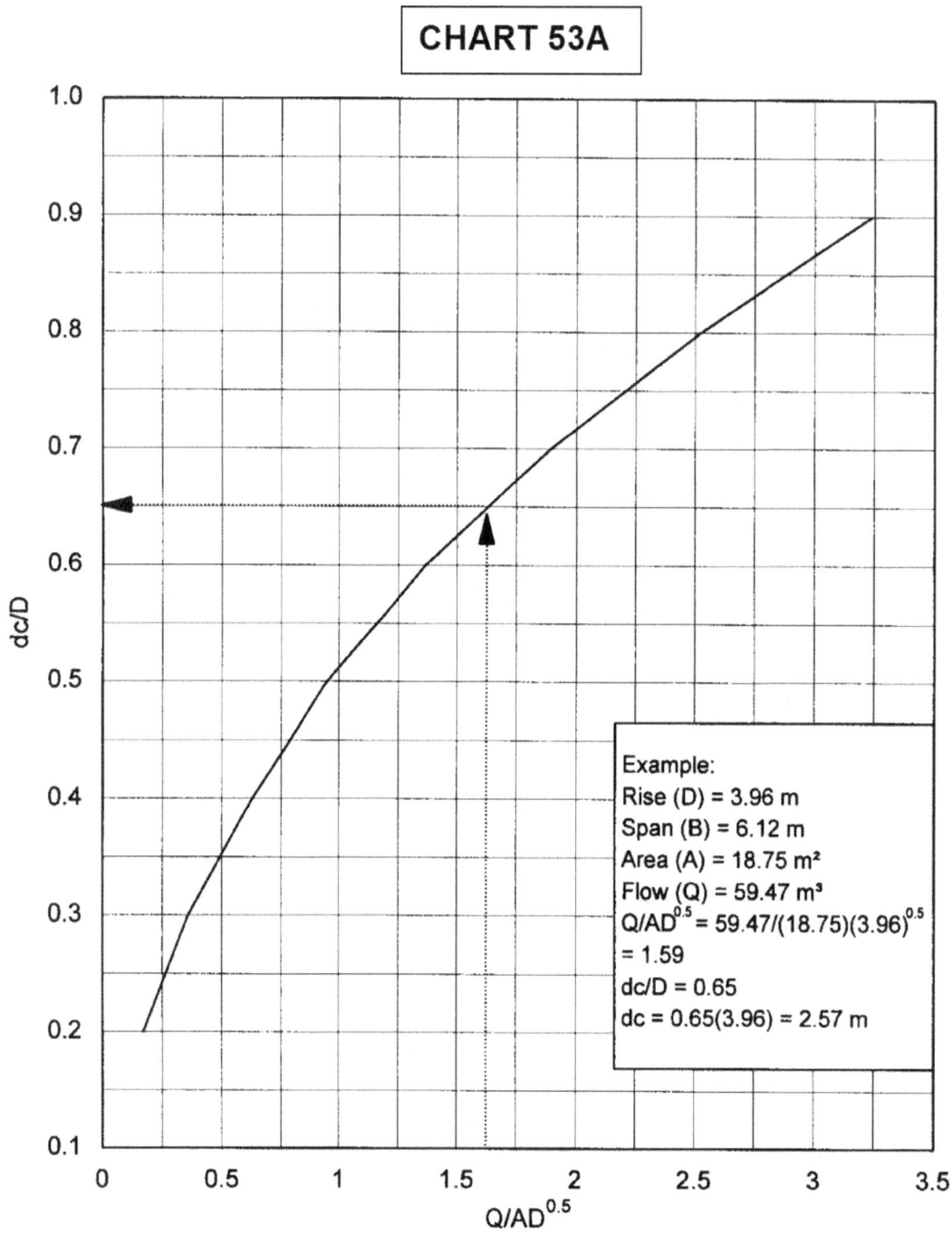

CHART 53A

Dimensionless Critical Depth Chart for Structural Plate Ellipse Long Axis Horizontal

Example:
Rise (D) = 3.96 m
Span (B) = 6.12 m
Area (A) = 18.75 m²
Flow (Q) = 59.47 m³
$Q/AD^{0.5} = 59.47/(18.75)(3.96)^{0.5}$
= 1.59
$dc/D = 0.65$
$dc = 0.65(3.96) = 2.57$ m

EXAMPLE :

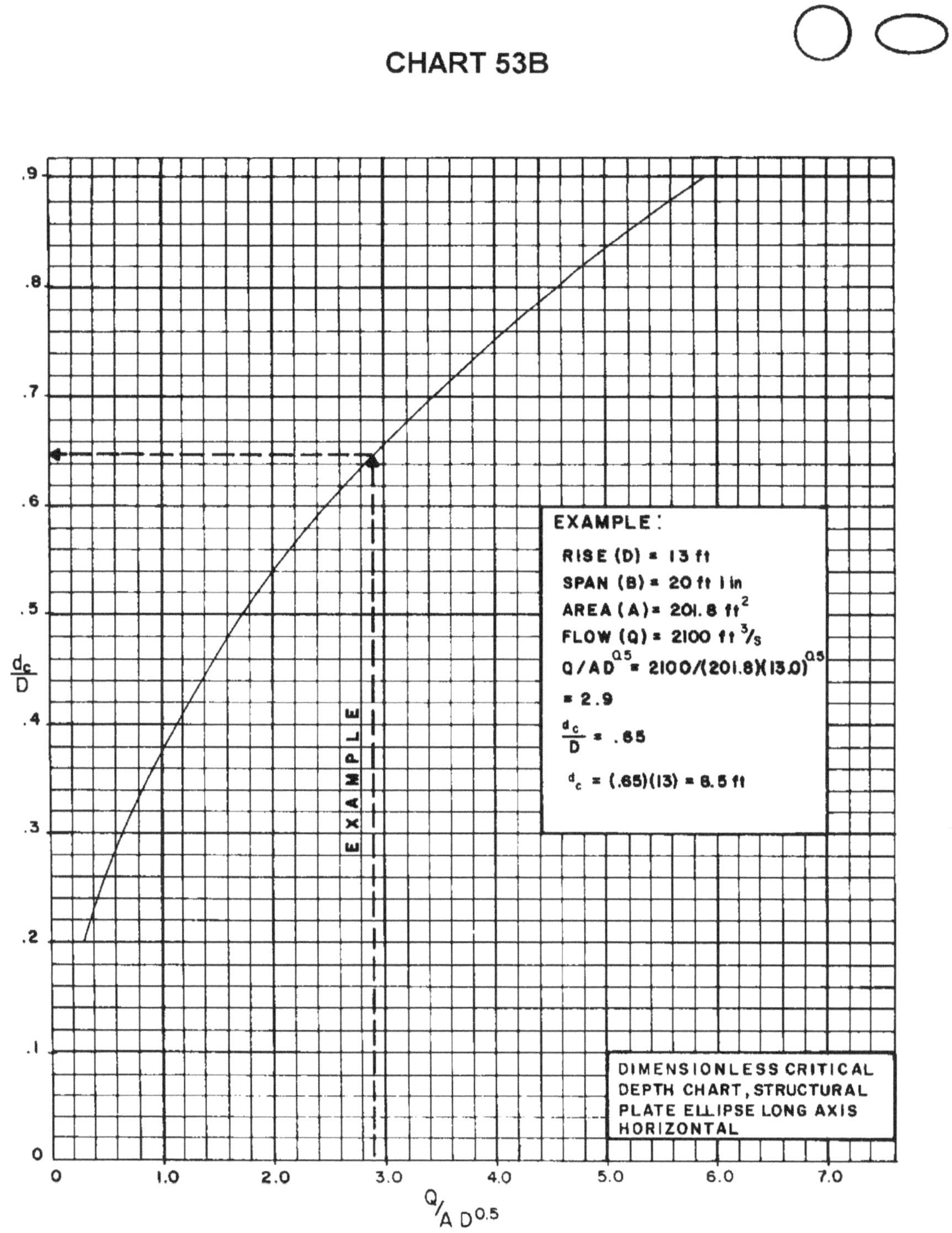

RISE (D) = 13 ft
SPAN (B) = 20 ft 1 in
AREA (A) = 201.8 ft^2
FLOW (Q) = 2100 ft^3/s
Q/AD$^{0.5}$ = 2100/(201.8)(13.0)$^{0.5}$
= 2.9

$\dfrac{d_c}{D}$ = .65

d_c = (.65)(13) = 8.5 ft

EXAMPLE

DIMENSIONLESS CRITICAL
DEPTH CHART, STRUCTURAL
PLATE ELLIPSE LONG AXIS
HORIZONTAL

$\dfrac{d_c}{D}$

Q/A D$^{0.5}$

CHART 54

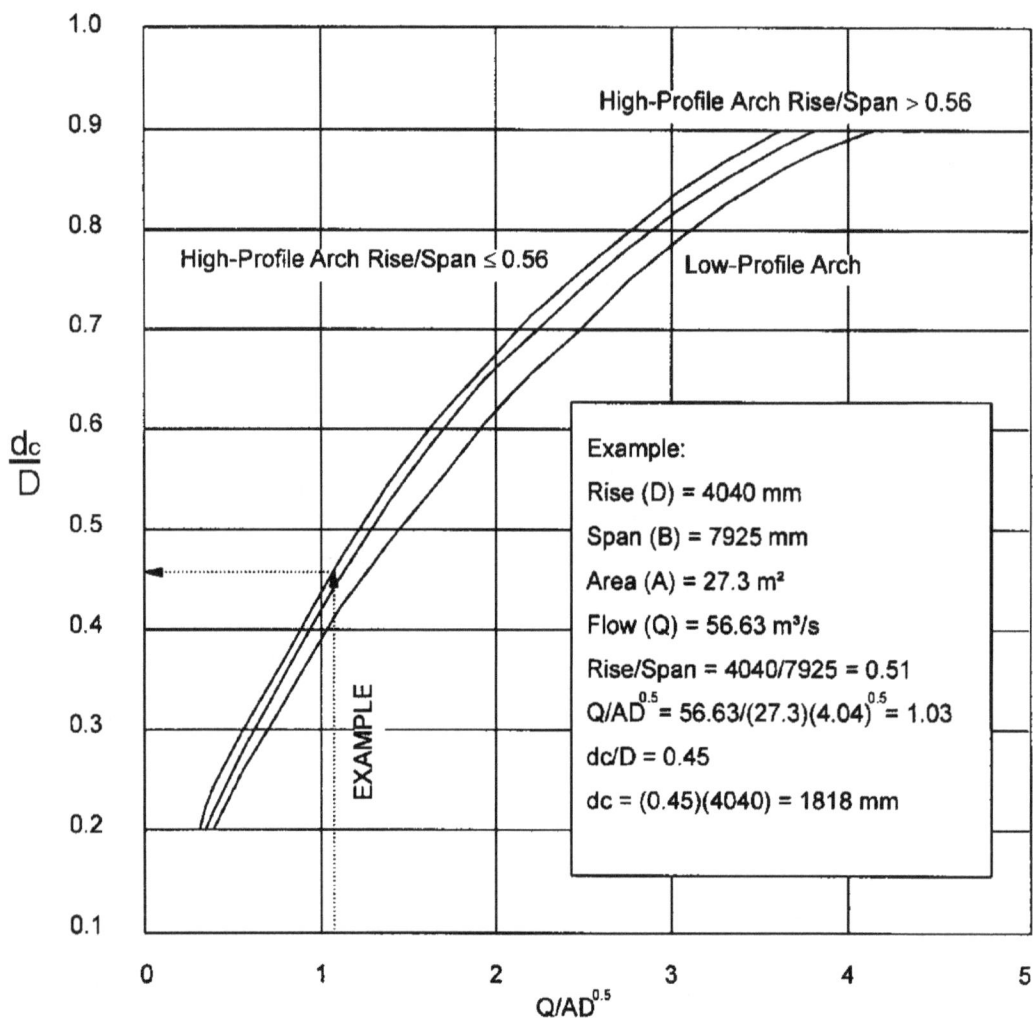

Dimensionless Critical Depth Chart for Structural Plate Low- and High-Profile Arches

$\frac{d_c}{D}$ axis (vertical): .9, .8, .7, .6, .5, .4, .3, .2, .1, 0

Horizontal axis: $\frac{Q}{A D^{0.5}}$ — 0, 1.0, 2.0, 3.0, 4.0, 5.0, 6.0, 7.0

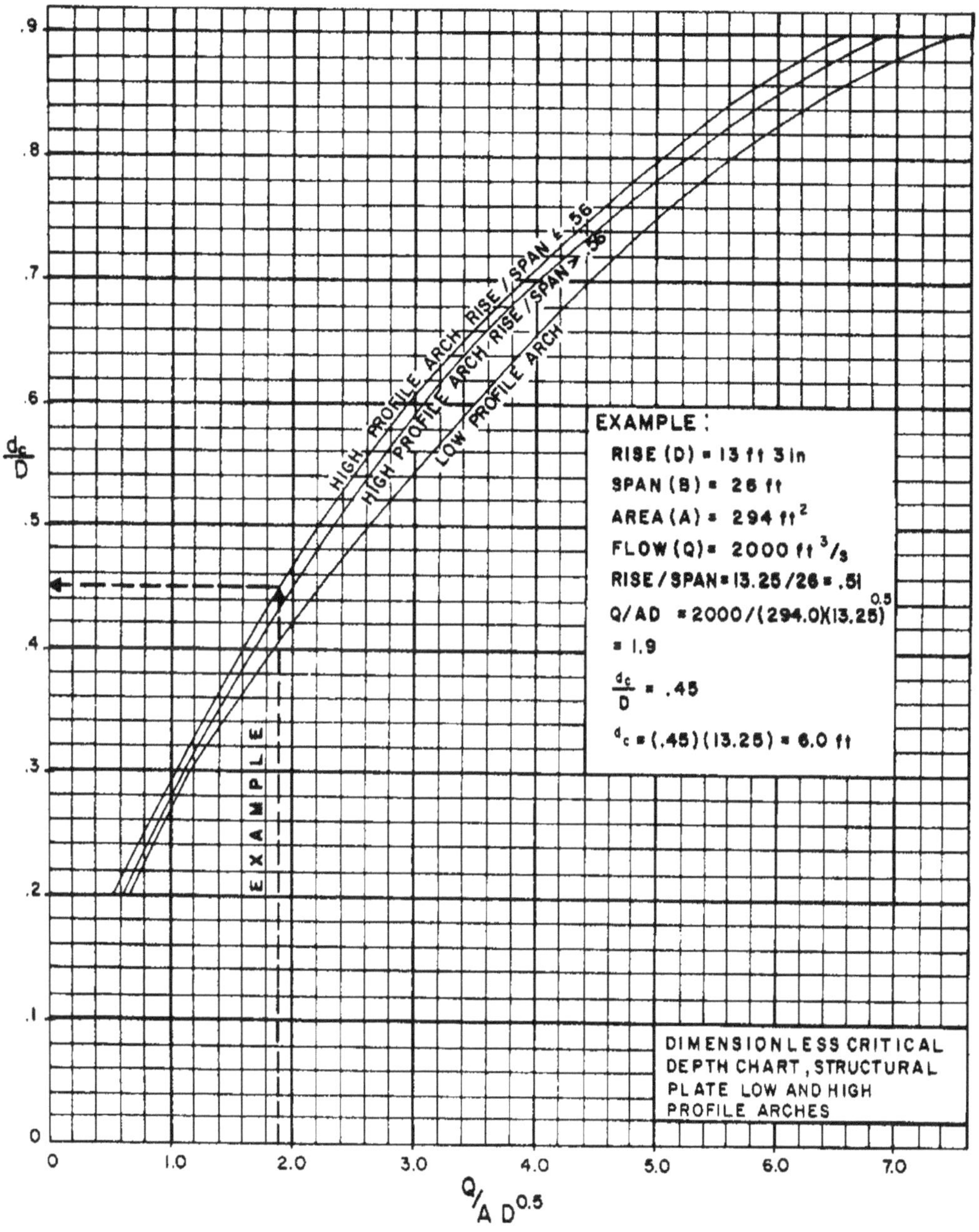

HIGH PROFILE ARCH RISE / SPAN ≤ .56

HIGH PROFILE ARCH RISE / SPAN > .56

LOW PROFILE ARCH

EXAMPLE

EXAMPLE :

RISE (D) = 13 ft 3 in

SPAN (B) = 26 ft

AREA (A) = 294 ft²

FLOW (Q) = 2000 ft³/s

RISE / SPAN = 13.25 / 26 = .51

$Q/AD^{0.5} = 2000/(294.0)(13.25)^{0.5}$

= 1.9

$\frac{d_c}{D}$ = .45

$d_c = (.45)(13.25) = 6.0$ ft

DIMENSIONLESS CRITICAL
DEPTH CHART, STRUCTURAL
PLATE LOW AND HIGH
PROFILE ARCHES

CHART 55A

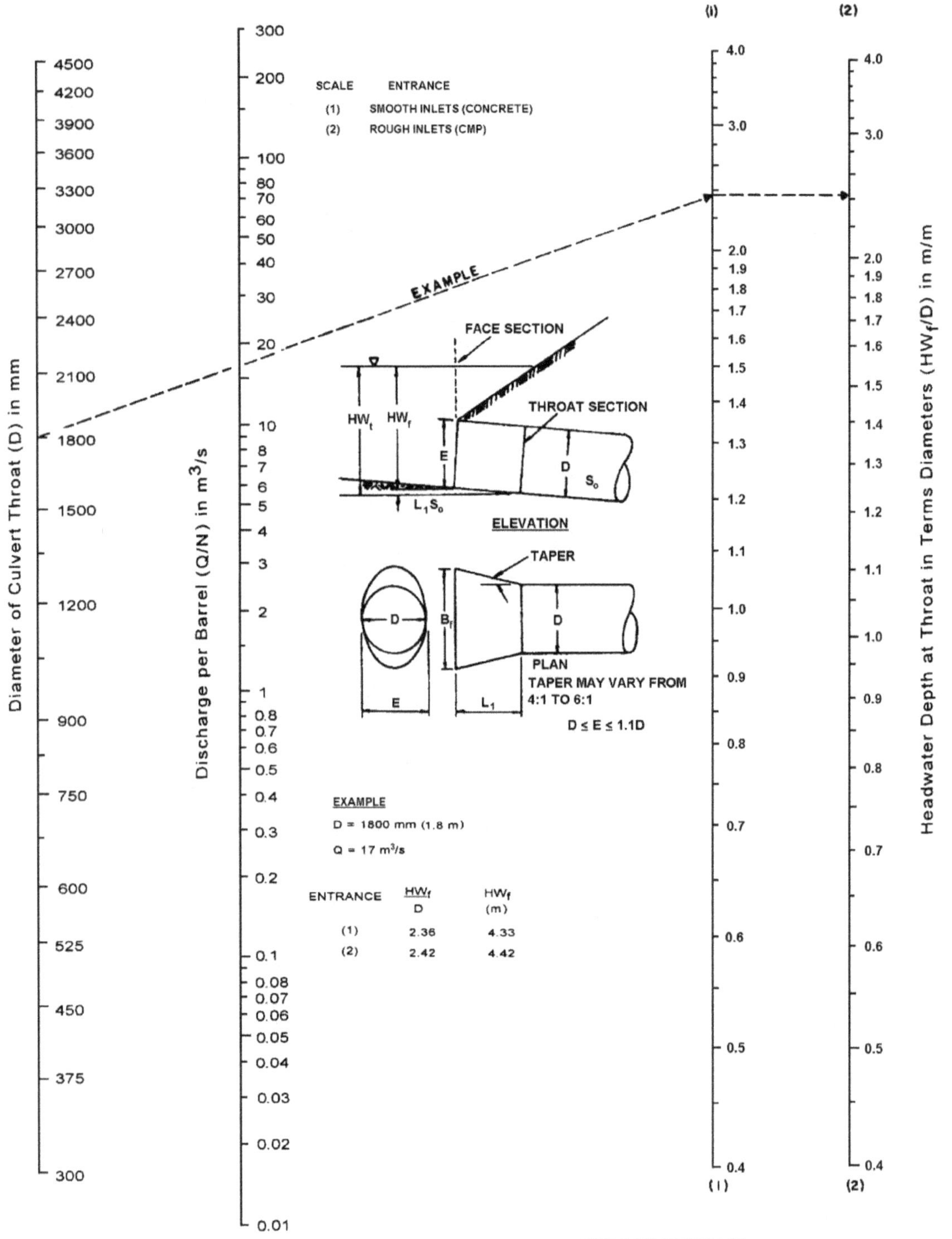

Diameter of Culvert Throat (D) in mm

4500
4200
3900
3600
3300
3000
2700
2400
2100
1800
1500
1200
900
750
600
525
450
375
300

Discharge per Barrel (Q/N) in m³/s

300
200
100
80
70
60
50
40
30
20
10
8
7
6
5
4
3
2
1
0.8
0.7
0.6
0.5
0.4
0.3
0.2
0.1
0.08
0.07
0.06
0.05
0.04
0.03
0.02
0.01

SCALE   ENTRANCE
(1)   SMOOTH INLETS (CONCRETE)
(2)   ROUGH INLETS (CMP)

EXAMPLE

FACE SECTION

THROAT SECTION

HW_t HW_f

E

D S_o

$L_1 S_o$

ELEVATION

TAPER

B_f D   D

E L_1

PLAN
TAPER MAY VARY FROM
4:1 TO 6:1

$D \leq E \leq 1.1D$

EXAMPLE

D = 1800 mm (1.8 m)

Q = 17 m³/s

ENTRANCE	$\dfrac{HW_f}{D}$	HW_f (m)
(1)	2.36	4.33
(2)	2.42	4.42

Headwater Depth at Throat in Terms Diameters (HW_f/D) in m/m

(1)
4.0
3.0
2.0
1.9
1.8
1.7
1.6
1.5
1.4
1.3
1.2
1.1
1.0
0.9
0.8
0.7
0.6
0.5
0.4
(1)

(2)
4.0
3.0
2.0
1.9
1.8
1.7
1.6
1.5
1.4
1.3
1.2
1.1
1.0
0.9
0.8
0.7
0.6
0.5
0.4
(2)

THROAT CONTROL
FOR SIDE-TAPERED INLETS TO PIPE CULVERT
(CIRCULAR SECTION ONLY)

CHART 55B

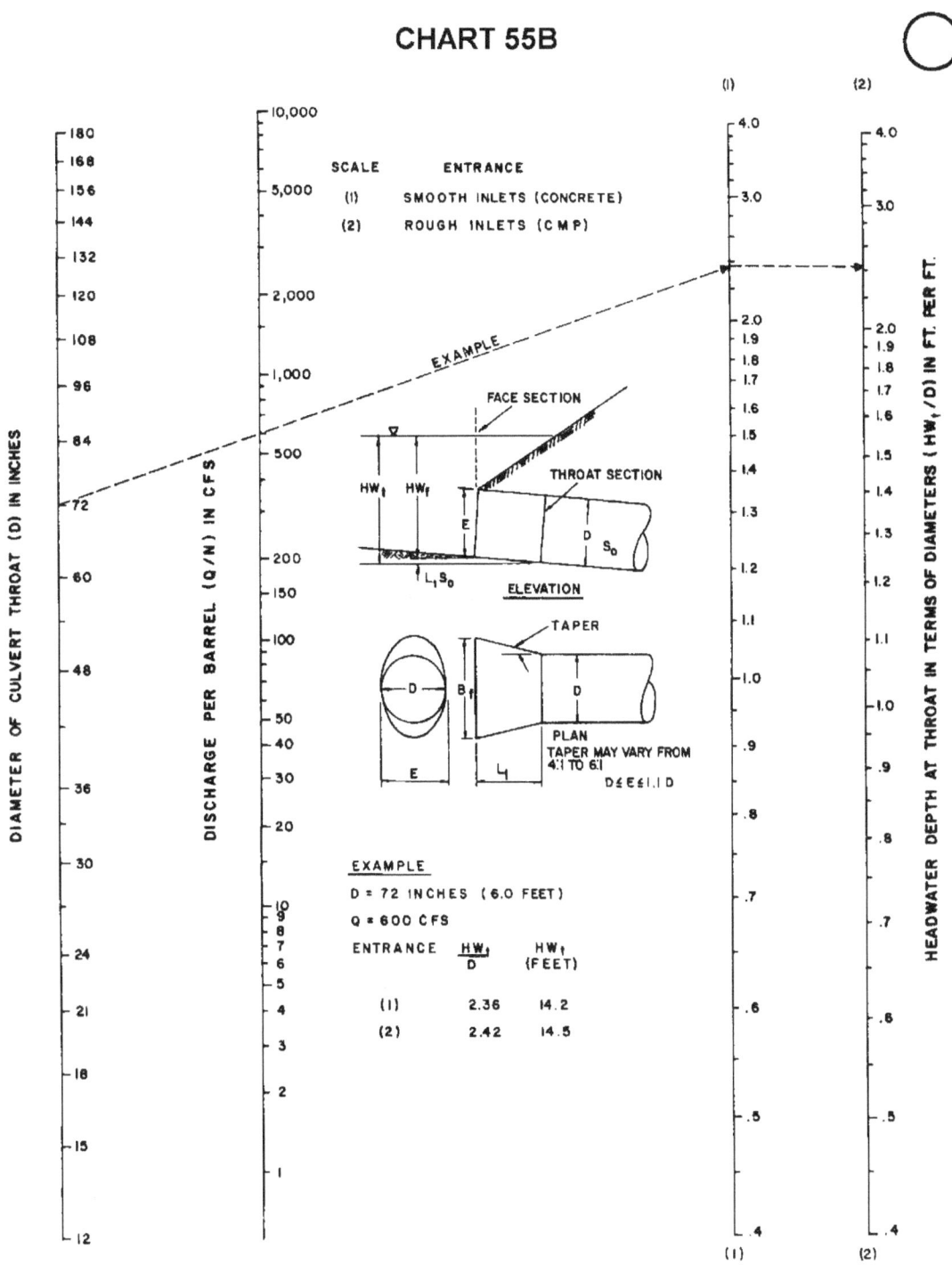

DIAMETER OF CULVERT THROAT (D) IN INCHES

DISCHARGE PER BARREL (Q/N) IN CFS

HEADWATER DEPTH AT THROAT IN TERMS OF DIAMETERS (HW$_t$/D) IN FT. PER FT.

SCALE	ENTRANCE
(1)	SMOOTH INLETS (CONCRETE)
(2)	ROUGH INLETS (C M P)

EXAMPLE

FACE SECTION

THROAT SECTION

ELEVATION

TAPER

PLAN
TAPER MAY VARY FROM
4:1 TO 6:1

$D \leq E \leq 1.1\, D$

EXAMPLE

D = 72 INCHES (6.0 FEET)

Q = 600 CFS

ENTRANCE	$\dfrac{HW_t}{D}$	HW_t (FEET)
(1)	2.36	14.2
(2)	2.42	14.5

THROAT CONTROL
FOR SIDE-TAPERED INLETS TO PIPE CULVERT
(CIRCULAR SECTION ONLY)

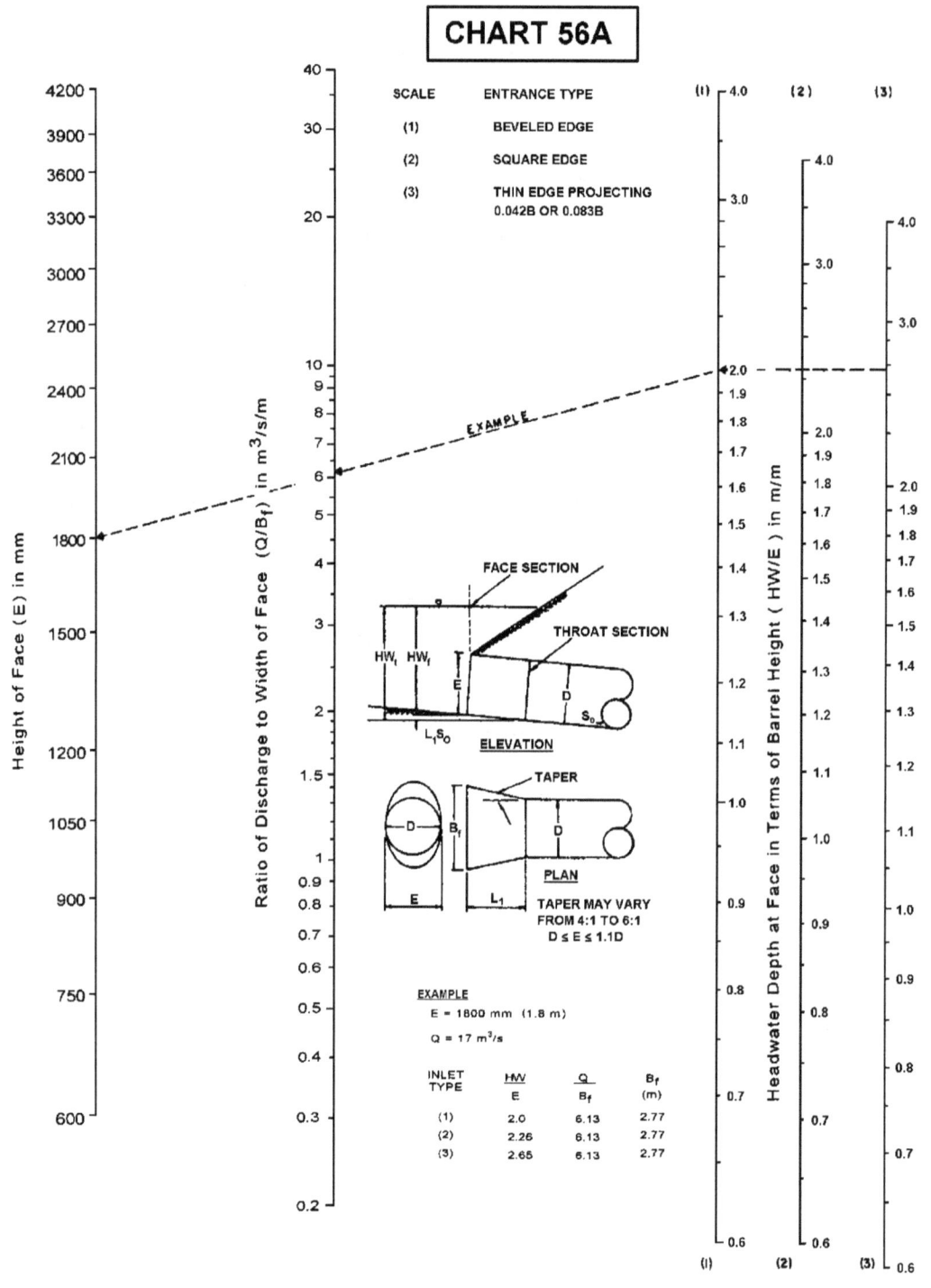

CHART 56A

Height of Face (E) in mm

4200
3900
3600
3300
3000
2700
2400
2100
1800
1500
1200
1050
900
750
600

Ratio of Discharge to Width of Face (Q/B_f) in m³/s/m

40
30
20

10
9
8
7
6
5
4
3
2
1.5
1
0.9
0.8
0.7
0.6
0.5
0.4
0.3
0.2

SCALE	ENTRANCE TYPE
(1)	BEVELED EDGE
(2)	SQUARE EDGE
(3)	THIN EDGE PROJECTING 0.042B OR 0.083B

EXAMPLE

FACE SECTION

THROAT SECTION

HW_t HW_f

E

L_1 S_o

ELEVATION

TAPER

D

PLAN

TAPER MAY VARY
FROM 4:1 TO 6:1
D ≤ E ≤ 1.1D

EXAMPLE

E = 1800 mm (1.8 m)

Q = 17 m³/s

INLET TYPE	$\frac{HW}{E}$	$\frac{Q}{B_f}$	B_f (m)
(1)	2.0	6.13	2.77
(2)	2.26	6.13	2.77
(3)	2.65	6.13	2.77

Headwater Depth at Face in Terms of Barrel Height (HW/E) in m/m

(1)
4.0
3.0
2.0
1.9
1.8
1.7
1.6
1.5
1.4
1.3
1.2
1.1
1.0
0.9
0.8
0.7
0.6

(2)
4.0
3.0
2.0
1.9
1.8
1.7
1.6
1.5
1.4
1.3
1.2
1.1
1.0
0.9
0.8
0.7
0.6

(3)
4.0
3.0
2.0
1.9
1.8
1.7
1.6
1.5
1.4
1.3
1.2
1.1
1.0
0.9
0.8
0.7
0.6

**FACE CONTROL FOR SIDE-TAPERED
INLETS TO PIPE CULVERTS
(NON-RECTANGULAR SECTIONS ONLY)**

CHART 56B

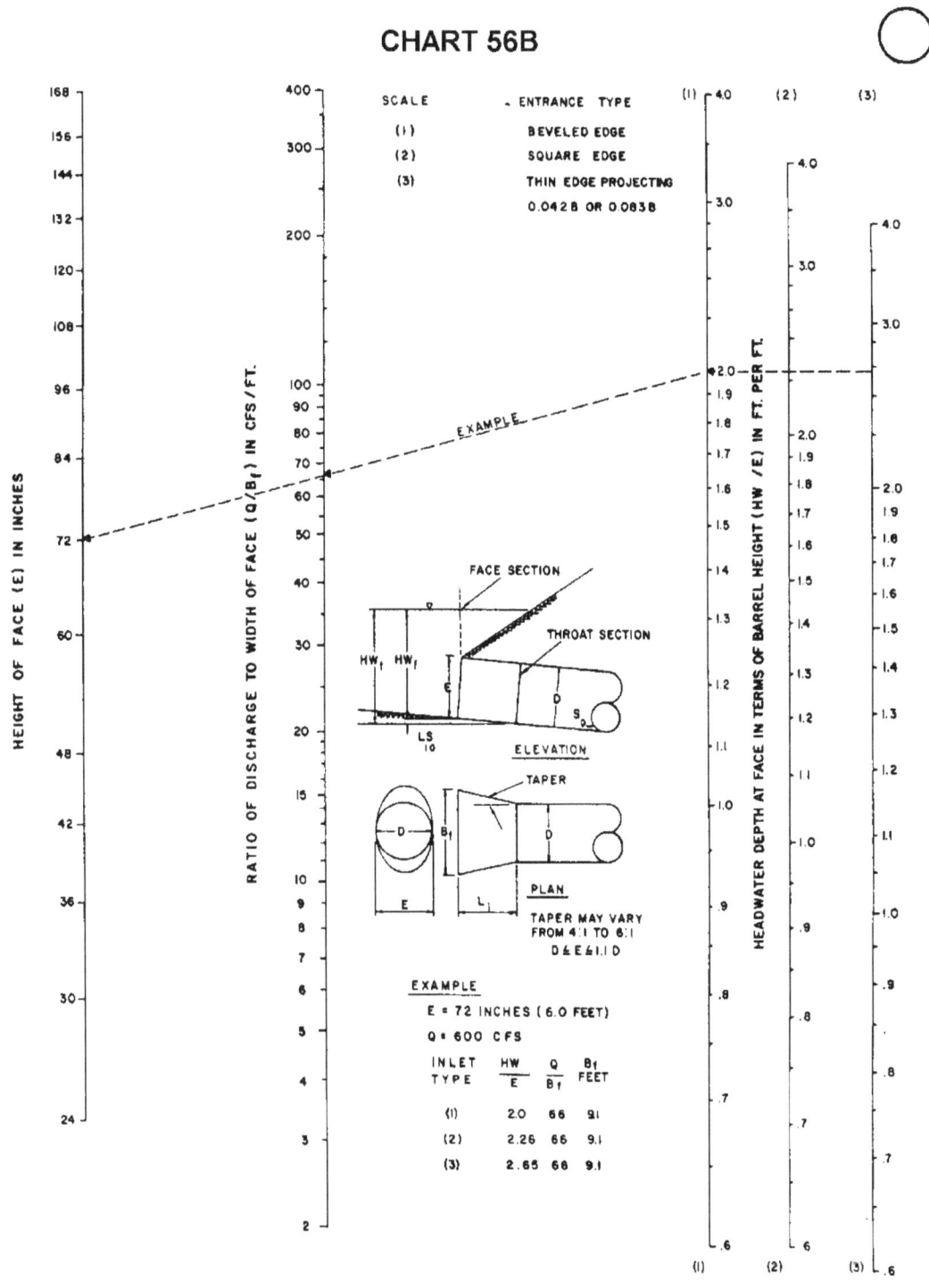

HEIGHT OF FACE (E) IN INCHES

RATIO OF DISCHARGE TO WIDTH OF FACE (Q/B_f) IN CFS/FT.

HEADWATER DEPTH AT FACE IN TERMS OF BARREL HEIGHT (HW /E) IN FT. PER FT.

SCALE — ENTRANCE TYPE

(1)   BEVELED EDGE

(2)   SQUARE EDGE

(3)   THIN EDGE PROJECTING

       0.042 B OR 0.083 B

EXAMPLE

FACE SECTION

THROAT SECTION

ELEVATION

TAPER

PLAN

TAPER MAY VARY
FROM 4:1 TO 6:1
D & E ≤ 1.1 D

EXAMPLE

E = 72 INCHES (6.0 FEET)

Q = 600 CFS

INLET TYPE	HW/E	Q/B_f	B_f FEET
(1)	2.0	66	9.1
(2)	2.26	66	9.1
(3)	2.65	66	9.1

FACE CONTROL FOR SIDE-TAPERED
INLETS TO PIPE CULVERTS
(NON-RECTANGULAR SECTIONS ONLY)

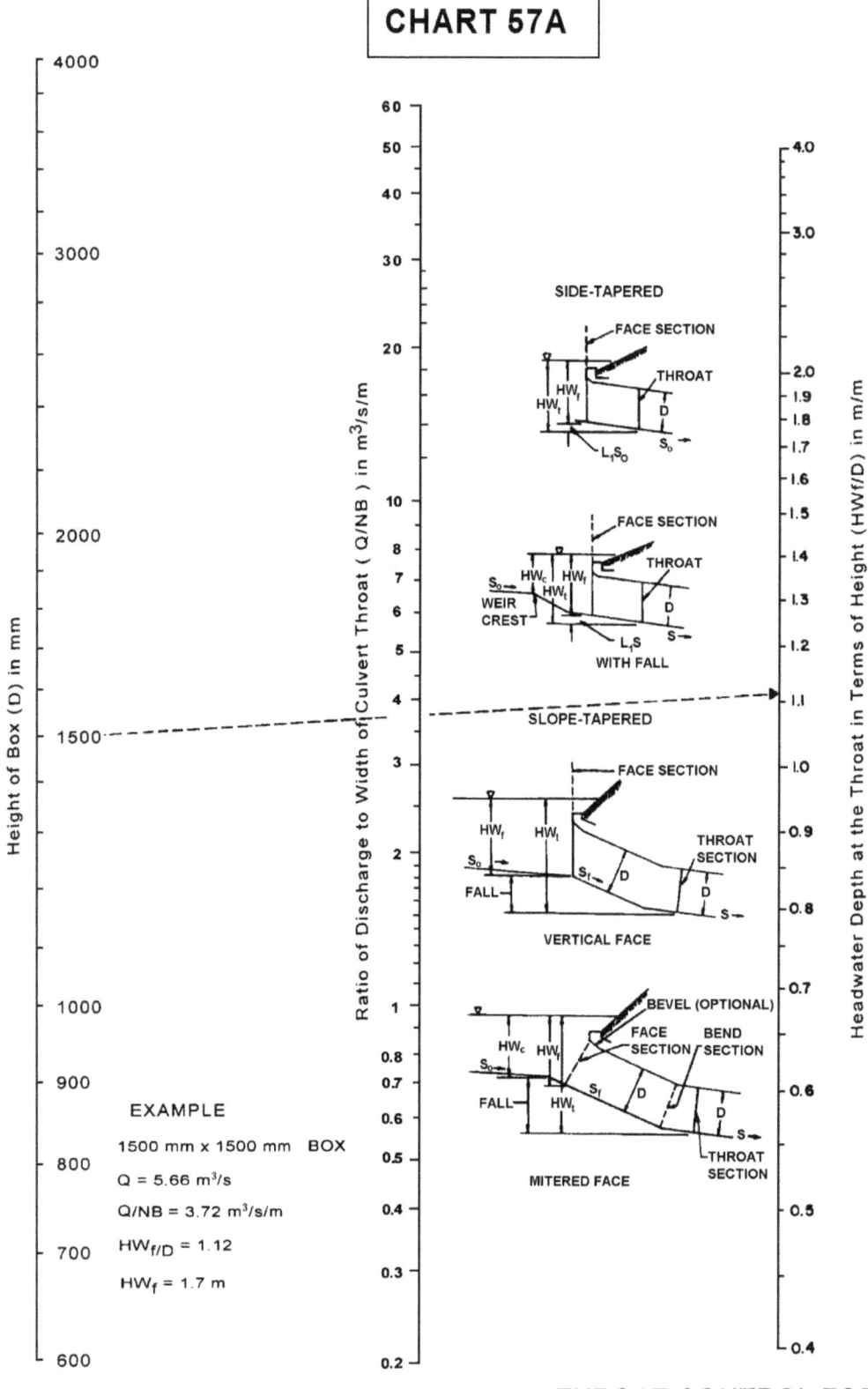

THROAT CONTROL FOR BOX CULVERTS WITH TAPERED INLETS

Height of Box (D) in mm

Ratio of Discharge to Width of Culvert Throat (Q/NB) in m³/s/m

Headwater Depth at the Throat in Terms of Height (HWf/D) in m/m

EXAMPLE

1500 mm x 1500 mm   BOX

Q = 5.66 m³/s

Q/NB = 3.72 m³/s/m

HW$_{f/D}$ = 1.12

HW$_f$ = 1.7 m

CHART 57B

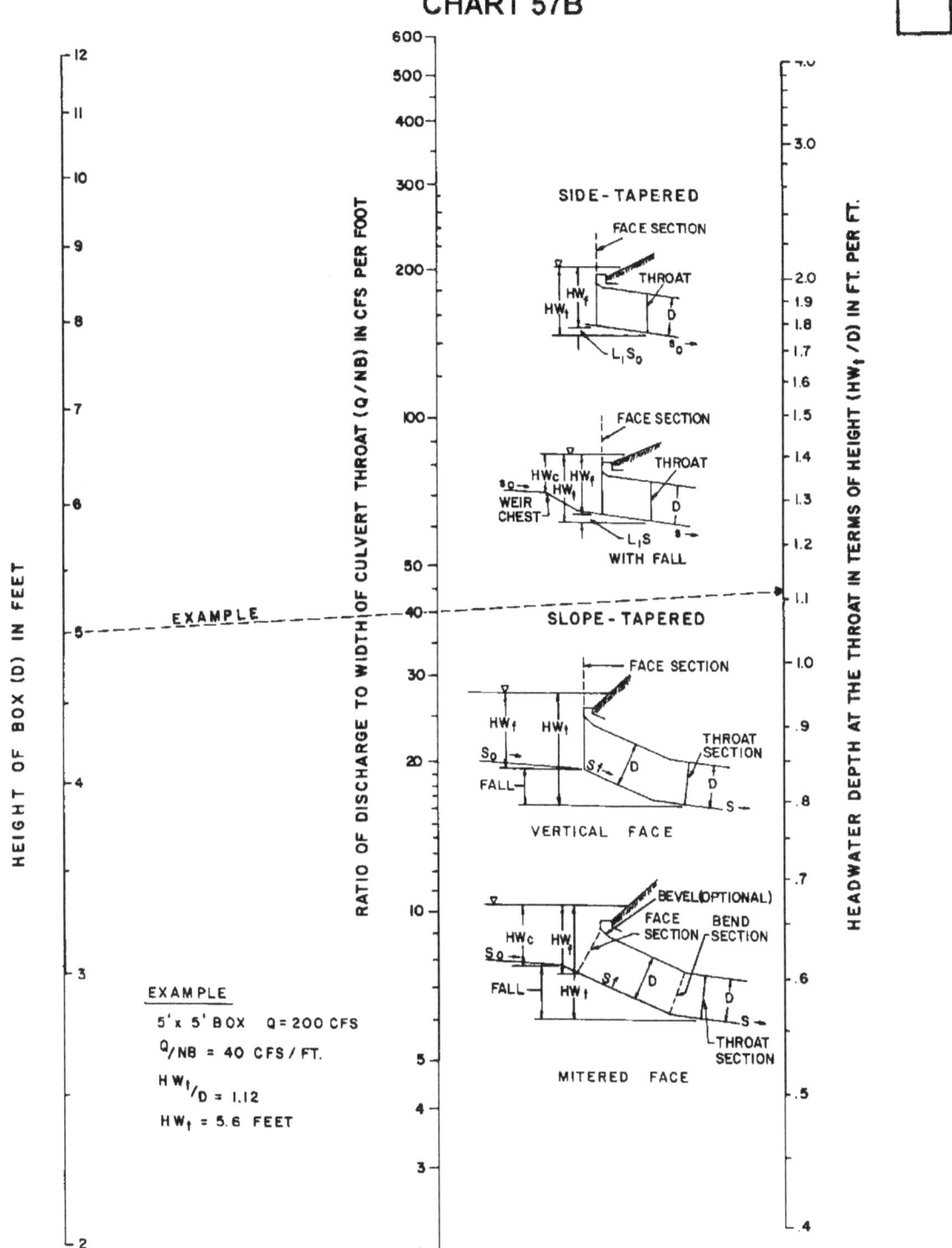

HEIGHT OF BOX (D) IN FEET

RATIO OF DISCHARGE TO WIDTH OF CULVERT THROAT (Q/NB) IN CFS PER FOOT

HEADWATER DEPTH AT THE THROAT IN TERMS OF HEIGHT (HW$_t$/D) IN FT. PER FT.

SIDE-TAPERED

FACE SECTION

THROAT

WITH FALL

SLOPE-TAPERED

FACE SECTION

THROAT SECTION

VERTICAL FACE

BEVEL (OPTIONAL)

FACE SECTION

BEND SECTION

THROAT SECTION

MITERED FACE

EXAMPLE

EXAMPLE

5' x 5' BOX   Q = 200 CFS

Q/NB = 40 CFS/FT.

HW$_t$/D = 1.12

HW$_t$ = 5.6 FEET

THROAT CONTROL FOR BOX
CULVERTS WITH TAPERED
INLETS

CHART 58A

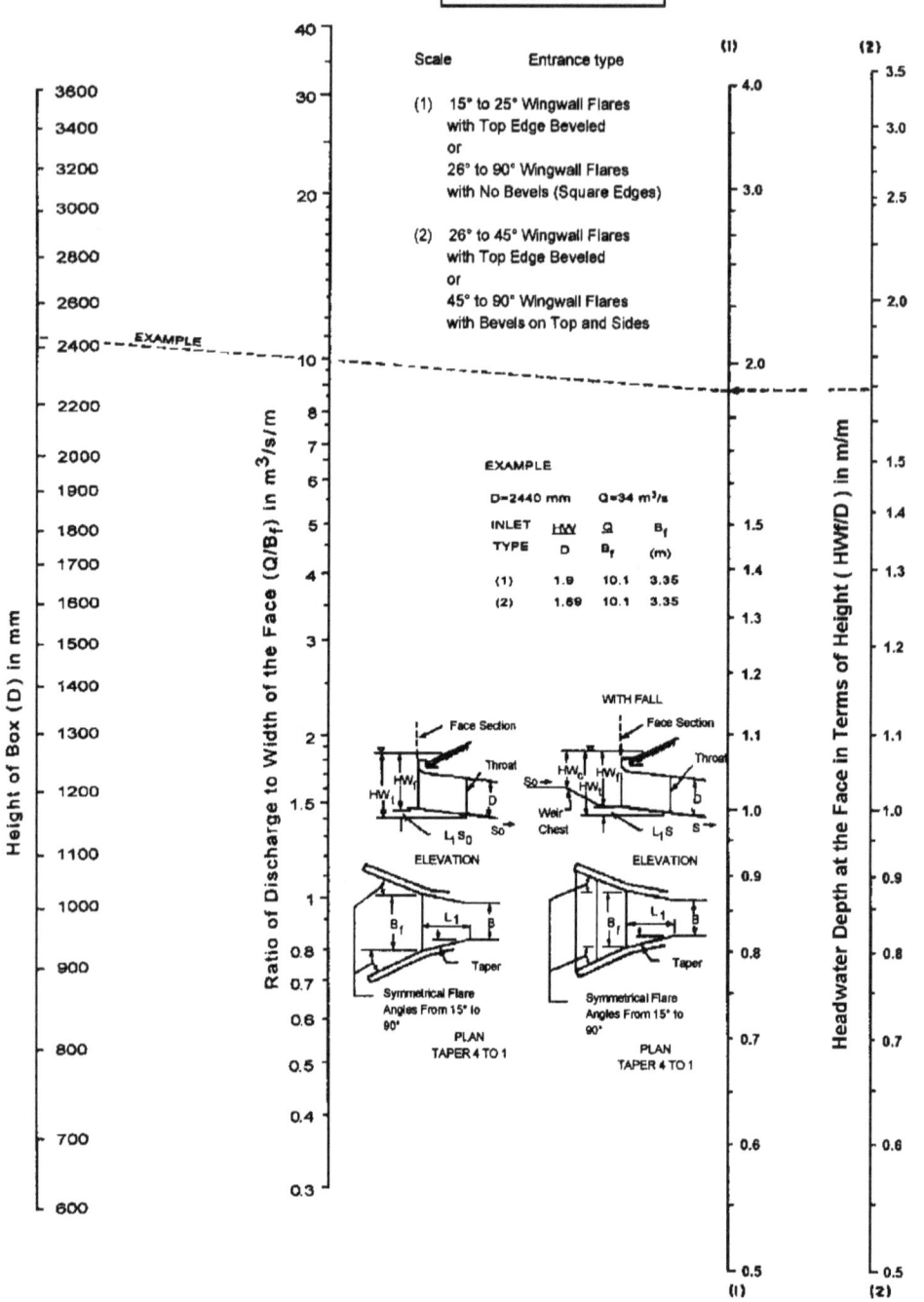

Height of Box (D) in mm (left scale): 3600, 3400, 3200, 3000, 2800, 2600, 2400, 2200, 2000, 1900, 1800, 1700, 1600, 1500, 1400, 1300, 1200, 1100, 1000, 900, 800, 700, 600

EXAMPLE

Ratio of Discharge to Width of the Face (Q/B_f) in $m^3/s/m$: 40, 30, 20, 10, 8, 7, 6, 5, 4, 3, 2, 1.5, 1, 0.8, 0.7, 0.6, 0.5, 0.4, 0.3

Scale   Entrance type

(1)   15° to 25° Wingwall Flares
       with Top Edge Beveled
       or
       26° to 90° Wingwall Flares
       with No Bevels (Square Edges)

(2)   26° to 45° Wingwall Flares
       with Top Edge Beveled
       or
       45° to 90° Wingwall Flares
       with Bevels on Top and Sides

EXAMPLE

D=2440 mm   Q=34 m^3/s

INLET TYPE	$\frac{HW}{D}$	$\frac{Q}{B_f}$	B_f (m)
(1)	1.9	10.1	3.35
(2)	1.69	10.1	3.35

(1) scale: 4.0, 3.0, 2.0, 1.5, 1.4, 1.3, 1.2, 1.1, 1.0, 0.9, 0.8, 0.7, 0.6, 0.5

(2) scale: 3.5, 3.0, 2.5, 2.0, 1.5, 1.4, 1.3, 1.2, 1.1, 1.0, 0.9, 0.8, 0.7, 0.6, 0.5

Headwater Depth at the Face in Terms of Height (HWf/D) in m/m

WITH FALL

ELEVATION

Symmetrical Flare
Angles From 15° to
90°

PLAN
TAPER 4 TO 1

FACE CONTROL FOR BOX CULVERTS
WITH SIDE-TAPERED INLETS

CHART 58B

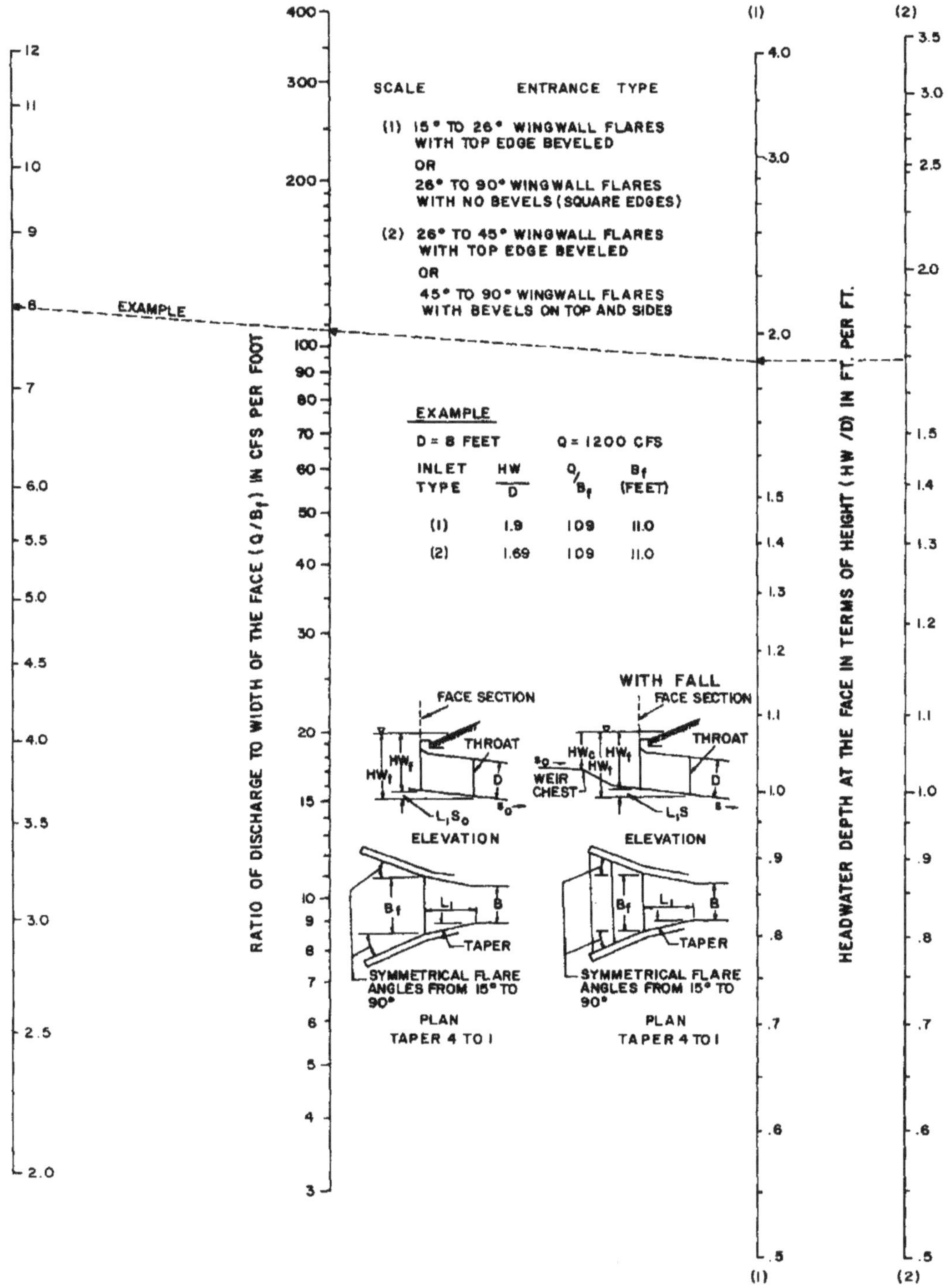

SCALE   ENTRANCE TYPE

(1) 15° TO 26° WINGWALL FLARES
    WITH TOP EDGE BEVELED
    OR
    26° TO 90° WINGWALL FLARES
    WITH NO BEVELS (SQUARE EDGES)

(2) 26° TO 45° WINGWALL FLARES
    WITH TOP EDGE BEVELED
    OR
    45° TO 90° WINGWALL FLARES
    WITH BEVELS ON TOP AND SIDES

EXAMPLE

<u>EXAMPLE</u>

D = 8 FEET   Q = 1200 CFS

INLET TYPE	$\frac{HW}{D}$	Q/B_f	B_f (FEET)
(1)	1.9	109	11.0
(2)	1.69	109	11.0

FACE SECTION   WITH FALL   FACE SECTION

THROAT   THROAT

HW_f HW_c HW_f
                     S_0 HW_f
                     WEIR
                     CHEST

$L_1 S_0$ S_0 $L_1 S$ S

ELEVATION   ELEVATION

B_f L_1 B B_f L_1 B

TAPER   TAPER

SYMMETRICAL FLARE   SYMMETRICAL FLARE
ANGLES FROM 15° TO   ANGLES FROM 15° TO
90°   90°

PLAN   PLAN
TAPER 4 TO 1   TAPER 4 TO 1

RATIO OF DISCHARGE TO WIDTH OF THE FACE (Q/B_f) IN CFS PER FOOT

HEADWATER DEPTH AT THE FACE IN TERMS OF HEIGHT (HW/D) IN FT. PER FT.

**FACE CONTROL FOR BOX CULVERTS
WITH SIDE TAPERED INLETS**

CHART 59 A

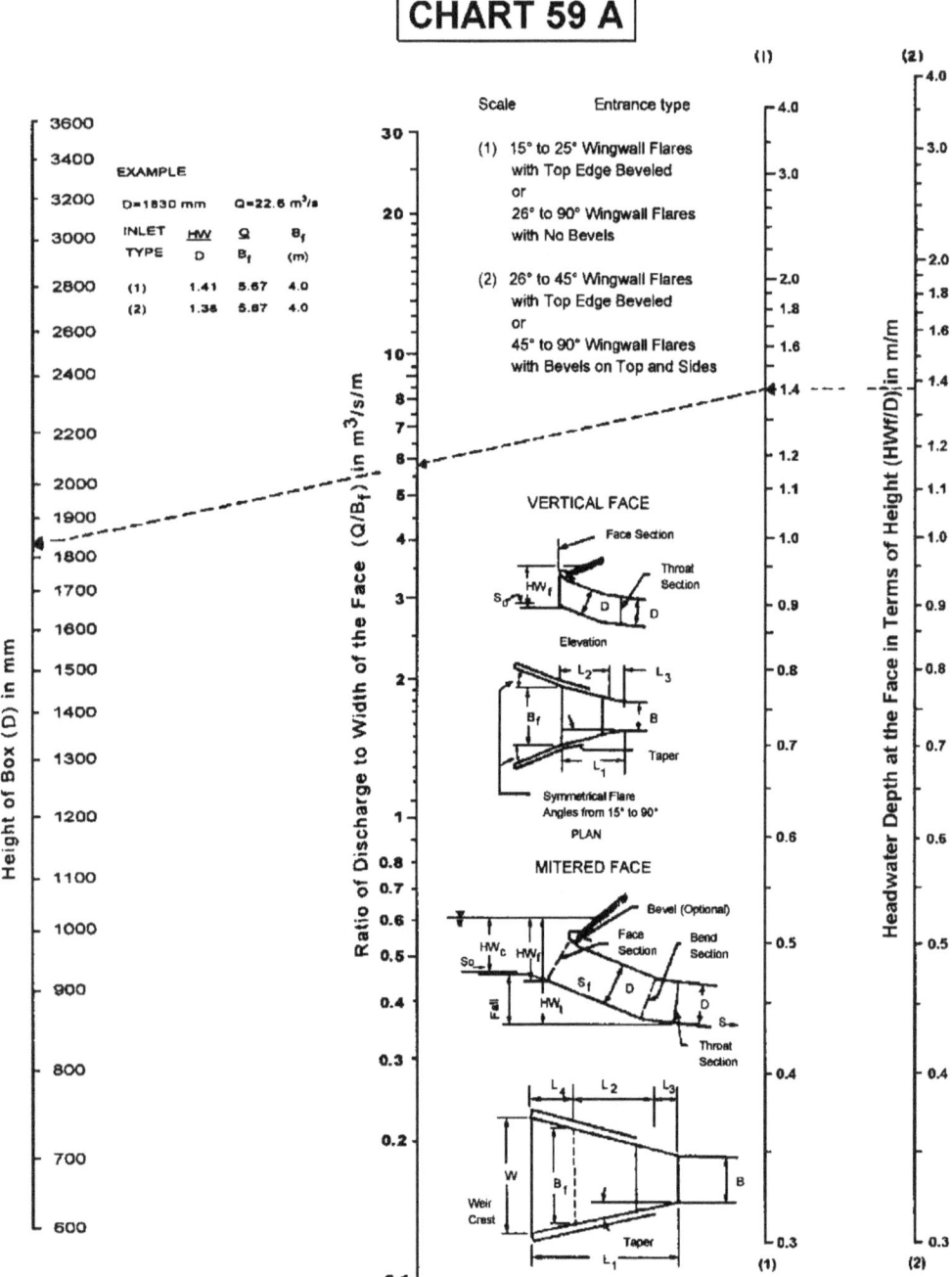

Height of Box (D) in mm

3600
3400
3200
3000
2800
2600
2400
2200
2000
1900
1800
1700
1600
1500
1400
1300
1200
1100
1000
900
800
700
600

EXAMPLE

D=1830 mm   Q=22.6 m³/s

INLET TYPE	$\frac{HW}{D}$	$\frac{Q}{B_f}$	B_f (m)
(1)	1.41	5.67	4.0
(2)	1.36	5.67	4.0

Ratio of Discharge to Width of the Face (Q/B_f) in m³/s/m

30
20
10
8
7
6
5
4
3
2
1
0.8
0.7
0.6
0.5
0.4
0.3
0.2
0.1

Scale   Entrance type

(1) 15° to 25° Wingwall Flares
with Top Edge Beveled
or
26° to 90° Wingwall Flares
with No Bevels

(2) 26° to 45° Wingwall Flares
with Top Edge Beveled
or
45° to 90° Wingwall Flares
with Bevels on Top and Sides

VERTICAL FACE

Face Section

Throat Section

HW_f S_o D   D

Elevation

L_2 L_3

B_f B

L_1 Taper

Symmetrical Flare Angles from 15° to 90°

PLAN

MITERED FACE

Bevel (Optional)

HW_c HW_f Face Section   Bend Section

S_o S_f D   D

Fall HW_i S

Throat Section

L_4 L_2 L_3

W B_f f   B

Weir Crest   Taper

L_1

Headwater Depth at the Face in Terms of Height (HWf/D) in m/m

(1)   (2)

4.0   4.0
3.0   3.0
2.0   2.0
1.8   1.8
1.6   1.6
1.4   1.4
1.2   1.2
1.1   1.1
1.0   1.0
0.9   0.9
0.8   0.8
0.7   0.7
0.6   0.6
0.5   0.5
0.4   0.4
0.3   0.3

(1)   (2)

FACE CONTROL FOR BOX
CULVERTS WITH SLOPE
TAPERED INLETS

CHART 59B

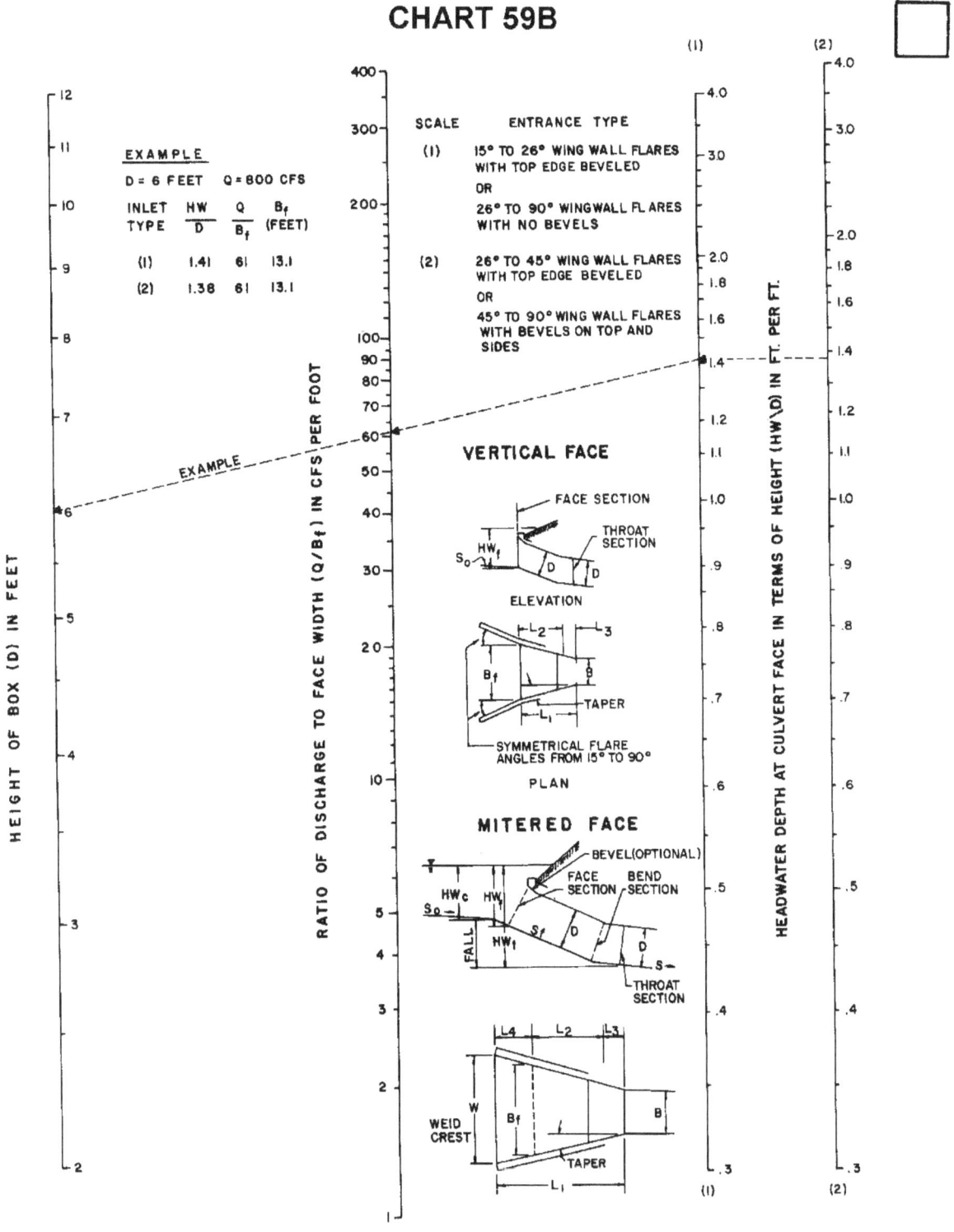

HEIGHT OF BOX (D) IN FEET

EXAMPLE

D = 6 FEET   Q = 800 CFS

INLET TYPE	$\frac{HW}{D}$	$\frac{Q}{B_f}$	B_f (FEET)
(1)	1.41	61	13.1
(2)	1.38	61	13.1

EXAMPLE

RATIO OF DISCHARGE TO FACE WIDTH (Q/B_f) IN CFS PER FOOT

SCALE   ENTRANCE TYPE

(1)   15° TO 26° WING WALL FLARES WITH TOP EDGE BEVELED
OR
26° TO 90° WINGWALL FLARES WITH NO BEVELS

(2)   26° TO 45° WING WALL FLARES WITH TOP EDGE BEVELED
OR
45° TO 90° WING WALL FLARES WITH BEVELS ON TOP AND SIDES

HEADWATER DEPTH AT CULVERT FACE IN TERMS OF HEIGHT (HW\D) IN FT. PER FT.

VERTICAL FACE

FACE SECTION

THROAT SECTION

S_0 HW_f D   D

ELEVATION

L_2 L_3

B_f B

TAPER

L_1

SYMMETRICAL FLARE ANGLES FROM 15° TO 90°

PLAN

MITERED FACE

BEVEL (OPTIONAL)

FACE SECTION   BEND SECTION

HW_c HW_f

S_0

S_f D

FALL HW_f D

S—

THROAT SECTION

L_4 L_2 L_3

W B_f B

WEID CREST

TAPER

L_1

FACE CONTROL FOR BOX
CULVERTS WITH SLOPE
TAPERED INLETS

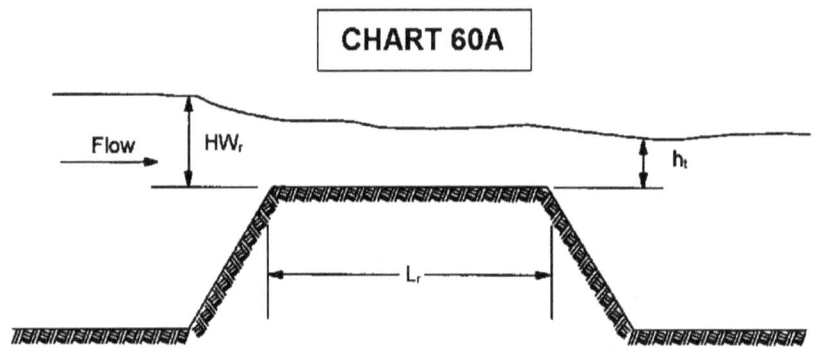

CHART 60A

$C_d = k_t C_r$

C_r = Coefficient of Free Discharge

k_t = Adjustment Factor for Submerged Weir Flow

(TW is Higher Than Roadway Elevation)

$Q_r = C_d \, L \, HW_r^{1.5}$

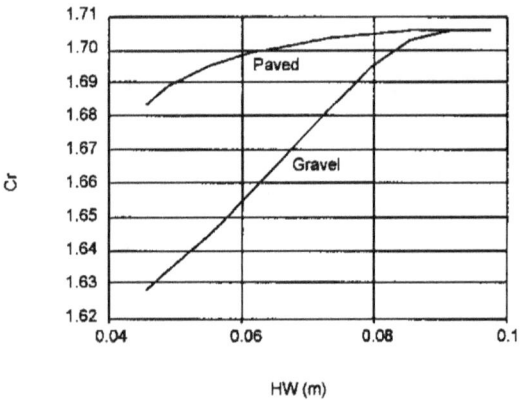

A) Discharge Coefficient for $HW_r/L_r > 0.15$

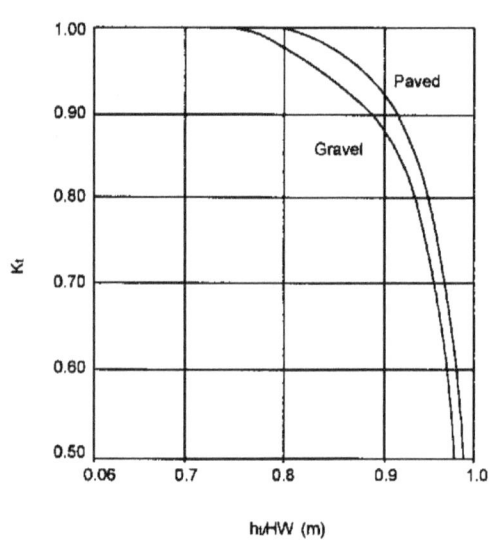

C) Submergence Factor

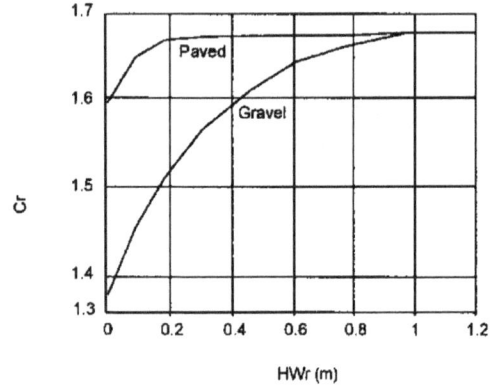

B) Discharge Coefficient for $HW_r/L_r \le 0.15$

Discharge Coefficients for Roadway Overtopping

CHART 60B

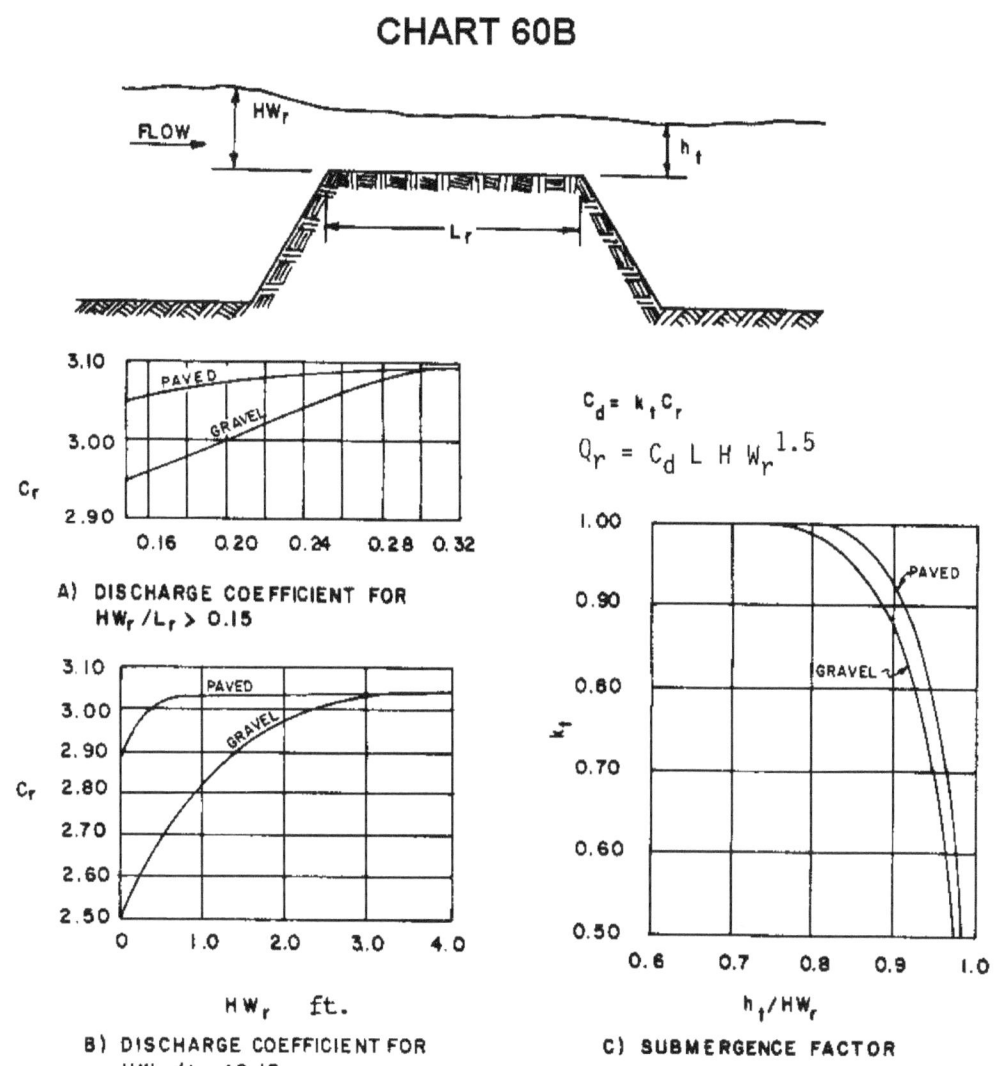

$c_d = k_t c_r$

$Q_r = C_d L H W_r^{1.5}$

A) DISCHARGE COEFFICIENT FOR $HW_r / L_r > 0.15$

B) DISCHARGE COEFFICIENT FOR $HW_r / L_r \leq 0.15$

C) SUBMERGENCE FACTOR

English Discharge Coefficients
for Roadway Overtopping

CULVERT DESIGN FORM

PROJECT: _____

STATION: _____

SHEET _____ OF _____

DESIGNER / DATE: _____ / _____

REVIEWER / DATE: _____ / _____

HYDROLOGICAL DATA

ROADWAY ELEVATION: _____ ( )

☐ METHOD: _____
☐ DRAINAGE AREA: _____
☐ CHANNEL SHAPE: _____
☐ ROUTING: _____

☐ STREAM SLOPE: _____
☐ OTHER: _____

EL_{hd}: _____ ( )

EL_{sf}: _____ ( ) S_o: _____

ORIGINAL STREAM BED

EL_o: _____ ( )

$S = S_o - T/L_o$

$S =$ _____

$L_o =$ _____

EL_i: _____ ( )

DESIGN FLOWS/TAILWATER

R.I. (YEARS)	FLOW (cfs)	TW (ft)
_____	_____	_____
_____	_____	_____

See Add'l Shts.

CULVERT DESCRIPTION:

HEADWATER CALCULATIONS

			INLET CONTROL			OUTLET CONTROL								
MATERIAL – SHAPE – SIZE – ENTRANCE	Total Flow Q (cfs)	Flow Per Barrel Q/N (1)	HW/D (2)	HW$_i$	T (3)	EL_{hi} (4)	TW (5)	d_c	$\frac{d_c+D}{2}$	h_o (6)	k_e	H (7)	EL_{ho} (8)	Control Headwater Elevation	Outlet Velocity	Comments

TECHNICAL FOOTNOTES:

(1) USE Q/NB FOR BOX CULVERTS

(2) HW$_i$/ D = HW / D OR HW$_i$/ D FROM DESIGN CHARTS

(3) T = HW – (EL_{hd} – EL_{sf})
T IS ZERO FOR CULVERTS ON GRADE

(4) EL_{hi} = HW$_i$ + EL_i (INVERT OF INLET CONTROL SECTION)

(5) TW BASED ON DOWN STREAM CONTROL OR FLOW DEPTH IN CHANNEL

(6) h_o = TW or (d_c + D) /2 (WHICHEVER IS GREATER)

(7) $H = [1 + k_e + (K_u \, n^2 \, L) / R^{1.33}] \, v^2 / 2g$ WHERE K_u = 19.63 (29 IN ENGLISH UNITS)

(8) EL_{ho} = EL_o + H + h_o

COMMENTS / DISCUSSION:

_____

SUBSCRIPT DEFINITIONS:

a. APPROXIMATE
f. CULVERT FACE
ha. ALLOWABLE HEADWATER
hi. HEADWATER IN INLET CONTROL
ho. HEADWATER IN OUTLET CONTROL
i. INLET CONTROL SECTION
o. OUTLET
sf. STREAMBED AT CULVERT FACE
tw. TAILWATER

CULVERT BARREL SELECTED:

SIZE: _____

SHAPE: _____

MATERIAL: _____ n _____

ENTRANCE: _____

PROJECT: _____

STATION: _____

SHEET _____ OF _____

TAPERED INLET DESIGN FORM

DESIGNER / DATE: _____ / _____

REVIEWER / DATE: _____ / _____

DESIGN DATA:

Q = _____ ( ); EL_{hi} _____ ( )

EL. THROAT INVERT _____ ( )

EL. STREAM BED AT FACE _____ ( )

T _____ TAPER _____ : 1 (4 : 1 TO 6 : 1)

STREAM SLOPE, S_o = _____ ( ) / ( )

SLOPE OF BARREL, S = _____ ( ) / ( )

S_D _____ : 1 (2 : 1 TO 3 : 1)

BARREL SHAPE AND MATERIAL: _____

N = _____ , B = _____ , D = _____

INLET EDGE DESCRIPTION _____

Symmetrical Wingwall Flare Angles From 15° to 90°

Taper (4H:1V to 6H:1V)

PLAN

ELEVATION
Side-taper

Symmetrical Wingwall Flare Angles From 15° to 90°

Taper (4H:1V to 6H:1V)

PLAN

Face of Section

Bend Section
Throat Section

ELEVATION
Slope-taper

COMMENTS

Q ( )	EL_{hi} ( )	EL. Throat Invert	EL. Face Invert (1)	HW$_i$ (2)	$\frac{HW_i}{E}$ (3)	$\frac{Q}{B_f}$ (4)	MIN. B_f (5)	Selected B_f	MIN. L_3 (6)	L_2 (7)	Check L_2 (8)	Adj. L_3 (9)	Adj. Taper (10)	L_1 (11)	EL. Crest Inv.	HW$_c$ (12)	MIN. W (13)

SLOPE-TAPERED ONLY spans columns (6)–(10). SIDE-TAPERED w/ depression spans columns EL. Crest Inv., (12), (13).

(1) SIDE-TAPERED : EL. FACE INVERT = EL. THROAT INVERT + 1 FT (0.3 M APPROX.)
    SLOPE-TAPERED : EL. FACE INVERT = EL. STREAM BED AT FACE

(2) HW$_i$ = EL_{hi} - EL. FACE INVERT

(3) 1.1 D ≥ E ≥ D; E = D FOR BOX CULVERTS

(4) FROM DESIGN CHARTS

(5) MIN. B_f = Q / (Q / B_f)

(6) MIN. L_3 = 0.5 NB

(7) L_2 = (EL. FACE INVERT - EL. THROAT INVERT) S_D

(8) CHECK $L_2 = \left[\frac{B_f - NB}{2}\right]$. TAPER - L_3

(9) If (8)>(7), ADJ. $L_3 = \left[\frac{B_f - NB}{2}\right]$. TAPER - L_2

(10) If (7) >(8), ADJ. TAPER = $(L_2 + L_3)$ / $\left[\frac{B_f - NB}{2}\right]$

(11) SIDE-TAPERED : L = $\left[\frac{B_f - NB}{2}\right]$. TAPER

     SLOPE-TAPERED : $L_1 = L_2 + L_3$

(12) HW$_C$ = EL_{hi} - EL. CREST INVERT

(13) MIN. W = K_u Q / HW$_c$[15] Where K_u = 0.35 (0.64 SI)

SELECTED DESIGN

B_f _____

L_1 _____

L_2 _____

L_3 _____

BEVELS ANGLE

b = _____ ( ) ; d = _____ ( )

TAPER _____ : 1

S_D = _____ : 1

DESIGN GUIDELINES

DESIGN GUIDELINE 1
CULVERT DESIGN USING NOMOGRAPHS

DG 1.1 BACKGROUND

Culvert design can be accomplished using design aids in this manual to manually determine the appropriate culvert size, shape (box or circle) and material that will accommodate a design flood at a given highway crossing. Section DG 1.2 provides the design procedure steps that should be followed. Section DG 1.3 applies the design steps to a circular shape. Section DG 1.4 applies the design procedure steps to a rectangular shape.

DG 1.2 DESIGN PROCEDURE

The following are the general steps that are followed to design a straight culvert (see Figure 3.18):

Step 1. Summarize hydrology data (Section 2.1) and site data (Section 2.2) for the culvert at the top of the Culvert Design Form. This information will have been collected or calculated prior to performing the actual culvert design. In addition, the site assessments (Section 2.3) have been completed.

Step 2. Select a preliminary culvert shape (Section 1.3.1), material (Section 1.3.2), size from standard plans, and entrance configuration from standard plans.

Step 3. Perform inlet control headwater (HW_i) calculations for the design flow rate (Section 3.3.2).

Step 4. Perform outlet control headwater (HW_o) calculations for the design flow rate (Section 3.3.3).

Step 5. The controlling headwater is the higher of HW_i and HW_o.

Step 6. Evaluate Results (Section 3.3.5) to determine if controlling HW is near allowable HW and less. If not close enough or higher, return to Step 2 and try another alternative.

Step 7. Calculate outlet velocity (V_o) for the controlling HW (Section 3.1.6) and compare with downstream channel velocity. If velocity is not acceptable, consider an energy dissipator (HEC-14) or a rougher material (return to Step 2).

Step 8. Check that culvert dimensions fit embankment and stream. If dimensions are satisfactory, repeat steps 3 through 5 for performance curve discharges (Section 3.2) and document the design.

DG 1.3 CIRCULAR CULVERT

Design a straight, circular culvert with no depression for a new rural roadway crossing for the 25-year flood. Use the standard practice of providing 2 ft (0.61 m) of freeboard below the subgrade shoulder.

The complete customary unit (CU) solution is detailed below. A summary of SI results is also provided based on the complete solution of the problem using the appropriate SI nomographs as detailed in HDS 5, second edition, Chapter III, Example Problem 1. Note that direct conversion of the CU solution to SI may yield slightly different results.

The following information is provided:

- DG 1.3.1 Design procedure steps based on Section DG 1.2 with a summary of results
- DG 1.3.2 Culvert Design Form (CDF)
- DG 1.3.3 Charts showing solutions

DG 1.3.1 Design Procedure

Step 1. Summarize hydrology data (Section 2.1) and site data (Section 2.2) for the culvert at the top of the CDF. The site assessments (Section 2.3) indicate that this is a stable stream reach without debris problems.

Description	Symbol	CU Units	SI Units
Design Discharge	Q_{25}	200 ft³/s	5.663 m³/s
Tailwater for Design Flood	TW	3.5 ft	1.067 m
Natural Stream Bed Slope	S_o	0.01 ft/ft	0.01 m/m
Approximate Culvert Length	L_a	200 ft	60.960 m
Elevation at Shoulder	EL_s	110 ft	33.528 m
Elevation of Allowable Headwater	El_{ha}	108 ft	32.918 m
Elevation at Culvert Face	EL_i	100 ft	30.480 m

Step 2. Select a preliminary culvert shape (Section 1.3.1), material (Section 1.3.2), size from standard plans, and entrance configuration from standard plans, enter on CDF. Try a corrugated metal pipe (CMP) in headwall with 45° bevel

72" CMP with 2-2/3 by 1/2 in corrugations (1800 mm with 68 x 13 mm)

Step 3. Perform inlet control headwater (HW$_i$) calculations for the design flow rate (Section 3.3.2) on CDF.

El_{hi} = 105.8 ft for 72" CMP (32.23 m for 1800 mm)

Step 4. Perform outlet control headwater (HW$_o$) calculations for the design flow rate (Section 3.3.3).

El_{ho} = 105.5 ft for 72" CMP (32.17 m for 1800 mm)

Step 5. The controlling headwater is the higher of HW$_i$ and HW$_o$.

El_{hi} = 105.8 ft for 72" CMP (32.23 m for 1800 mm) is < El_{ha} = 108 ft (32.918 m)

Step 6. Evaluate Results (Section 3.3.5) to determine if controlling HW is near allowable HW and less. If not close enough or higher, return to Step 2 and try another alternative. Since HW is not near El_{ha} = 108 ft, the following alternates with controlling HW were tried.

El_{ho} = 108.9 ft for 60" CMP > El_{ha} (try smoother pipe)
El_{hi} = 106.8 ft for 60" RCP < El_{ha} = 108 ft (try smaller)
El_{hi} = 108.0 ft for 54" RCP = El_{ha} = 108 ft (ok)

(For the SI solution the acceptable alternative is a 1500 mm RCP, groove end)

Step 7. Calculate outlet velocity (V_o) for the controlling HW (Section 3.1.6) and compare with downstream channel velocity. If velocity is not acceptable, consider an energy dissipator (HEC-14) or a rougher material (return to Step 2).

54" RCP V_o = 15.3 ft/s

(For the 1500 mm RCP SI solution the outlet velocity is 4.8 m/s)

Step 8. Check that culvert dimensions fit embankment and stream. If dimensions are satisfactory, repeat steps 3 through 5 for performance curve discharges (Section 3.2) and document the design.

54" (4.5 ft) provides 5.5 ft of cover over the inlet crown
1350 mm provides 1.7 m of cover over the inlet crown

DG 1.3.2 HY-8 Solution

The hand solution shown in Section 1.3.1 can be duplicated using current version of HY-8:

- Enter Crossing Data shown in Step 1 and on the CDF.

- Select the following options: U.S. Customary Units, Outlet Control Profiles and Exit Loss Standard Method.

- Analyze the crossing which brings up Crossing Summary Table that shows no overtopping.

- Select Culvert Summary Table which shows that inlet control governs: HW_i = 7.9 ft and El_{hi} = 107.9 ft ≈ 108 ft of the nomograph solution.

- The outlet velocity is 14.9 ft/s which is less than 15.3 ft/s, because normal depth has not been reached at the end of the culvert.

- Similar results are obtained when the SI solution is verified with HY-8.

DG 1.3.3 Culvert Design Form

PROJECT: Example Problem No. 1

STATION: 1 + 00

SHEET 1 **OF** 1

CULVERT DESIGN FORM

DESIGNER / DATE: WJJ / 7/18

REVIEWER / DATE: JMN / 7/19

HYDROLOGICAL DATA

- ☐ METHOD: Rational
- ☐ DRAINAGE AREA: 125 Ac. ☐ STREAM SLOPE: 1.0%
- ☐ CHANNEL SHAPE: Trapezoidal
- ☐ ROUTING: N/A ☐ OTHER:

DESIGN FLOWS/TAILWATER

R.I. (YEARS)	FLOW (cfs)	TW (ft)
25	200	3.5

ROADWAY ELEVATION: ( )

EL_{ha}: **108.0** (ft)

EL_i : **100.0** (ft)

EL_{sf}: **100.0** (ft) S_o : **0.01**

ORIGINAL STREAM BED

EL_o: **98.0** (ft)

$S = S_o - T/L_a$ $S = $ **0.01** $L_a = $ **200**

See Add'l Shts.

HEADWATER CALCULATIONS

CULVERT DESCRIPTION:
MATERIAL – SHAPE – SIZE – ENTRANCE

	Total Flow Q (cfs)	Flow Per Barrel Q/N (1)	INLET CONTROL				OUTLET CONTROL							Control Headwater Elevation	Outlet Velocity	Comments
			HW/D (2)	HW_i	T (3)	EL_{hi} (4)	TW (5)	d_c	$\frac{d_c+D}{2}$	h_o (6)	k_e	H (7)	EL_{ho} (8)			
C.M.P. - Circ. - 72 in. - bevel 45° in headwall	200	200	0.96	5.76	--	105.8	3.5	3.8	4.9	4.9	0.2	2.6	105.5	105.8	9.0	Try 60" CMP
C.M.P. - Circ. - 60 in. - bevel 45° in headwall	200	200	1.40	7.10	--	107.0	3.5	4.1	4.6	4.6	0.2	6.3	108.9	108.9	11.9	Try 60" Conc.
Conc. Circ. - 60 in. - groove end	200	200	1.36	6.80	--	106.8	3.5	4.1	4.6	4.6	0.2	2.9	105.5	106.8	15.6	Try 54" Conc.
Conc. Circ. - 54 in. - groove end	200	200	1.77	7.97	--	108.0	3.5	4.1	4.3	4.3	0.2	4.7	107.0	108.0	15.3	OK

TECHNICAL FOOTNOTES:

(1) USE Q/NB FOR BOX CULVERTS

(2) HW_i / D = HW / D OR HW_i / D FROM DESIGN CHARTS

(3) T = HW – (EL_{hi} – EL_o)
    T IS ZERO FOR CULVERTS ON GRADE

(4) EL_{hi} = HW_i + EL_i (INVERT OF INLET CONTROL SECTION)

(5) TW BASED ON DOWN STREAM CONTROL OR FLOW DEPTH IN CHANNEL

(6) h_o = TW or $(d_c + D)/2$ (WHICHEVER IS GREATER)

(7) $H = [1 + k_e + (K_u n^2 L) / R^{1.33}] \frac{v^2}{2g}$ WHERE K_u = 19.63 (29 IN ENGLISH UNITS)

(8) EL_{ho} = EL_o + H + h_o

COMMENTS / DISCUSSION:

High outlet velocity - outlet protection or larger conduit may be necessary

SUBSCRIPT DEFINITIONS:

- a. APPROXIMATE
- f. CULVERT FACE
- ha. ALLOWABLE HEADWATER
- hi. HEADWATER IN INLET CONTROL
- ho. HEADWATER IN OUTLET CONTROL
- i. INLET CONTROL SECTION
- o. OUTLET
- sf. STREAMBED AT CULVERT FACE
- tw. TAILWATER

CULVERT BARREL SELECTED:

SIZE: 54 in.

SHAPE: Circular

MATERIAL: Conc. **n** .012

ENTRANCE: Groove End

DG 1.3.4 Charts (Circular)

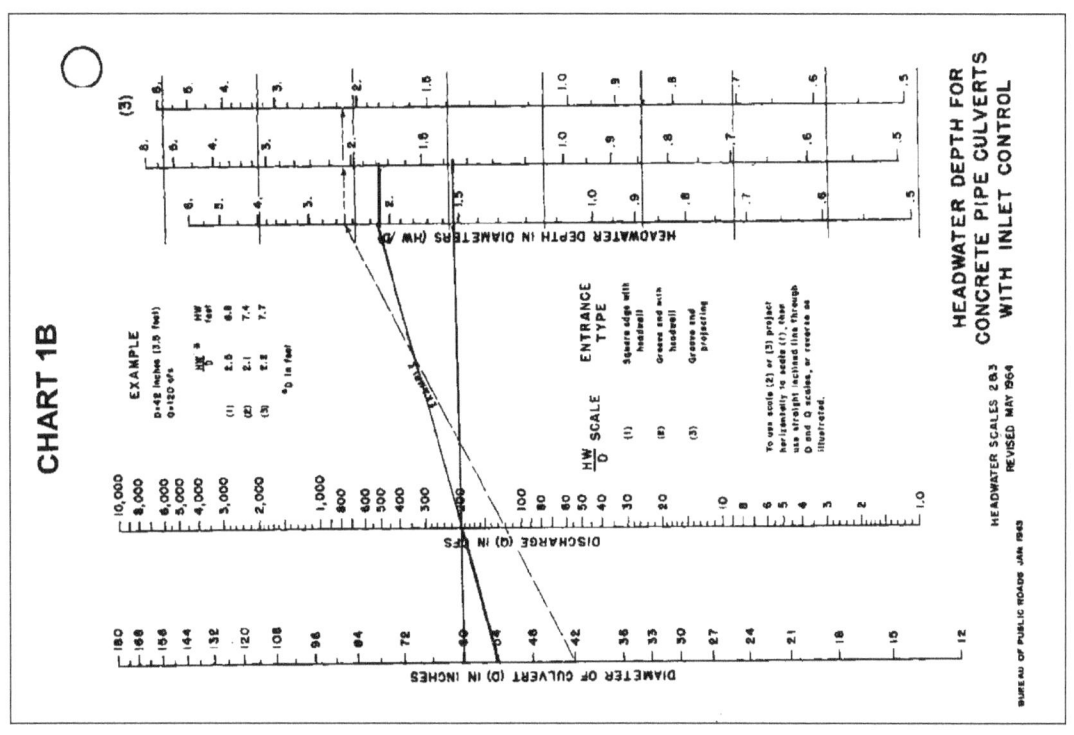

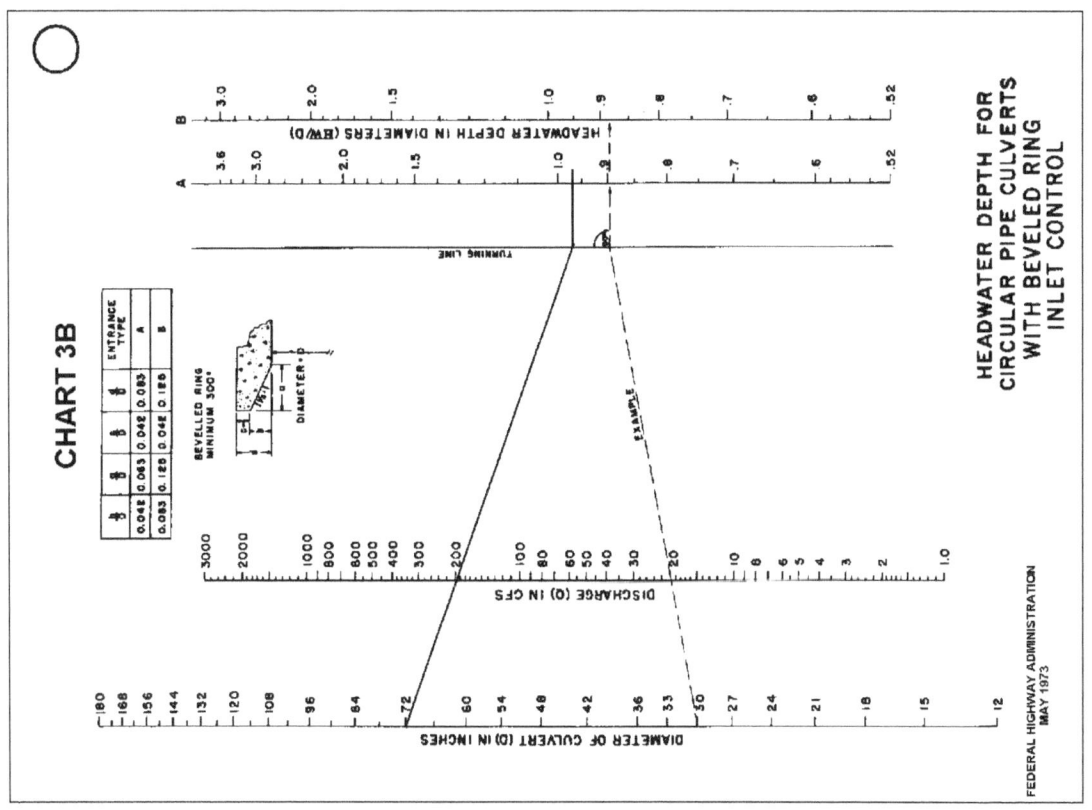

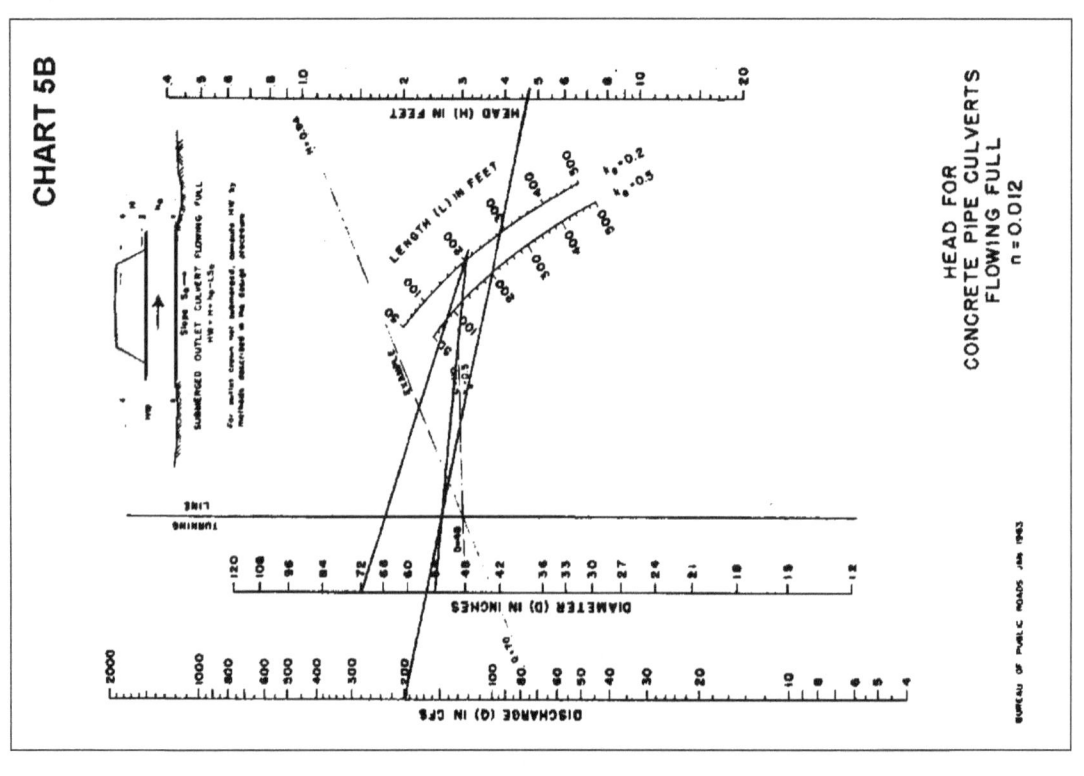

CHART 5B

HEAD FOR
CONCRETE PIPE CULVERTS
FLOWING FULL
n = 0.012

BUREAU OF PUBLIC ROADS JAN. 1963

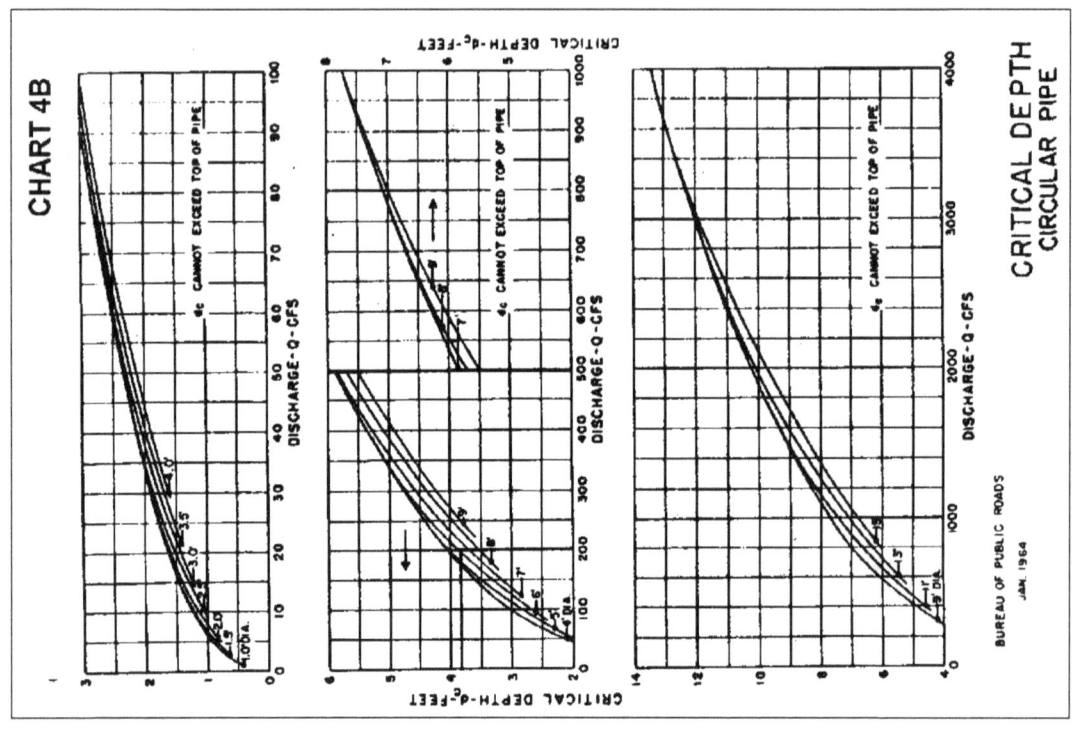

CHART 4B

CRITICAL DEPTH
CIRCULAR PIPE

BUREAU OF PUBLIC ROADS
JAN. 1964

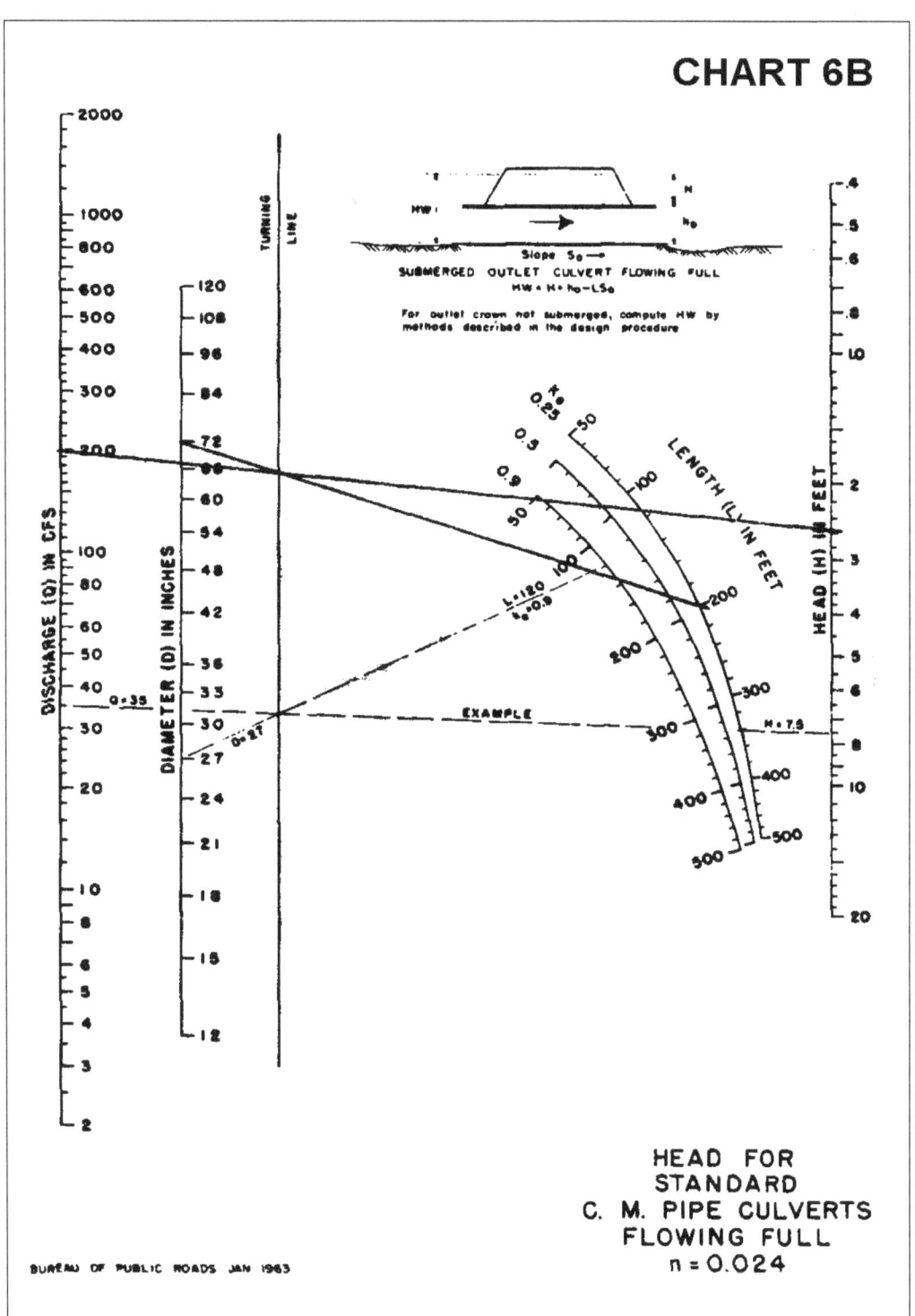

CHART 6B

**HEAD FOR
STANDARD
C. M. PIPE CULVERTS
FLOWING FULL
n = 0.024**

BUREAU OF PUBLIC ROADS JAN 1963

DG 1.4 BOX CULVERT

Design a straight, reinforced concrete box culvert (RCB) culvert with no depression for a new primary roadway crossing for the 50-year flood. Use the standard practice of providing 2 ft (0.61 m) of freeboard below the subgrade shoulder.

The complete customary unit (CU) solution is detailed below. A summary of SI results is also provided based on the complete solution of the problem using the appropriate SI nomographs as detailed in HDS 5, second edition, Chapter III, Example Problem 2. Note that direct conversion of the CU solution to SI may yield slightly different results.

The following information is provided:

- DG 1.4.1 Design procedure steps based on Section DG 1.2 with a summary of results
- DG 1.4.2 Culvert Design Form (CDF)
- Design Charts 8B, 10B, 14B, and 15B are used in this solution, but not provided.

DG 1.4.1 Design Procedure

Step 1. Summarize hydrology data (Section 2.1) and site data (Section 2.2) for the culvert at the top of the CDF. The site assessments (Section 2.3) indicate that this is a stable stream reach without debris problems.

Description	Symbol	CU Units	SI Units
Design Discharge	Q_{50}	300 ft³/s	8.5 m³/s
Tailwater for Design Flood	TW	4.0 ft	1.219 m
Natural Stream Bed Slope	S_o	0.02 ft/ft	0.02 m/m
Approximate Culvert Length	L_a	250 ft	76.2 m
Elevation at Shoulder	EL_s	113.5 ft	34.595 m
Elevation of Allowable Headwater	El_{ha}	110 ft	33.528 m
Elevation at Culvert Invert	EL_i	100 ft	30.480 m

Step 2. Select a preliminary culvert shape (Section 1.3.1), material (Section 1.3.2), size from standard plans, and entrance configuration from standard plans, enter on CDF. Design a RCB culvert for this installation with both square edges and 45° bevel in a headwall.

6' x 5' RCB (1829 x 1424 mm) with square edges

Step 3. Perform inlet control headwater (HW$_i$) calculations for the design flow rate (Section 3.3.2) on CDF.

El_{hi} = 107.9 ft (32.90 m)

Step 4. Perform outlet control headwater (HW$_o$) calculations for the design flow rate (Section 3.3.3).

El_{ho} = 103.2 ft (31.418 m)

Step 5. The controlling headwater is the higher of HW$_i$ and HW$_o$.

El_{hi} = 107.9 ft (32.90 m) < El_{ha} = 110 ft (33.528 m)

Step 6. Evaluate Results (Section 3.3.5) to determine if controlling HW below allowable HW. If not close enough or higher, return to Step 2 and try another alternative. Since HW is not near El_{ha} = 110 ft, the following alternative with controlling HW were tried: 5x5' with square edges and with 45° bevel.

El_{hi} = 109.6 ft for 5x5' RCB (33.45 m) ≈ El_{ha} (ok, but try adding bevels)
El_{hi} = 108.6 ft for 5x5' RCB (33.10 m) < El_{ha} (bevels provide additional capacity)

Step 7. Calculate outlet velocity (V_o) for the controlling HW (Section 3.1.6) and compare with downstream channel velocity. If velocity is not acceptable, consider an energy dissipator (HEC-14) or a rougher material (return to Step 2).

V_o = 20.8 ft/s for 5x5' RCB (6.47 m/s for 1524x1524 mm)

Step 8. Check that culvert dimensions fit embankment and stream. If dimensions are satisfactory, repeat steps 3 through 5 for performance curve discharges (Section 3.2) and document the design.

5' high RCB provides 8.5' of cover over the inlet crown
1524 mm (1.52 m) high RCB provides 2.6 m of cover over the inlet crown

DG 1.4.2 HY-8 Solution

The hand solution shown in Section 1.4.1 can be duplicated using the current version of HY-8:

- Enter Crossing Data shown in Step 1 and on the CDF.

- Select the following options: U.S. Customary Units, Outlet Control Fullflow and Exit Loss Standard Method.

- Analyze Crossing brings up Crossing Summary Table that shows no overtopping.

- Select Culvert Summary Table which shows that inlet control governs: HW_i = 9.65 ft and El_{hi} = 109.65 ft ≈ 109.6 ft of the nomograph solution for square edges.

- The outlet velocity is 19.61 ft/s which is less than 20.8 ft/s. The HY-8 velocity is lower, because normal depth has not been reached at the end of the culvert.

- Similar results are obtained when the SI solution is verified with HY-8.

DG 1.4.3 Culvert Design Form

PROJECT: Example Problem No. 2

STATION: 1 + 00

SHEET 1 **OF** 1

ROADWAY ELEVATION: _____ ( )

EL_{ha}: 110.0 (ft)

EL_{sf}: 100.0 (ft) S_o : 0.02

ORIGINAL STREAM BED

EL_i : 100.0 (ft)

$S = S_o - T/L_a$ S = 0.02 L_a = 250

EL_o: 95.0 (ft)

HYDROLOGICAL DATA

- ☐ METHOD: SCS
- ☐ DRAINAGE AREA: 200 Ac. ☐ STREAM SLOPE: 2.0%
- ☐ CHANNEL SHAPE: Trapezoidal
- ☐ ROUTING: N/A ☐ OTHER: _____

DESIGN FLOWS/TAILWATER

R.I. (YEARS)	FLOW (cfs)	TW (ft)
50	300	4.0

CULVERT DESCRIPTION:

MATERIAL – SHAPE – SIZE – ENTRANCE

		HEADWATER CALCULATIONS																

	Total Flow Q (cfs)	Flow Per Barrel Q/N (1)	INLET CONTROL				OUTLET CONTROL								Control Headwater Elevation	Outlet Velocity	Comments
			HW/D (2)	HW$_i$ (3)	T (3)	EL$_{hi}$ (4)	TW (5)	d$_c$	$\frac{d_c + D}{2}$	h$_o$ (6)	k$_o$	H (7)	EL$_{ho}$ (8)		
Concrete - Box - 6' x 5' - Sq. edge	300	50	1.57	7.9	--	107.9	4.0	4.2	4.6	4.6	0.5	3.55	103.2	107.9	21.1	OK try Sm. box
Concrete - Box - 5' x 5' - Sq. edge	300	60	1.91	9.6	--	109.6	4.0	4.8	4.9	4.9	0.5	5.2	105.1	109.6	20.8	Check bevels
Concrete - Box - 5' x 5' - 45° bevel	300	60	1.71	8.55	--	108.6	4.0	4.8	4.9	4.9	0.2	5.0	104.9	108.6	20.8	OK

TECHNICAL FOOTNOTES:

(1) USE Q/NB FOR BOX CULVERTS

(2) HW / D = HW / D OR HW$_i$ / D FROM DESIGN CHARTS

(3) T = HW$_i$ – (EL$_{ha}$ – EL$_{sf}$)
T IS ZERO FOR CULVERTS ON GRADE

(4) EL$_{hi}$ = HW$_i$ + EL$_i$ (INVERT OF INLET CONTROL SECTION)

(5) TW BASED ON DOWN STREAM CONTROL OR FLOW DEPTH IN CHANNEL

(6) h$_o$ = TW or (d$_c$ + D) /2 (WHICHEVER IS GREATER)

(7) H = [1 + k$_e$ + (K$_u$ n^2 L) / R$^{1.33}$]v^2 / 2g WHERE K$_u$ = 19.63 (29 IN ENGLISH UNITS)

(8) EL$_{ho}$ = EL$_o$ + H + h$_o$

SUBSCRIPT DEFINITIONS:

- a. APPROXIMATE
- f. CULVERT FACE
- ha. ALLOWABLE HEADWATER
- hi. HEADWATER IN INLET CONTROL
- ho. HEADWATER IN OUTLET CONTROL
- i. INLET CONTROL SECTION
- o. OUTLET
- sf. STREAMBED AT CULVERT FACE
- tw. TAILWATER

COMMENTS / DISCUSSION:

5' x 5' box will work with or without bevels. Bevels provide additional flow capacity.

CULVERT BARREL SELECTED:

SIZE: 5 ft x 5 ft

SHAPE: Rectangular

MATERIAL: Conc. **n** .012

ENTRANCE: Sq. edge - 90° headwall

See Add'l Shts.

DG 1.5 RCB Performance Curve

An existing 7 ft by 7 ft concrete box culvert with a headwall and square edges was designed for a 50-year flood of 600 ft³/s and a design headwater elevation of 114 ft (34.747 m). Upstream development has increased the 50-year runoff to 1,000 ft³/s. Prepare a performance curve for this installation, including any roadway overtopping, up to a flow rate of 1,000 ft³/s.

The roadway is gravel with a width of 40 ft. The roadway profile may be approximated as a broad crested weir 200 ft long. Use Figure 3.11 to calculate overtopping flows,

The complete customary unit (CU) solution is detailed below. The solution of the SI problem is detailed in HDS 5, second edition, Chapter III, Example Problem 4.

The following information is provided:

- DG 1.5.1 Design procedure steps based on Section DG 1.2 with a summary of results
- DG 1.5.2 Culvert Design Form (CDF)
- Design Charts 8B, 14B, 15B and Figure 3,11 are used in this solution, but not provided.

DG 1.5.1 Design Procedure

Step 1. Summarize hydrology data (Section 2.1) and site data (Section 2.2) for the culvert at the top of the CDF. The site assessments (Section 2.3) indicate that this is a stable stream reach without debris problems.

Description	Symbol	CU Units	SI Units
Design Discharge	Q_{50}	1000 ft³/s	28.317 m³/s
Tailwater for Design Flood	TW	4.1 ft	1.250 m
Natural Stream Bed Slope	S_o	0.05 ft/ft	0.05 m/m
Approximate Culvert Length	L_a	200 ft	60.96 m
Elevation at Overtopping	EL_o	116 ft	35.357 m
Elevation of Allowable Headwater	EL_{ha}	114 ft	34.747 m
Elevation at Culvert Invert	EL_i	100 ft	30.480 m

Flow (ft³/s)	TW (ft)	Flow (m³/s)	TW (m)
400	2.6	11.327	0.793
600	3.1	16.99	0.945
800	3.8	22.654	1.158
1000	4.1	28.317	1.25

Step 2. Through Step 7 are summarized on the CDF (Section DG 1.5.2).

Step 8. Check that culvert dimensions fit embankment and stream. If dimensions are satisfactory, repeat steps 3 through 5 for performance curve discharges (Section 3.2) and document the design. These steps were repeated for a range of discharges up to 1000 cfs. The following figure is used for overtopping calculations shown in table and the performance curve plot. A similar solution for the SI problem is shown in HDS 5, second edition, Example Problem 4.

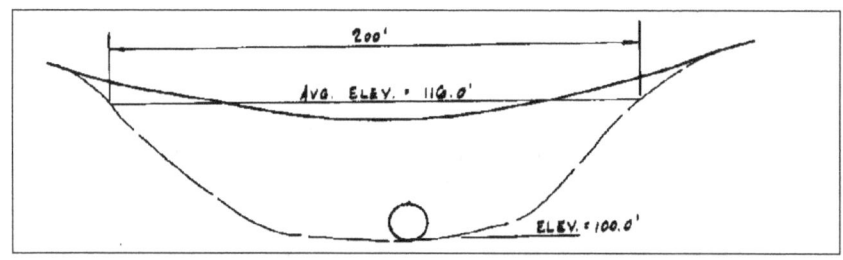

| Culvert Q_c | EL_{hi} | HW_r | k_t | C_r | Overtopping Q_o | Total Q |
(ft^3/s)	(ft)	(ft)			(ft^3/s)	(ft^3/s)
400	108.1	-			-	400
600	111.6	-			-	600
700	113.7	-			-	700
800	116.5	0.5	1	2.7	191	991
850	117.9	1.5	1	2.92	1073	1923
1000	122.5	6.5	1	3.04	10075	11075

$Q_o = C_d L(HW_r)^{1.5}$ and $C_d = k_t C_r$

At Q_c = 800 cfs, HW_r = 0.5 ft, $Q_o = 2.7(200)(0.5)^{1.5} = 191$ ft^3/s

At Q_c = 850 cfs, HW_r = 1.5 ft, $Q_o = 2.92(200)(1.5)^{1.5} = 1073$ ft^3/s

At Q_c = 1000 cfs, HW_r = 6.5 ft, $Q_o = 3.04(200)(6.5)^{1.5} = 10075$ ft^3/s

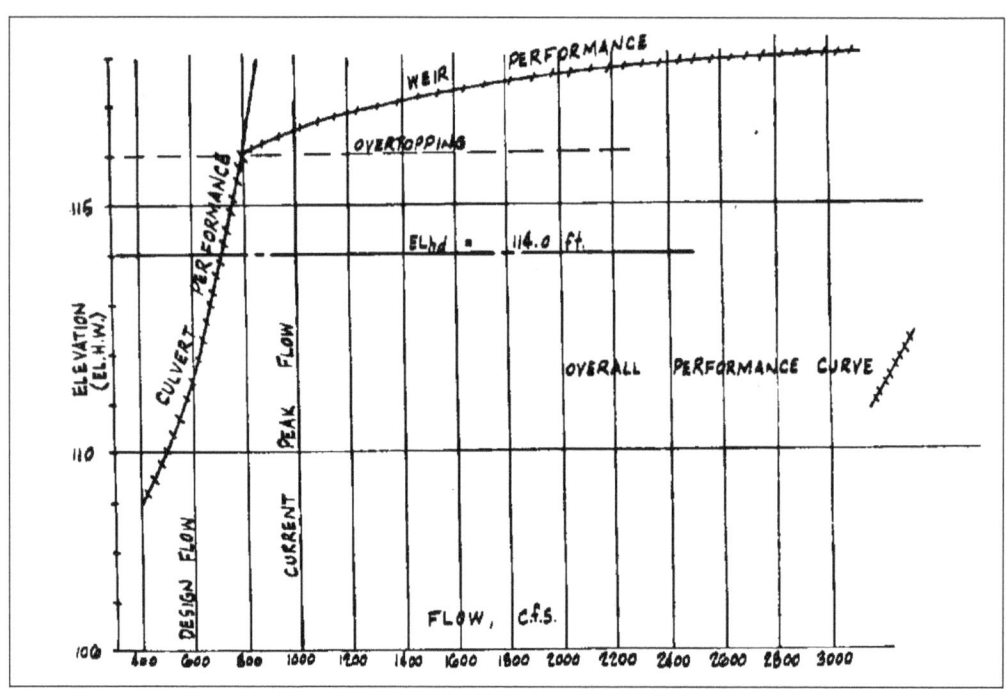

DG 1.5.2 HY-8 Solution

The hand solution shown in Section 1.5.1 can be duplicated using the current version of HY-8:

- Enter Crossing Data shown in Step 1 and on the CDF. The TW has to be input as a rating curve with interpolation used for missing discharges.

- Select the following options: U.S. Customary Units, Outlet Control Profiles and Exit Loss Standard Method.

- Analyze Crossing brings up Crossing Summary Table that shows overtopping begins at 798.5 ft^3/s. At a combined flow of 1000 ft^3/s, the culvert is discharging 816.23 ft^3/s and 183.52 ft^3/s goes over the roadway.

- Select Culvert Summary Table which shows that inlet control governs: HW$_i$ = 16.49 ft and El$_{hi}$ = 116.49 ft ≈ 116.5 ft of the performance curve for 1000 ft^3/s.

- Similar results are obtained when the SI solution is verified with HY-8.

DG 1.5.3 Culvert Design Form

CULVERT DESIGN FORM

PROJECT: Example Problem No. 4	STATION: 4 + 50
	SHEET 1 OF 3
	DESIGNER / DATE: WJJ / 7/18
	REVIEWER / DATE: JMN / 7/19

HYDROLOGICAL DATA

☐ METHOD: SCS
☐ DRAINAGE AREA: 400 Ac.   ☐ STREAM SLOPE: 5.0%
☐ CHANNEL SHAPE: Trapezoidal
☐ ROUTING: N/A   ☐ OTHER:

ROADWAY ELEVATION: _____ (   )

EL_{ha}: 114.0 (ft)

HW_i

EL_i: 100.0 (ft)

EL_{sf}: 100.0 (ft) S_o: 0.05

ORIGINAL STREAM BED

H   h_o

EL_o: 90.0 (ft)

$S = S_o - T/L_a$ $S = 0.05$ $L_a = 200$

DESIGN FLOWS/TAILWATER

R.I. (YEARS)	FLOW (cfs)	TW (ft)
50 (Old)	600	3.1
50 (New)	1,000	4.1

HEADWATER CALCULATIONS

CULVERT DESCRIPTION: MATERIAL – SHAPE – SIZE – ENTRANCE	Total Flow Q (cfs)	Flow Per Barrel Q/N (1)	INLET CONTROL				OUTLET CONTROL							Control Headwater Elevation	Outlet Velocity	Comments
			HW/D (2)	HW_i	T (3)	EL_{hi} (4)	TW (5)	d_c	$\frac{d_c+D}{2}$	h_o (6)	k_e	H (7)	EL_{ho} (8)			
Concrete - Box - 7' x 7' - Sq. edge	400	57.1	1.15	8.1	--	108.1	2.6	4.6	5.8	5.8	0.5	1.95	97.8	108.1	--	--
Concrete - Box - 7' x 7' - Sq. edge	600	85.7	1.65	11.6	--	111.6	3.1	6.1	6.6	6.6	0.5	4.4	101.0	111.6	--	--
Concrete - Box - 7' x 7' - Sq. edge	700	100.0	1.95	13.7	--	113.7	3.5	6.8	6.9	6.9	0.5	6.0	102.9	113.7	--	--
Concrete - Box - 7' x 7' - Sq. edge	800	114.3	2.35	16.5	--	116.5	3.8	≥7	7.0	7.0	0.5	7.9	104.9	116.5	--	--
Concrete - Box - 7' x 7' - Sq. edge	850	121.4	2.55	17.9	--	117.9	3.9	≥7	7.0	7.0	0.5	9.0	106.0	117.9	--	--
Concrete - Box - 7' x 7' - Sq. edge	1000	142.9	3.21	22.5	--	122.5	4.1	≥7	7.0	7.0	0.5	12.6	109.6	122.5	--	--

TECHNICAL FOOTNOTES:

(1) USE Q/NB FOR BOX CULVERTS

(2) HW_i / D = HW_i / D OR HW_i / D FROM DESIGN CHARTS

(3) $T = HW_i - (EL_{ha} - EL_{sf})$ T IS ZERO FOR CULVERTS ON GRADE

(4) $EL_{hi} = HW_i + EL_i$ (INVERT OF INLET CONTROL SECTION)

(5) TW BASED ON DOWN STREAM CONTROL OR FLOW DEPTH IN CHANNEL

(6) $h_o = TW$ or $(d_c + D)/2$ (WHICHEVER IS GREATER)

(7) $H = [1 + k_e + (K_u n^2 L)/R^{1.33}] V^2/2g$ WHERE $K_u = 19.63$ (29 IN ENGLISH UNITS)

(8) $EL_{ho} = EL_o + H + h_o$

SUBSCRIPT DEFINITIONS:

a. APPROXIMATE
f. CULVERT FACE
ha. ALLOWABLE HEADWATER
hi. HEADWATER IN INLET CONTROL
ho. HEADWATER IN OUTLET CONTROL
i. INLET CONTROL SECTION
o. OUTLET
sf. STREAMBED AT CULVERT FACE
tw. TAILWATER

COMMENTS / DISCUSSION:

New Q_{50} results in roadway overtopping. 2.5' above DL_{hd}

CULVERT BARREL SELECTED:

SIZE: 7' x 7'

SHAPE: Rectangular

MATERIAL: Conc.   n .012

ENTRANCE: Sq. edge - w/headwall

DESIGN GUIDELINE 2
CULVERTS WITHOUT DESIGN CHARTS

DG 2.1 BACKGROUND

A culvert without design charts is a culvert shape that does not have laboratory determined orifice and weir coefficients (Appendix A). Therefore, the shape does not have inlet control or outlet control nomographs in Appendix C and cannot be designed using Design Guideline 1. These shapes can be designed using this guideline and generic design aids (Charts 51 through 54) in Appendix C to manually determine the appropriate culvert size, shape and material that will accommodate a design flood at a given highway crossing. Section DG 2.2 provides the design procedure steps that should be followed. These steps are similar to Design Guideline 1, but use different design aids. Section DG 2.3 applies the design steps to a long span shape. Section DG 2.4 outlines the solution procedure using HY-8.

DG 2.2 DESIGN PROCEDURE

The following general steps are used to design a straight culvert (see Figure 3.18):

Step 1. Summarize hydrology data (Section 2.1) and site data (Section 2.2) for the culvert. This information will have been collected or calculated prior to performing the actual culvert design. In addition, the site assessments (Section 2.3) have been completed.

Step 2. Select a preliminary culvert shape (Section 1.3.1), material (Section 1.3.2), size and inlet configuration from manufacturers' information. Obtain the area, A, and the interior height, D, for the selected barrel.

Step 3. Perform inlet control headwater depth (HW_i) calculations for the design flow rate (Section 3.6.1). Use Charts 51 for circular or elliptical conduits with the long horizontal axis at the mid-point of the barrel. Use Chart 52 for high and low profile structural plate arches.

    a.   Calculate $Q/AD^{0.5}$

    b.   Enter the appropriate design chart with $Q/AD^{0.5}$ and the selected edge condition, read HW/D.

    c.   Multiply HW/D by D to obtain the inlet control headwater depth, HW_i.

    d.   If approach velocity (V_u) is known, deduct the approach velocity head ($V_u^2/2g$) from HW_i to obtain the inlet control headwater depth. If V_u is neglected, use HW_i.

    e.   If $HW_i > HW_{ha}$ or if the culvert is oversized, select another size and/or inlet edge condition and return to step 3a.

Step 4. Perform outlet control headwater (HW_o) calculations for the design flow rate (Section 3.6.2) using either backwater calculations or Equation 3.6b.

      a.   **Partly Full Flow**. Large conduits, such as long span culverts, usually flow partly full throughout their lengths. In addition, the invert of the culvert is often natural. In these situations, it is advisable to perform backwater calculations to determine the headwater elevation.

      b.   **Full Flow**. If the conduit flows full or nearly full throughout its length, Equation 3.6b ($HW_o = TW + H_L - SL$) may be used to calculate the outlet control headwater depth.

Step 5. The controlling headwater depth is the higher of HW_i and HW_o.

Step 6. Evaluate Results (Section 3.6.3) to determine if controlling HW is near allowable HW and less. If not close enough or higher, return to Step 2 and try another alternative.

DG 2.3 LONG SPAN EXAMPLE

Design a long span, structural plate, corrugated metal elliptical culvert with a headwall to pass the Q_{25} under a high roadway fill. The design discharge should be below the crown of the conduit at the inlet, but the Q_{100} check flow may exceed the crown by no more than 5 ft (1.524 m).

The complete customary unit (CU) solution is detailed below. A summary of SI results is also provided based on the complete solution of the problem using the appropriate SI nomographs as detailed in HDS 5, second edition, Chapter III, Example Problem 5. Note that direct conversion of the CU solution to SI may yield slightly different results.

Step 1. Summarize hydrology data (Section 2.1) and site data (Section 2.2) for the culvert. This information will have been collected or calculated prior to performing the actual culvert design. In addition, the site assessments (Section 2.3) have been completed.

Description	Symbol	CU Units	SI Units
Design Discharge	Q_{25}	5,500 ft³/s	155.744 m³/s
Check Discharge	Q_{100}	7,500 ft³/s	212.378 m³/s
Tailwater for Design Flood	TW_{25}	16.0 ft	4.877 m
Tailwater for Check Flood	TW_{100}	19.0 ft	5.791 m
Natural Stream Bed Slope	S_o	0.01 ft/ft	0.01 m/m
Approximate Culvert Length	L_a	200 ft	60.690 m
Elevation at Shoulder	EL_s	260 ft	79.248 m
Elevation of Allowable Headwater	El_{ha}	240 ft	73.152 m
Elevation at Inlet Invert	EL_i	220 ft	67.056 m
Elevation at Outlet Invert	EL_o	218 ft	66.446 m

Step 2. Select a preliminary culvert shape (Section 1.3.1), material (Section 1.3.2), size and inlet configuration from manufacturers' information. Obtain the full barrel area, A, and the interior height, D, for the selected barrel.

Barrel Selected	Symbol	CU Units	SI Units
Width	B	30 ft	9144 mm
Height	D	20 ft	6096 mm
Barrel Area	A	487.5 ft^2	45.289 m^2
Corrugations		6 x 2 in	150 x 50 mm

Step 3. Perform inlet control headwater depth (HW_i) calculations for the design flow rate (Section 3.6.1). Use Charts 51 for circular or elliptical conduits with the long horizontal axis at the mid-point of the barrel. Headwall provides a square edge condition.

a. Calculate $Q/AD^{0.5}$

$$AD^{0.5} = (487.5)(20)^{0.5} = 2180$$

$$Q/AD^{0.5} = 5500/2180 = 2.52$$

b. Enter the appropriate design chart with $Q/AD^{0.5}$ and the selected edge condition, read HW/D.

Chart 51b, HW/D = 0.90

c. Multiply HW/D by D to obtain the inlet control headwater depth, HW_i.

HW_i = (0.90)(20) = 18 ft (5.486 m)
EL_{hi} = 220 + 18 = 238 ft (72.542 m)

d. V_u is neglected, use HW_i.

e. $HW_i < HW_{ha}$ (18 ft < 20 ft), size is ok

Step 3. Perform inlet control headwater (HW_i) calculations for the check flow rate.

a. $Q/AD^{0.5} = 3.44$

b. Chart 51b, HW/D = 1.13

c. HW_i = (1.13)(20) = 22.6 ft (6.888 m)
EL_{hi} = 220 + 22.6 = 242.6 ft (73.944 m)

d. V_u is neglected, use HW_i

e. HW_i < D + 5 ft (242.6 ft < 240 + 5 ft), size is ok

Step 4. Perform outlet control headwater (HW_o) calculations for the design flow rate (Section 3.6.2) using either backwater calculations or Equation 3.6b. Backwater calculations will be necessary to check Outlet Control.

From hydraulic tables for elliptical conduits (FHWA, 1962):

Barrel Selected	Symbol	CU Units	SI Units
Design Discharge	Q	5,500 ft³/s	155.744 m³/s
Manning's Roughness	n	0.034	0.034
Critical Depth	d_c	12.4 ft	3.780 m
Normal Depth	d_n	13.1 ft	3.993 m
Tailwater	TW	16.0 ft	4.877 m

Barrel Selected	Symbol	CU Units	SI Units
Design Discharge	Q	7,500 ft³/s	212.378 m³/s
Manning's Roughness	n	0.034	0.034
Critical Depth	d_c	14.6 ft	4.450 m
Normal Depth	d_n	16.7 ft	5.090 m
Tailwater	TW	19.0 ft	5.791 m

Since $d_n > d_c$, flow is subcritical. Since TW > d_n, water surface has an M1 profile. Backwater calculations should start at TW depth downstream and progress upstream.

Area and Hydraulic Radius vs. depth from data obtained from tables.

d/D	d	A/BD	A	R/D	R
0.65	13.0	0.5537	332.2	0.3642	7.28
0.70	14.0	0.6013	360.8	0.3781	7.56
0.75	15.0	0.6472	388.3	0.3886	7.77
0.80	16.0	0.6908	414.5	0.3950	7.90
0.85	17.0	0.7313	438.8	0.3959	7.92
0.90	18.0	0.7671	460.3	0.3870	7.74
0.95	19.0	0.7953	477.2	0.3649	7.30
1.00	20.0	0.8108	486.5	0.3060	6.12

Complete Water Surface Computations (see attached calculation sheet).

for Q = 5,500 ft³/s:
HW_o = specific head (H) + k_e ($V^2/2g$)
HW_o = 18.004 + (0.5)(3.208) = 19.6 ft (5.974 m)
EL_{ho} = 220 + 19.6 = 239.6 ft (73.030 m)

for Q = 7,500 ft³/s:
HW_o = specific head (H) + k_e ($V^2/2g$)
HW_o = 22.627 + (0.5)(3.89) = 24.6 ft (7.49 m)
EL_{ho} = 220 + 24.6 = 244.6 ft (74.554 m)

Step 5.   The controlling headwater is the higher of HW_i and HW_o: HW_o

Description	Symbol	CU Units	SI Units
Design Discharge	Q_{25}	5,500 ft³/s	155.744 m³/s
Elevation of Allowable Headwater	El_{ha}	240.0 ft	73.152 m
Elevation at Inlet Control	EL_{hi}	238.0 ft	72.542 m
Elevation at Outlet Control	EL_0	239.6 ft	73.030 m

Description	Symbol	CU Units	SI Units
Check Discharge	Q_{100}	7,500 ft³/s	212.378 m³/s
Elevation of Allowable Headwater	El_{ha}	245.0 ft	74.676 m
Elevation at Inlet Control	EL_{hi}	242.6 ft	73.944 m
Elevation at Outlet Control	EL_0	244.6 ft	74.554 m

Step 6.   Evaluate Results (Section 3.6.3) to determine if controlling HW is near allowable HW and less. If not close enough or higher, return to Step 2 and try another alternative.

This culvert design meets the requirements stated in the problem.

WATER SURFACE PROFILE COMPUTATIONS
(ENGLISH UNITS)

Identification: _Example Problem No.5, HDS No.5_   By: _JMN_   Date: _3/14_

Channel Shape: _Elliptical C.M.P., B=30 ft, D=20 ft, d_n=13.1 ft, TW=16 ft, S_0=0.01_
_M_1 Profile, Barrel Length=200 ft, Start at TW=16 ft_

Manning n= 0.034
Q= 5500 c.f.s.

d	A	R	$V=\frac{Q}{A}$	$\frac{V^2}{2g}$	$H=d+\frac{V^2}{2g}$	ΔH	$R^{2/3}$	$AR^{2/3}$	$S_f=\left(\frac{K_n}{AR^{2/3}}\right)^2$	$\bar{S}_f=\left(S_{f11}-S_{f12}\right)$	$S_0-\bar{S}_f$	$\Delta L=\frac{\Delta H}{S_0-\bar{S}_f}$	$L=\Sigma\Delta L$
16	414.5	7.90	13.27	2.73	18.734		3.969	1645.3	0.00682				0.0
15.5	401.6	7.85	13.76	2.91	18.412	0.322	3.953	1587.35	.00630	0.00656	0.00359	89.7	89.7
15.0	388.3	7.77	14.16	3.115	18.115	0.297	3.926	1521.33	.00618	.00648	.00352	84.4	165.1
14.8	382.9	7.34	14.36	3.208	18.004	0.111	3.779	1447.12	.00752	.00714	.00286	38.8	203.9
													ok

(1) SUBTRACT SECOND H FROM FIRST H VALUE

(2) $K_n=\frac{Qn}{1.49}=\frac{(5500)(.034)}{1.49}=125.50$

(3) IF ΔL IS +, PROFILE IS PROGRESSING UPSTREAM.
  - ΔL DENOTES DOWNSTREAM PROGRESSION.

WATER SURFACE PROFILE COMPUTATIONS
(ENGLISH UNITS)

Identification: _Example Problem No.5, HDS No.5_   By: _WJJ_   Date: _3/14_

Channel Shape: _Elliptical C.M.P., B=30 ft, D=20 ft, d_n=16.7 ft, TW=19.0 ft, S_0=0.01_
_M_2 Profile, Barrel Length=200 ft, Start at TW=19.0 ft_

Manning n= 0.034
Q= 7500 ft³/s

d	A	R	$V=\frac{Q}{A}$	$\frac{V^2}{2g}$	$H=d+\frac{V^2}{2g}$	ΔH	$R^{2/3}$	$AR^{2/3}$	$S_f=\left(\frac{K_n}{AR^{2/3}}\right)^2$	$\bar{S}_f=\left(S_{f11}-S_{f12}\right)$	$S_0-\bar{S}_f$	$\Delta L=\frac{\Delta H}{S_0-\bar{S}_f}$	$L=\Sigma\Delta L$
19.0	477.20	7.30	15.72	3.836	22.835		3.765	1796.6	0.00907				0.0
18.8	474.24	7.268	15.82	3.833	22.684	0.151	3.803	1803.4	.08901	0.00904	0.00096	157.6	157.6
18.74	472.65	7.457	15.88	3.9098	23.610	.074	3.820	1805.3	.00899	.00900	.00100	74.1	231.7
18.35	473.80	7.440	15.83	3.8911	23.621	.057	3.814	1800.9	.00897	.00899	.00101	56.46	214.1
													ok

(1) SUBTRACT SECOND H FROM FIRST H VALUE

(2) $K_n=\frac{Qn}{1.49}=\frac{(7500)(0.034)}{1.49}=171.14$

(3) IF ΔL IS +, PROFILE IS PROGRESSING UPSTREAM.
  - ΔL DENOTES DOWNSTREAM PROGRESSION.

DG 2.4 LONG SPAN HY-8 SOLUTION

The hand solution shown in Section 2.3 can be duplicated using the current version of HY-8:

- Select the following options: U.S. Customary Units, Outlet Control Profiles and Exit Loss Standard Method.

- Enter Crossing Data shown in Step 1, select Elliptical Shape, Size 22 (361" by 242"), change n to 0.034, select headwall for inlet configuration

- Analyze Crossing brings up Crossing Summary Table that shows no overtopping.

- Select Culvert Summary Table which shows that outlet control governs:

  - For 5500 ft^3/s - HW_o = 19.9 ft and El_{ho} = 239.9 ft ≈ 239.6 ft for hand solution for square edges with M1 profile. The hand solution full barrel area is 487.5 ft^2 versus the smaller 471 ft^2 of shape 22.

  - For 7500 ft^3/s: HW_o = 25.47 ft and El_{ho} = 245.47 ft ≈ 244.6 ft for hand solution.

  - The inlet control depths computed using Chart 52 ≈ Chart 51 solutions of the hand solution. The charts are similar for square edges.

- Similar results are obtained when the SI solution is verified with HY-8.

(page intentionally left blank)

DESIGN GUIDELINE 3

TAPERED INLET DESIGN USING NOMOGRAPHS

DG 3.1 BACKGROUND

Tapered inlet design begins with the selection of the culvert barrel size, shape, and material. These calculations are performed using Design Guideline 1 and the Culvert Design Form (CDF). If the culvert operates in inlet control, the Tapered Inlet Design Form (TIDF) and the design charts contained in Appendix C are used to design the tapered inlet. The result will be one or more culvert designs, with and without tapered inlets, all of which meet the site design criteria. The designer must select the best design for the site under consideration.

In the design of tapered inlets, the goal is to select a barrel and depression that provides inlet control at the efficient throat section in the design range of headwater and discharge. This is because the throat section has the same geometry as the barrel, and the barrel is the most costly part of the culvert. The inlet face is then sized large enough to pass the design flow without acting as a control section in the design discharge range. Some slight oversizing of the face is beneficial because the cost of constructing the tapered inlet is usually minor compared with the cost of the barrel. The required size of the face can be reduced by use of beveled edges. Design charts are provided for the face with and without bevels.

DG 3.2 DESIGN PROCEDURE

The following steps outline the design process for culverts with tapered inlets. Step 1 is the same for all culverts, with and without tapered inlets.

Step 1. Determine Culvert Barrel Size and Throat Depression (T)

> Barrel - The culvert barrel size is determined using Design Guideline 1 and the Culvert Design Form (CDF). The CDF should be completed for all barrels of interest. Since a tapered inlet is being designed, the appropriate Throat Control Chart 55 (pipes) and Chart 57 (boxes) should be used for inlet control. Chart 55 (pipes) has two scales: one for smooth inlets (RCP) and one for rough inlets (CMP).

> Depression - The CDF provides the required Throat Depression (T) at a straight culvert inlet, at a side-tapered inlet throat and at a slope tapered inlet throat.

> Performance Curve Data - Plot outlet control performance curves for the barrels of interest and inlet control performance curves for the faces of culverts with nonenlarged inlets and for the throats of tapered inlets.

Step 2. Select Tapered Inlet Alternatives

> Side-tapered should be used If T < D/4 (see Section DG 3.3).
> Side-tapered or Slope-tapered can be used if D/4 $\leq$ T $\leq$ 1.5D.

Step 3. Determine Tapered Inlet Dimensions

Face Width - Use the Tapered Inlet Design Form (TIDF), Figure 3.24, and Chart 56 (side-tapered pipe), Chart 58 (side-tapered RCB) or Chart 59 (slope-tapered RCB) in Appendix C to determine the face width (B_f) which is then used to calculate the inlet dimensions.

Side-tapered inlet - Check to assure that the drop between the face section and the throat section is one foot or less. If not, revise face invert elevation

Side-tapered inlet with depression upstream of the face - Calculate the minimum crest width (W) and compare to the proposed crest width. The wingwall flare angle may need to be increased to obtain the necessary crest width.

Step 4. Select Tapered Inlet

Evaluate the design alternatives that were dimensioned in Step 3 and select the alternative that is the most cost effective to construct.

DG 3.3 DIMENSIONAL LIMITATIONS

The following dimensional limitations must be observed when designing tapered inlets using the design charts of this publication. Tapered inlets can only be used where the culvert width is less than three times its height, (B < 3 D).

DG 3.3.1 Side-Tapered Inlets

Taper: 4:1 $\leq$ TAPER $\leq$ 6:1 (Tapers less divergent than 6:1 may be used but performance will be underestimated. (NBS 6th and 7th Progress Report)

Wingwall Flare Angle: (a) 15-degrees to 26-degrees with top edge beveled or (b) 26-degrees to 90-degrees with or without bevels.

Throat Depression (T): If a depression is used upstream of the face, extend the barrel invert slope upstream from the face a distance of D/2 before sloping upward more steeply. The maximum vertical slope of the apron is 1V:2H.

Face Height (E): $D \leq E \leq 1.1D$ for circular barrels

DG 3.3.2 Slope-Tapered Inlets.

Taper: 4:1 $\leq$ TAPER $\leq$ 6:1 (Tapers > 6:1 may be used, but performance will be underestimated.)

Wingwall Flare Angle: (a) 15-degrees to 26-degrees with top edge beveled or (b) 26-degrees to 90-degrees with or without bevels.

Depression Slope (S_D) : IV:3H $\geq$ S_D $\geq$ IV:2H, If S_D > IV:3H, use side-tapered design.

Length to Bend (L_3): Minimum L_3 = 0.5B

Thoat Depression (T): $D/4 \leq T \leq 1.5D$ If $T < D/4$, use side-tapered design. If $T > 1.5D$, estimate friction losses (H_1) between the face and the throat by using Equation (DG 3.1) and add the additional losses to HW_t.

$$H_1 = \left[\frac{K_U\, n^2 L_i}{R^{1.33}} \right] \frac{Q^2}{2gA^2}$$ (DG 3.1)

where:

K_U	is 29 (19.63 in SI Units)
H_1	is the friction head loss in the tapered inlet, ft (m)
n	is the Manning's n for the tapered inlet material
L_1	is the length of the tapered inlet, ft (m)
R	is the average hydraulic radius of the tapered inlet = $(A_f + A_t))/(P_f + P_t)$, ft (m)
Q	is the flow rate, ft³/s (m³/s)
g	is the gravitational acceleration, ft/s² (mt/s²)
A	is the average cross sectional area of the tapered inlet = $(A_f + A_t)/2$, ft² (m²)

DG 3.4 TAPERED INLETS FOR BOX CULVERT

The design procedure of Section DG 3.2 and dimensional limitations in Section DG 3.3 should be followed for side-tapered and slope-tapered inlets for box culverts. Section DG 3.4.1 provides the only inlet configurations that can be used with tapered inlets for box culverts. Section DG 3.4.2 should be reviewed if the proposed culvert has more than one barrel. A hand worked design example is provided in Sections DG 3.4.3 to 3.4.8.

DG 3.4.1 Alternative Inlet Configurations

For determining the required face width, there are two nomographs in Appendix C, Chart 58 for side-tapered inlets and Chart 59 for slope-tapered inlets. Each Chart has two scales, and each scale refers to a specific inlet edge condition. The edge conditions are depicted in Figure DG 3.1. Both the inlet edge condition and the wingwall flare angle affect the performance of the face section for box culverts:

- Scale 1 applies to the less favorable edge conditions, defined as either:
    - wingwall flares of 15-degrees to 26-degrees and a 1:1 top edge bevel, or
    - wingwall flares of 26-degrees to 90-degrees and square edges (no bevels). A 90-degree wingwall flare is a straight headwall.

- Scale 2 applies to the more favorable edge conditions, defined as either:

    - wingwall flares of 26-degrees to 45-degrees with 1:1 top edge bevel, or

    - wingwall flares of 45-degrees to 90-degrees with a 1:1 bevel on the side and top edges.

Undesirable wingwall flare angles less than 15-degrees, or 26-degrees without a top bevel, are not covered by the charts. Although the large 33.7 degree bevels can be used, the smaller 45 degree bevels are preferred due to structural considerations.

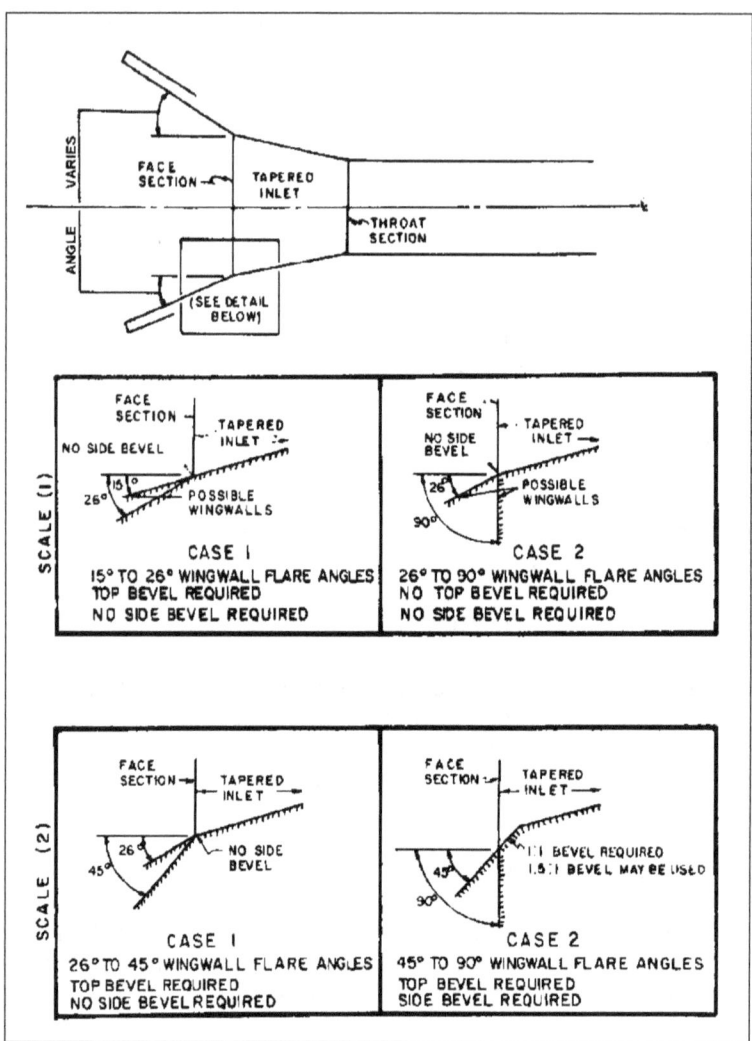

Figure DG 3.1. Inlet edge conditions, face section, rectangular tapered Inlets.

DG 3.4.2 Multiple Barrel Designs

When designing side-tapered or slope-tapered inlets for box culverts with double barrels, the required face width derived from the design procedures is the total clear width of the face. The thickness of the center wall must be added to this clear width to obtain the total face width. No design procedures are available for tapered inlets on box culverts with more than two barrels.

DG 3.4.3 RCB Culvert Example

Design the smallest possible barrel and depression to pass the Q_{50} without exceeding the EL_{ha}. The culvert will be located in a rural area with a low risk of damage. Underground utilities limit the available depression to 2.5 ft (0.762 m) below the standard streambed elevation at the inlet. Use a reinforced concrete box (RCB) culvert with n=0.012.

The complete customary unit (CU) solution is detailed below. A summary of SI results is also provided based on the complete solution of the problem using the appropriate SI nomographs as detailed in HDS 5, second edition, Chapter IV, Example Problem 1.   Note that direct conversion of the CU solution to SI may yield slightly different results.

Note: Charts 14B, 15B, 57B, 58B, and 59B are used in this solution (see Section DG 3.4.7).

Step 1. Determine Culvert Barrel Size & Throat Depression (T) - The following data was used in Design Guideline 1 and the CDF (Section DG 3.4.4) to select the barrel size and depression (T).

Site Description	Symbol	CU Units	SI Units
Design Discharge	Q_{50}	400 ft³/s	11.327 m³/s
Natural Stream Bed Slope	S_o	0.05 ft/ft	0.05 m/m
Approximate Culvert Length	L_a	300 ft	91.440 m
Elevation at Shoulder	EL_s	196 ft	59.741 m
Elevation at Allowable Headwater	El_{ha}	195 ft	59.436 m
Elevation at Streambed Face	EL_{sf}	187.5 ft	57.150 m
Elevation at Outlet Invert	EL_o	172.5 ft	52.578 m

Flow (ft³/s)	TW (ft)	Flow (m³/s)	TW (m)
300	4.4	8.495	1.341
400	4.9	11.327	1.494
500	5.3	14.159	1.615

Barrel Selected	Symbol	CU Units	SI Units
Width	B	5 ft	1524 mm
Height	D	5 ft	1524 mm
Throat Depression	T	2 ft	0.45 m
Elevation at Inlet Invert	EL_i	185.5 ft	56.70 m
Culvert Slope	S	0.043 ft/ft	0.045 m/m
Outlet Velocity	V_o	29.1 ft/s	8.87 m/s

Step 2. Select Tapered Inlet Alternatives

Both side-tapered and slope-tapered inlets will be designed:
Depression (T) = 2 ft is greater than the minimum D/4 = 5/4 = 1.25 ft.

Step 3. Determine Tapered Inlet Dimensions (Side-tapered)

The TIDF (see Section DG 3.4.5) is used to determine the following:

Description	Symbol	CU Units	SI Units
Face Width	B_f	9 ft	3.05 m
Side-taper	taper	4:1	4:1
Throat Depression	T	2 ft	0.45 m
Depression Slope	S_D	IV:2H	IV:2H
Length (Face to Throat)	L_1	8 ft	2.95 m

Step 3. Determine Tapered Inlet Dimensions (Slope-tapered)

The TIDF (see Section DG 3.4.5) is used to determine the following:

Description	Symbol	CU Units	SI Units
Face Width	B_f	8 ft	2.44 m
Side-taper	taper	4.33:1[1]	4.1
Throat Depression	T	2 ft	0.45 m
Depression Slope	S_D	IV:2H	IV:2H
Length (Face to Throat)	L_1	6.5 ft	1.83 m
Length (Face to Bend)	L_2	4.0 ft	0.90 m
Length (Bend to Face)	L_3	2.5 ft	0.93 m

[1]The taper as adjusted in CU solution while L_3 was adjusted in SI solution.

Step 4. Select Tapered Inlet

A slope-tapered inlet with a vertical face is selected since it is the smaller inlet. The inlet will have 26-degree to 90-degree wingwalls with no bevels. The performance curves are shown in the following figure:

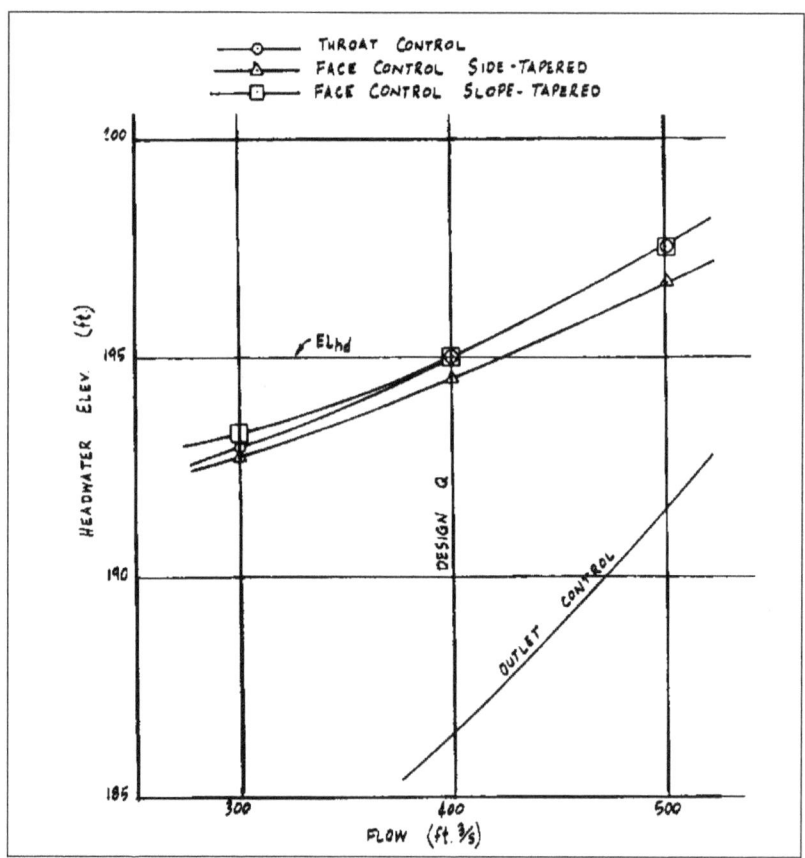

Performance curves for 5 ft by 5 ft RCB with a tapered inlet.

DG 3.4.4 RCB HY-8 Solution

The hand solution shown in Section 3.4.3 can be duplicated using the current version of HY-8:

- Enter Crossing Data shown in Step 1 and CDF. For Tailwater, enter rating curve.

- For Culvert Type, select Side-Tapered. For Inlet Edge, select square edge top. For Inlet depression, select yes and enter 2 ft, slope of 2:1 and crest width of 9 ft.

- Analyze Crossing brings up Crossing Summary Table that shows some overtopping if 196 was used for the crest elevation.

- Select Culvert Summary Table which shows that at 400 ft^3/s inlet control governs: HW_i = 9.15 ft and El_{hi} = 194.92 ft $\approx$ 195 ft of the nomograph solution for square edges.

- Select Improved Inlet Table which shows at 400 ft^3/s that throat control governs and that the face control headwater is 193.58 $\approx$ 193.8 of nomograph solution.

- The Slope-Tapered results can be obtained by changing Culvert Type, Face Width to 8 ft, slope to 2:1 and depression to 2 ft. The improved inlet table shows that the throat control headwater is the same El_{hi} = 194.92 ft, but the face control headwater is 195.16 $\approx$ 195 of nomograph solution. The smaller face slightly controls.

- Similar results are obtained when the SI solution is verified with HY-8.

DG 3.4.5 RCB Culvert CDF

PROJECT: Example Problem No. 1 - English

STATION: 5 + 00
SHEET 1 OF 4

CULVERT DESIGN FORM
DESIGNER / DATE: WJJ / 7/16
REVIEWER / DATE: JMN / 7/17

HYDROLOGICAL DATA

See Add'l Shts

☐ METHOD: Rational
☐ DRAINAGE AREA: 150 Ac. ☐ STREAM SLOPE: 5.0%
☐ CHANNEL SHAPE: Trapezoidal
☐ ROUTING: N/A ☐ OTHER: --

DESIGN FLOWS/TAILWATER

R.I. (YEARS)	FLOW (cfs)	TW (ft)
50	400	4.9

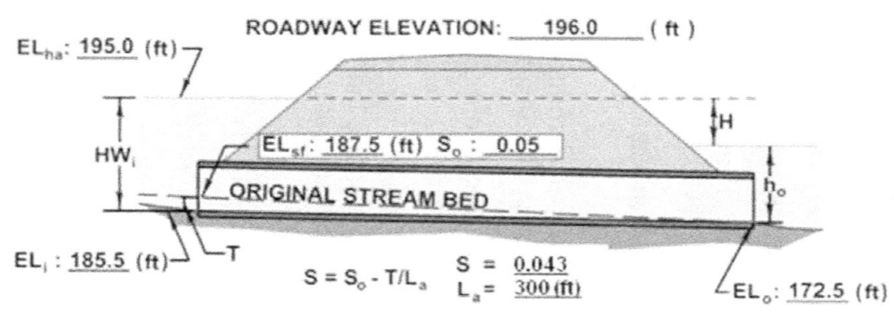
ROADWAY ELEVATION: 196.0 (ft)
EL$_{ha}$: 195.0 (ft)
HW$_i$
EL$_{sf}$: 187.5 (ft) S$_o$: 0.05
ORIGINAL STREAM BED
H
h$_o$
EL$_i$: 185.5 (ft)
T
S = S$_o$ - T/L$_a$
S = 0.043
L$_a$ = 300 (ft)
EL$_o$: 172.5 (ft)

CULVERT DESCRIPTION:

HEADWATER CALCULATIONS

MATERIAL – SHAPE – SIZE – ENTRANCE	Total Flow Q (cfs)	Flow Per Barrel Q / N (1)	INLET CONTROL: HW$_i$/D (2)	HW$_i$	T (3)	EL$_{hi}$ (4)	OUTLET CONTROL: TW (5)	d_c	$\frac{d_c+D}{2}$	h_o (6)	k_e	H (7)	EL$_{ho}$ (8)	Control Headwater Elevation	Outlet Velocity	Comments
Conc. Box - 5' x 5' - 45" bevel	400	80	2.4	12.0	4.5	195.0	4.9	≥ 5	5.0	5.0	0.2	8.9	186.4	195.0	--	Excess T reqd
Conc. Box - 5' x 5' - tapered inlet throat	400	80	1.9	9.5	2.0	195.0	4.9	≥ 5	5.0	5.0	0.2	8.9	186.4	195.0	29.1	Inlet and T OK
Conc. Box - 5' x 5' - tapered inlet throat	300	60	1.48	7.4	2.0	192.9	4.4	4.8	4.9	4.9	0.2	5.0	182.4	192.9	--	--
Conc. Box - 5' x 5' - tapered inlet throat	500	100	2.4	12.0	2.0	197.5	5.3	≥ 5	5.0	5.0	0.2	14.1	191.6	197.5	--	--

TECHNICAL FOOTNOTES:

(1) USE Q/NB FOR BOX CULVERTS

(2) HW_i / D = HW / D OR HW_i / D FROM DESIGN CHARTS

(3) $T = HW_i - (EL_{ha} - EL_{sf})$
T IS ZERO FOR CULVERTS ON GRADE

(4) $EL_{hi} = HW_i + EL_i$ (INVERT OF INLET CONTROL SECTION)

(5) TW BASED ON DOWN STREAM CONTROL OR FLOW DEPTH IN CHANNEL

(6) h_o = TW or $(d_c + D)/2$ (WHICHEVER IS GREATER)

(7) $H = [1 + k_e + (K_u n^2 L) / R^{1.33}] v^2 / 2g$ WHERE K_u = 19.63 (29 IN ENGLISH UNITS)

(8) $EL_{ho} = EL_o + H + h_o$

SUBSCRIPT DEFINITIONS:

a. APPROXIMATE
f. CULVERT FACE
ha. ALLOWABLE HEADWATER
hi. HEADWATER IN INLET CONTROL
ho. HEADWATER IN OUTLET CONTROL
i. INLET CONTROL SECTION
o. OUTLET
sf. STREAMBED AT CULVERT FACE
tw. TAILWATER

COMMENTS / DISCUSSION:

Use 5' x 5' concrete box with tapered inlet. Check side- and slope-tapered entrances.

CULVERT BARREL SELECTED:

SIZE: 5' x 5'
SHAPE: Rectangular
MATERIAL: Conc. n .012
ENTRANCE: Tapered Inlet

DG 3.4.6 RCB Side-Tapered TIDF

PROJECT: Example Problem No. 1 - English	STATION: 5 + 00
	SHEET 2 OF 4

TAPERED INLET DESIGN FORM
DESIGNER / DATE: WJJ / 7/16
REVIEWER / DATE: JMN / 7/17

DESIGN DATA:

- Q 50 = 400 (cfs) ; EL_{hi} 195.0 (ft)
- EL. THROAT INVERT 185.5 (ft)
- EL. STREAM BED AT FACE 187.5 (ft)
- T 2.0 TAPER 4 : 1 (4:1 TO 6:1)
- STREAM SLOPE, S_o, = 0.05 ft/ft
- SLOPE OF BARREL, S = 0.043 ft/ft
- S_D ---- : 1 (2 : 1 TO 3 : 1)
- BARREL SHAPE AND MATERIAL: Box - Conc.
- N = 1 B = 5 ft D = 5 ft
- INLET EDGE DESCRIPTION Tapered Inlet Throat

COMMENTS

Use 26° to 90° wingwalls with no bevels on a side tapered inlet.

							Side-Taper		Slope-Taper	SLOPE-TAPERED ONLY					SIDE-TAPERED w/depression		
Q (cfs)	EL_{hi} (ft)	EL. Throat Invert	EL. Face Invert (1)	HW_i (2)	HW_i/E (3)	Q/B_f (4)	MIN. B_f (5)	Selected B_f	MIN. L_3 (6)	L_2 (7)	Check L_2 (8)	Adj. L_3 (9)	Adj. Taper (10)	L_1 (11)	EL. (Crest Inv)	HW_c (12)	MIN. W (13)
400	195.0	185.5	186.5	8.5	1.7	47	8.51	9.0	--					8.0	188.5	6.5	8.5
400	193.8	185.5	185.8*	7.95	1.59	44	--	9.0	--	FACE CONTROL PERFORMANCE DATA					--	--	--
300	192.1	185.5	185.8	6.25	1.25	33	--	9.0	--	FACE CONTROL PERFORMANCE DATA					--	--	--
500	196.0	185.5	185.8	10.2	2.04	56	--	9.0	--	FACE CONTROL PERFORMANCE DATA					--	--	--

(1) SIDE-TAPERED : EL. FACE INVERT = EL. THROAT INVERT + 1 FT (0.3 M APPROX.)
SLOPE-TAPERED : EL. FACE INVERT = EL. STREAM BED AT FACE

(2) HW_i = EL_{hi} - EL. FACE INVERT

(3) 1.1 D ≥ E ≥ D; E = D FOR BOX CULVERTS

(4) FROM DESIGN CHARTS

(5) MIN. B_f = Q / (Q / B_f)

(6) MIN. L_3 = 0.5 NB

(7) L_2 = (EL. FACE INVERT - EL. THROAT INVERT) S_D

(8) CHECK $L_2 = \left[\dfrac{B_f - NB}{2}\right]$ TAPER $- L_3$

(9) IF (8)>(7), ADJ. $L_3 = \left[\dfrac{B_f - NB}{2}\right]$. TAPER $- L_2$

(10) IF (7) >(8), ADJ. TAPER = $(L_2 + L_3)$ / $\left[\dfrac{B_f - NB}{2}\right]$

(11) SIDE-TAPERED : L = $\left[\dfrac{B_f - NB}{2}\right]$ TAPER
SLOPE-TAPERED : $L_1 = L_2 + L_3$

(12) $HW_c = EL_{hi}$ - EL. CREST INVERT

(13) MIN. W = K_u Q / $HW_c^{1.5}$ Where K_u = 0.35 (0.64 SI)

*ACTUAL EL. FACE INVERT - EL_i + L_1 S = 185.5 + (8.0) (.043) = 185.8 ft

SELECTED DESIGN

- B_f ____9.0____
- L_1 ____8.0____
- L_2 _____
- L_3 _____
- BEVELS ANGLE ___0°___
- b = 0 ( ) ; d = 0 ( )
- TAPER ___4___ : 1
- S_D _____ : 1

DG 3.4.7 RCB Slope-Tapered TIDF

DESIGN DATA:

Q 50 = 400 ( ); EL_H 195.0 (ft)

EL. THROAT INVERT 185.5 (ft)

EL. STREAM BED AT FACE 187.5 (ft)

T 2.0 TAPER 4 : 1 (4:1 TO 6:1)

STREAM SLOPE, S_o = 0.05 ft/ft

SLOPE OF BARREL, S = 0.044 ft/ft

S_i 2 : 1 (2:1 TO 3:1)

BARREL SHAPE AND MATERIAL: Box - Conc.

N = 1 B = 5 ft D = 5 ft

INLET EDGE DESCRIPTION Tapered Inlet Throat

COMMENTS

Use 26° to 90° wingwalls flares with no bevels on a slope-tapered inlet.

Side-Taper

Slope-Taper

Q ( )	EL_{hi} ( )	EL. Throat Invert	EL. Face Invert (1)	HW_i (2)	$\frac{HW_i}{E}$ (3)	$\frac{Q}{B_f}$ (4)	MIN. B_f (5)	Selected B_f	MIN. L_3 (6)	L_2 (7)	Check L_2 (8)	Adj. L_3 (9)	Adj. Taper (10)	L_1 (11)	EL. Crest Inv.	HW_c (12)	W (13)
											SLOPE-TAPERED ONLY				SIDE-TAPERED w/depression		
400	195.0	185.5	187.5	7.5	1.5	50	8.0	8.0	2.5	4.0	3.5	--	4.33	6.5	--	--	--
300	193.2	185.5	187.5	5.7	1.14	37.5	--	8.0			FACE CONTROL PERFORMANCE DATA				--	--	--
500	197.5	185.5	187.5	10.0	2.0	62.5	--	8.0			FACE CONTROL PERFORMANCE DATA				--	--	--

(1) SIDE-TAPERED : EL. FACE INVERT = EL. THROAT INVERT = EL. THROAT INVERT + 1 FT (0.3 M APPROX.)

SLOPE-TAPERED : EL. FACE INVERT = EL. STREAM BED AT FACE

(2) $HW_i = EL_{hi} -$ EL. FACE INVERT

(3) 1.1 D ≥ E ≥ D; E = D FOR BOX CULVERTS

(4) FROM DESIGN CHARTS

(5) MIN. $B_f = Q / (Q / B_f)$

(6) MIN. $L_3 = 0.5$ NB

(7) $L_2 =$ (EL. FACE INVERT- EL. THROAT INVERT) S_D

(8) CHECK $L_2 = \left[\dfrac{B_f - NB}{2} \right] \cdot$ TAPER $- L_3$

(9) If (8)>(7), ADJ. $L_3 = \left[\dfrac{B_f - NB}{2} \right] \cdot$ TAPER $- L_2$

(10) If (7) >(8), ADJ. TAPER $= (L_2 + L_3) / \left[\dfrac{B_f - NB}{2} \right]$

(11) SIDE-TAPERED : $L_1 = \left[\dfrac{B_f - NB}{2} \right] \cdot$ TAPER

SLOPE-TAPERED : $L_1 = L_2 + L_3$

(12) $HW_C = EL_{hi} -$ EL. CREST INVERT

(13) MIN. $W = K_U Q / HW_c{}^{1.5}$ Where $K_U = 0.35$ (0.64 SI)

SELECTED DESIGN

B_f _____ 8.0 _____

L_1 _____ 6.5 _____

L_2 _____ 4.0 _____

L_3 _____ 2.5 _____

BEVELS ANGLE _____ 0° _____

b = 0 ( );d = 0 ( )

TAPER 4.33 : 1

S_D 2.0 : 1

DG 3.4.8 RCB Charts

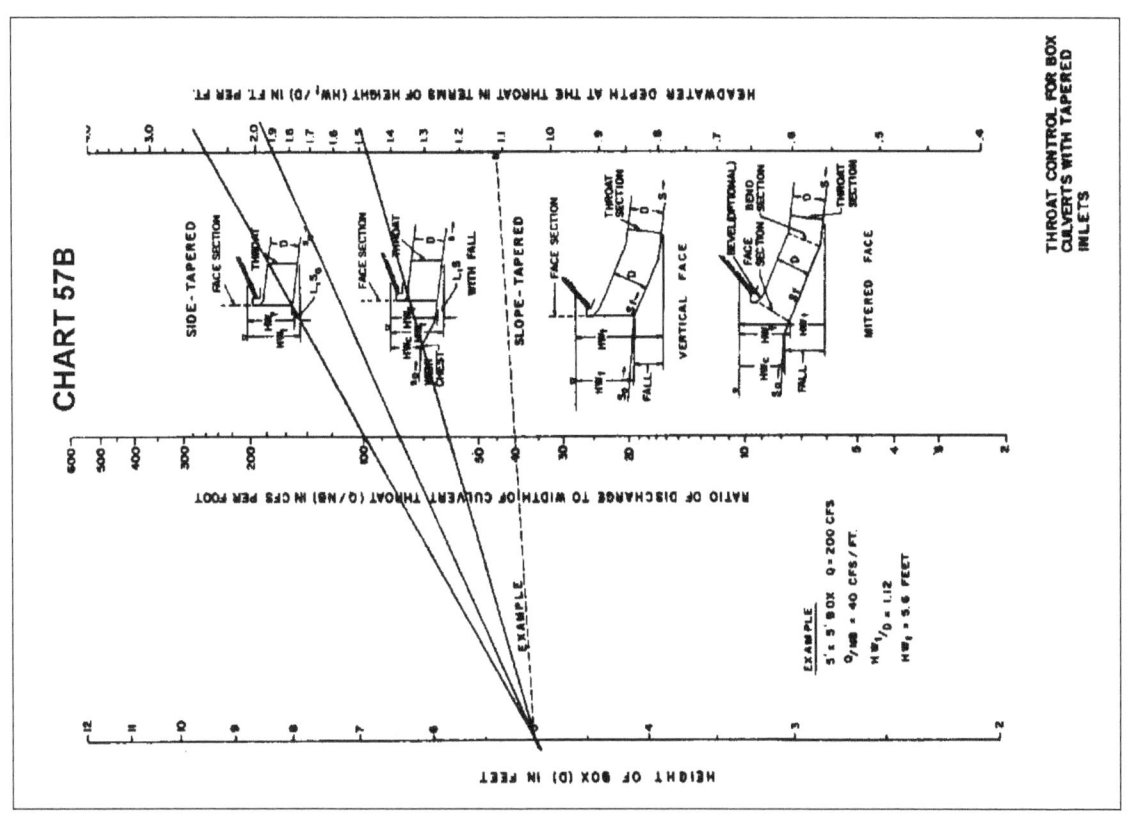

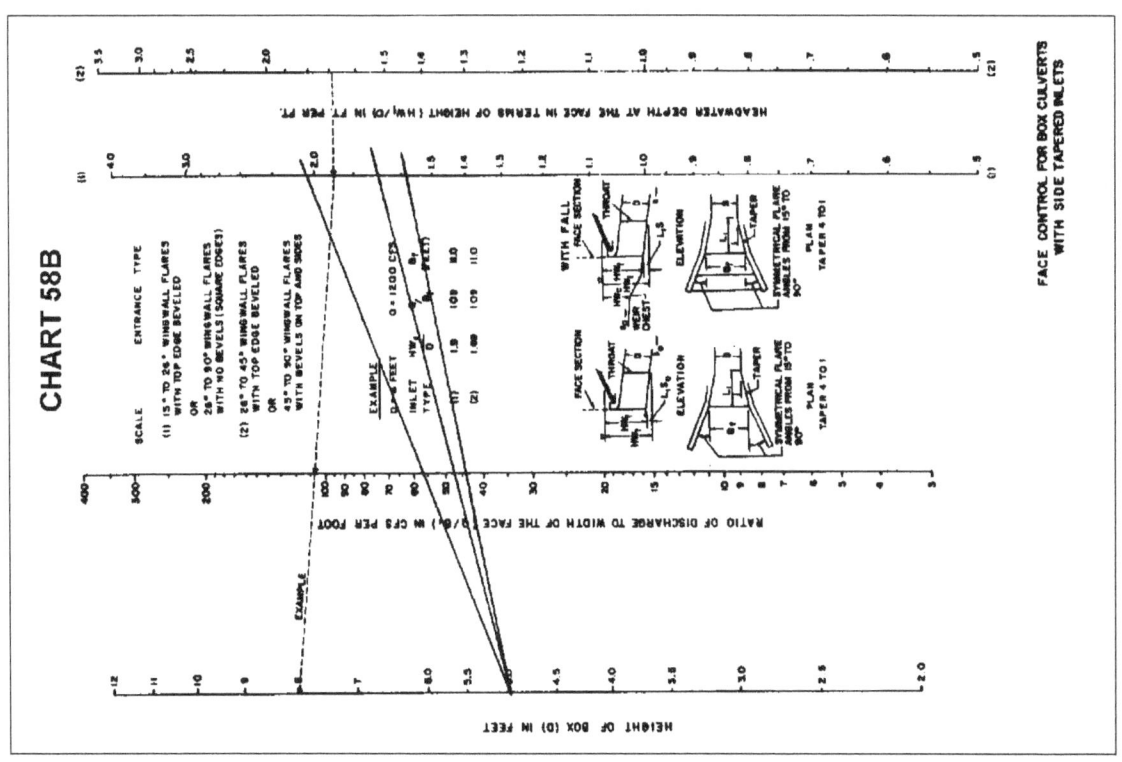

CHART 59B

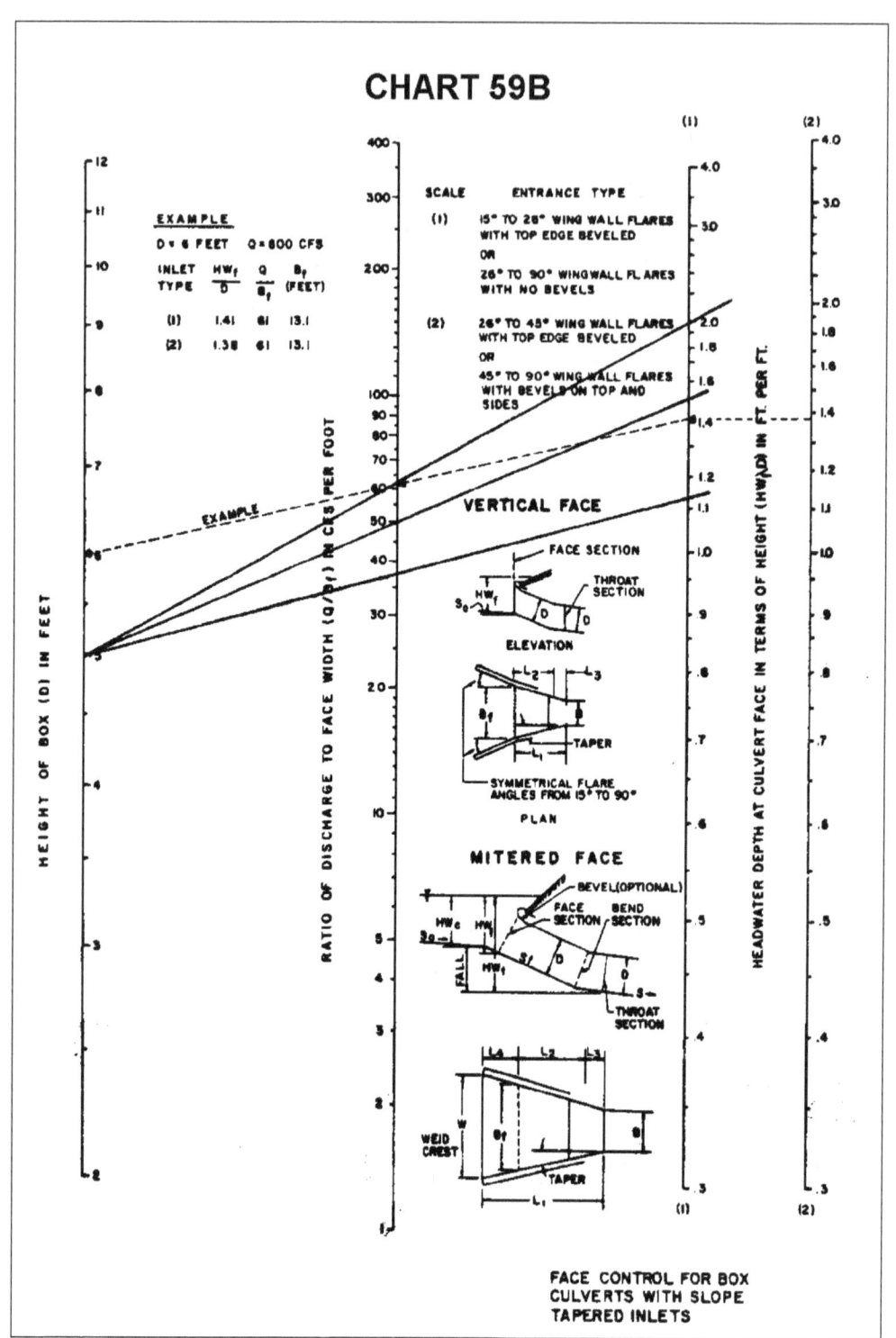

FACE CONTROL FOR BOX
CULVERTS WITH SLOPE
TAPERED INLETS

DG 3.5 TAPERED INLETS FOR CIRCULAR PIPE CULVERTS

The design procedure of Section DG 3.2 and dimensional limitations in Section DG 3.3 should be followed for side-tapered and slope-tapered inlets for circular pipe culverts. Section DG 3.5.1 provides the inlet configurations that can be used with tapered inlets for pipe culverts. Section DG 3.5.3 should be reviewed if the proposed culvert has more than one barrel. A hand worked design example is provided in Sections DG 3.5.4 through 3.5.8.

DG 3.5.1 Alternative Inlet Configurations

Tapered inlet configurations are shown in Figure DG 3.1. For the side-tapered inlet, either prefabricated inlets with nonrectangular cross sections or cast-in-place rectangular inlets are used. The rectangular inlets are joined to the circular pipe using a square to circular throat transition section (see Section 3.5.2).

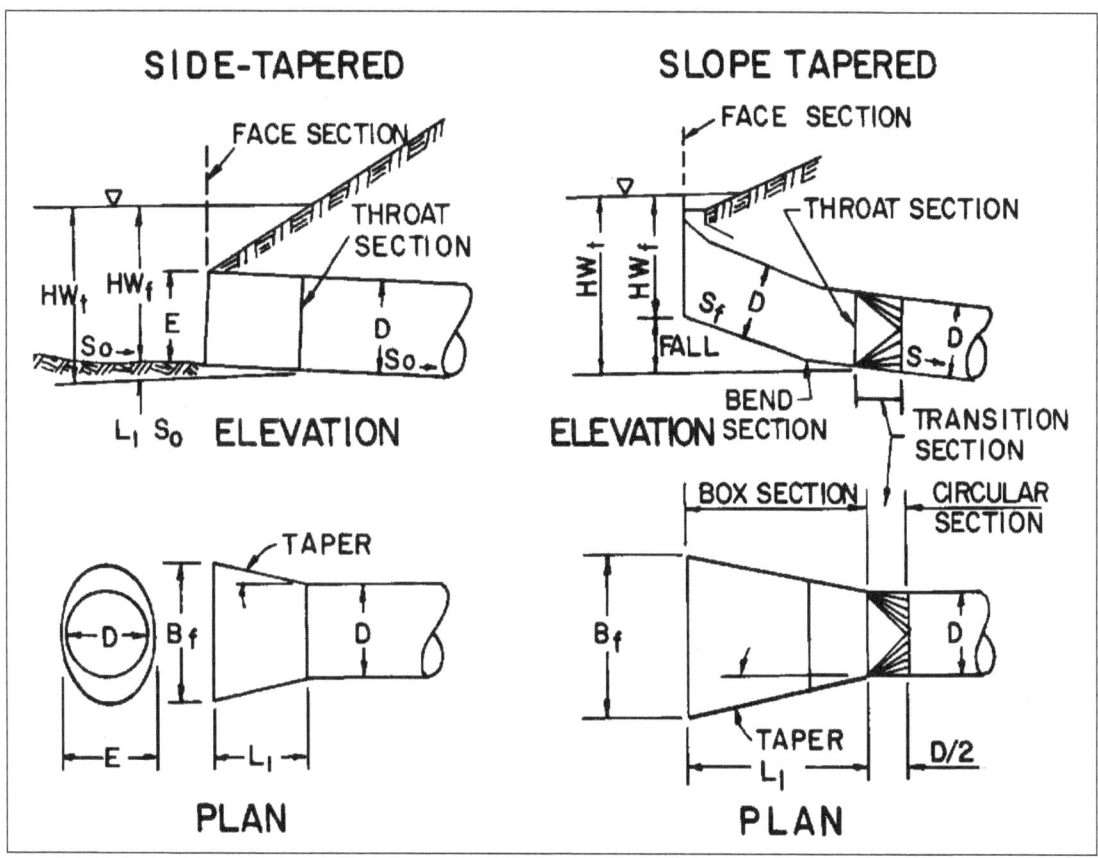

Figure DG 3.2. Tapered inlets for pipe culverts.

For slope-tapered inlets, the rectangular inlet is the only option for which design charts are available. The square to circular transition section is used to join the slope-tapered inlet to the circular pipe.

The design chart for sizing the face of a side-tapered inlet with a nonrectangular face includes three different edge condition scales: thin-edge projecting, square edge and bevel edged. The face area is larger than the barrel area and may be any nonrectangular shape, including an oval, a circle, a circular segment, or a pipe-arch. To design a rectangular side-tapered inlet for a circular pipe culvert, use the design charts in Appendix C for rectangular side-tapered inlets. Additional head can be provided on the throat control section of a side-tapered inlet by constructing a depression upstream of the face section. The depression designs are the same as for box culverts.

DG 3.5.2 RCB Tapered Inlets for Pipes

Rectangular inlets are adapted to pipe culverts as shown in Figure DG 3.3. The slope-tapered inlet is connected to the pipe culvert by use of a square to circular transition. The design of the slope-tapered inlet is the same as for box culverts. There are two throat sections, one square and one circular, but the circular throat section will control the flow because the area is much smaller than the square throat section.

In addition to the dimensional limitations given previously for all tapered inlets (Section DG 3.3), the following criteria apply to the application of rectangular side-tapered and slope-tapered inlets to circular pipe culverts:

- The transition from the square throat section to the circular throat section must be $\geq$ D/2. If excessive lengths are used, the frictional loss within the transition section of the culvert must be considered in the design using Equation (DG 3.1).
- The square throat dimension must equal the diameter of the circular pipe culvert

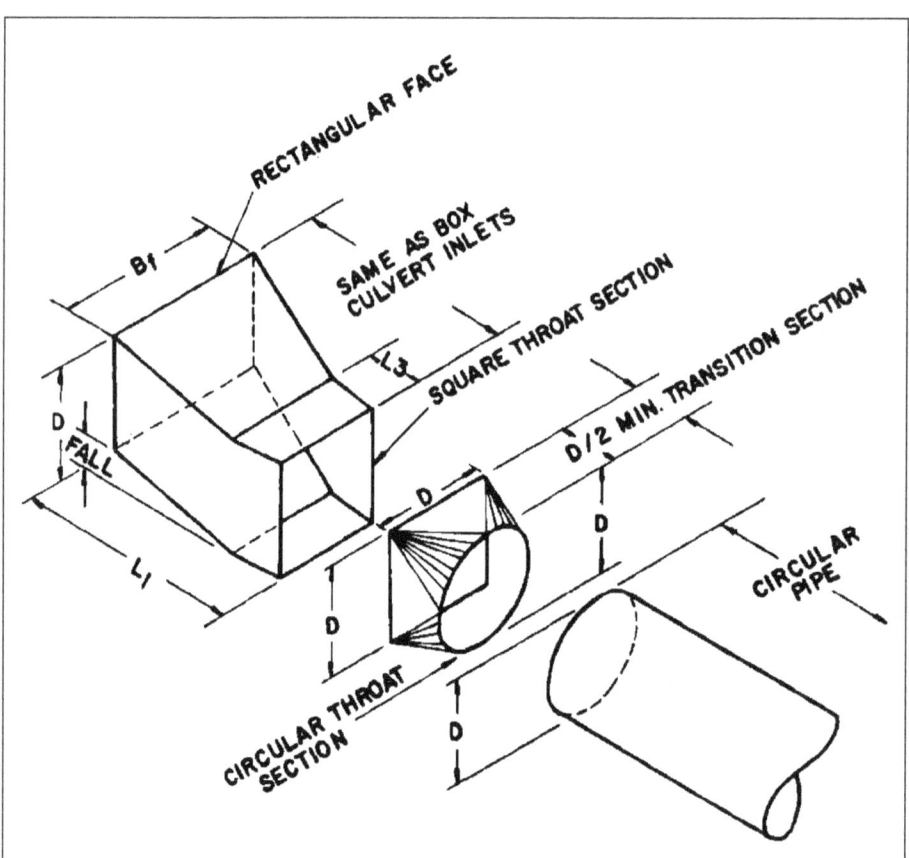

Figure DG 3.3. Slope-Tapered Inlet for Circular Pipe Culvert.

DG 3.5.3 Multiple Barrel Designs

Each barrel of the culvert must have an individual side-tapered inlet with a non-rectangular face design. For rectangular side-tapered inlets with a square to round transition, double barrel designs are the same as for box culverts. However, the center wall at the transition must be flared to provide adequate space between the pipes for proper backfill and compaction. The amount of flare required will depend on the size of the pipes and the construction techniques used. No more than two circular barrels may be feed from the throat section of a rectangular side-tapered inlet.

Double barrel slope-tapered inlets may be designed in the same manner as for rectangular side-tapered designs. Again, no more than two barrels may be feed from a single inlet structure.

DG 3.5.4 CMP Culvert Example

Design a corrugated metal pipe (CMP) with standard (2-2/3 by 1/2 in) corrugations (68 mm x 13 mm) with a tapered inlet. Investigate both a corrugated side-tapered inlet and a concrete slope-tapered inlet. Use normal depth in the natural channel as the tailwater depth.

The complete customary unit (CU) solution is detailed below. A summary of SI results is also provided based on the complete solution of the problem using the appropriate SI nomographs as detailed in HDS 5, second edition, Chapter IV, Example Problem 3. Note that direct conversion of the CU solution to SI may yield slightly different results.

NOTE: Charts 4B, 6B, 55B, 56B, and 59B are used in this solution, but are not provided.

Step 1. Determine Culvert Barrel Size & Throat Depression (T) - The following data was used in Design Guideline 1 and the CDF (Section DG 3.5.5) to select the barrel size and depression (T).

Description	Symbol	CU Units	SI Units
Design Discharge	Q_{50}	150 ft³/s	4.25 m³/s
Natural Stream Bed Slope	S_o	0.05 ft/ft	0.05 m/m
Approximate Culvert Length	L_a	350 ft	106.680 m
Elevation at Shoulder	EL_s	102.0 ft	31.090 m
Elevation of Allowable Headwater	El_{ha}	96.0 ft	29.261 m
Elevation at Culvert Face	EL_{sf}	92.5 ft	28.194 m
Elevation at Outlet Invert	EL_o	75.0 ft	22.860 m

The downstream channel approximates a trapezoid shape that has a 5 ft (1.524 m) bottom width, 2H:1V side slopes and a Manning's n = 0.03. The tailwater values are:

Flow (ft³/s)	TW (ft)	Flow (m³/s)	TW (m)
100	1.4	2.83	0.43
150	1.6	4.25	0.49
200	1.9	5.66	0.58

Barrel Selected	Symbol	CU Units	SI Units
Diameter	D	4 ft	1200 mm
Throat Depression (smooth)	T	2.8 ft	0.82 m
Elevation at Inlet Invert (smooth)	EL_i	89.5 ft	27.37 m
Throat Depression (rough)[1]	T	3.0 ft	0.87 m
Elevation at Inlet Invert (rough)[1]	EL_i	89.5 ft	27.32 m
Culvert Slope	S	0.042 ft/ft	0.042 m/m
Outlet Velocity	V_o	29.0 ft/s	8.84 m/s
[1]The rough values were used for both inlets in SI units (revise slope-tapered).			

Step 2. Select Tapered Inlet Alternatives

Both side-tapered and slope-tapered inlets will be designed.
Depression (T) = 3.0 ft is greater than the minimum depression D/4 = 4/4 = 1.0 ft.

Step 3. Determine Tapered Inlet Dimensions (Side-tapered)

The TIDF (see Section DG 3.5.6) is used to determine the following:

Rough CMP Side-tapered Inlet Dimensions			
Description	Symbol	CU Units	SI Units
Culvert Diameter	D	4 ft	1200 mm
Face Width	B_f	6 ft	1830 mm
Side-taper	taper	4:1	4:1
Throat Depression	T	3 ft	0.87 m
Depression Slope	S_D	IV:2H	IV:2H
Length (Face to Throat)	L_1	4.0 ft	1.26 m
Minimum Crest Width	W	10 ft	3.10 m

Step 3. Determine Tapered Inlet Dimensions (Slope-tapered)

The TIDF (see Section DG 3.5.7) is used to determine the following:

Smooth RCB Slope-tapered Inlet Dimensions			
Description	Symbol	CU Units	SI Units
Culvert Diameter	D	4 ft	1200 mm
Face Width	B_f	8 ft	2440 mm
Side-taper	taper	4:1	4:1
Throat Depression	T	2.8 ft	0.87 m[*]
Depression Slope	S_D	IV:2H	IV:2H
Length (Face to Throat)	L_1	8.0 ft	2.48 m
Length (Face to Bend)	L_2	5.6 ft	1.74 m
Length (Bend to Face)	L_3	2.4 ft	0.74 m
[*]Rough value used in SI solution.			

Step 4. Select Tapered Inlet

The side-tapered inlet has a smaller face, but requires a paved depression. The slope-tapered inlet has the depression enclosed in the inlet. Since either design will pass the required Q at the EL_{ha} of 96 ft (29.261 m), alternates should be bid to determine which is the less expensive to construct. Both inlets have 45 to 90 degree wingwalls with bevels on top and sides. The performance curves are shown in the following figure:

Performance Curves for 4 ft CMP with Tapered Inlet

DG 3.5.5 CMP HY-8 Solution

The hand solution shown in Section 3.5.4 can be duplicated using the current version of HY-8:

- Enter Crossing Data shown in Step 1 and CDF. For Tailwater, use trap channel.
- For Culvert Type, select Side-Tapered. For Inlet Edge, select bevels. For Inlet depression, select yes and enter 3 ft, slope of 2:1 and crest width of 10 ft.
- Analyze Crossing brings up Crossing Summary Table that shows some overtopping if 102 was used for the crest elevation.
- Select Culvert Summary Table which shows that at 150 ft^3/s inlet control governs: HW_i = 6.45 ft and El_{hi} = 96.12 ft ≈ 96 ft of the nomograph solution for bevel edges.
- Select Improved Inlet Table which shows that throat control governs at 150 ft^3/s and that the face control headwater is 95.28 ≈ 95.2 of nomograph solution with the adjusted face invert elevation.
- The Slope-Tapered results can be obtained by changing Culvert Type, Face Width to 8 ft, slope to 2:1 and depression to 2.8 ft (smooth inlet). The improved inlet table shows that the throat control headwater is El_{hi} = 95.03 ft for the conservative square throat and that the face control headwater is 96.03 ≈ 96 of nomograph solution. Face control is shown as the controlling headwater, but essentially matches the circular throat of side-tapered solution.
- Similar results are obtained when the SI solution is verified with HY-8.

DG 3.5.6 CMP Culvert CDF

PROJECT:	Example Problem No. 3 - English	STATION:	6 + 00
		DESIGNER / DATE:	WJJ / 7/18
		REVIEWER / DATE:	JMN / 7/19
		SHEET	1 OF 4

ROADWAY ELEVATION: ___102.0___ (ft)

EL_{ha}: __96.0__ (ft)

EL_{sf}: __92.5__ (ft) S_o : __0.05__

ORIGINAL STREAM BED

EL_i : __89.5__ (ft) (CMP)
         89.7 (RCP)

EL_o = __75.0__ (ft)

$S = S_o - T/L_a$ $S = \underline{0.042}$ $L_a = \underline{350}$

HYDROLOGICAL DATA

- METHOD: __S.C.S.__ ☐ STREAM SLOPE: __5.0%__
- DRAINAGE AREA: 100 Ac. ☐
- CHANNEL SHAPE: __Trapezoidal__
- ROUTING: __N/A__ ☐ OTHER: __--__

DESIGN FLOWS/TAILWATER

R.I. (YEARS)	FLOW (cfs)	TW (ft)
50	150	1.6

CULVERT DESCRIPTION:

HEADWATER CALCULATIONS

MATERIAL – SHAPE – SIZE – ENTRANCE	Total Flow Q (cfs)	Flow Per Barrel Q/N (1)	INLET CONTROL				OUTLET CONTROL							Control Headwater Elevation	Outlet Velocity	Comments
			HW/D (2)	HW_i	T (3)	EL_{hi} (4)	TW (5)	d_c	$\frac{d_c+D}{2}$	h_o (6)	k_a	H (7)	EL_{ho} (8)			
C.M.P. - Circ. - 48" - Rough Inlet Throat	150	--	1.62	6.5	3.0	96.0	1.6	3.6	3.8	3.8	0.25	16.0	94.8	96.0	29.0	OK
C.M.P. - Circ. - 48" - Rough Inlet Throat	100	--	1.22	4.9	3.0	94.4	1.4	3.1	3.5	3.5	0.25	6.9	85.4	94.4	--	Perform. Data
C.M.P. - Circ. - 48" - Rough Inlet Throat	200	--	2.22	8.9	3.0	98.4	1.9	≥4	4.0	4.0	0.25	25.0	104.0	104.0	--	Perform. Data
C.M.P. - Circ. - 48" - Smooth Inlet Throat	150	--	1.57	6.3	2.8	96.0	1.6	3.6	3.8	3.8	0.25	16.0	94.8	96.0	--	OK

TECHNICAL FOOTNOTES:

(1) USE Q/NB FOR BOX CULVERTS

(2) HW$_i$/D = HW/ D OR HW$_i$/ D FROM DESIGN CHARTS

(3) T = HW$_i$ – (EL$_{hi}$ – EL$_d$);
T IS ZERO FOR CULVERTS ON GRADE

(4) EL$_{hi}$ = HW$_i$ + EL (INVERT OF INLET CONTROL SECTION)

(5) TW BASED ON DOWN STREAM CONTROL OR FLOW DEPTH IN CHANNEL

(6) h$_o$ = TW or (d$_c$ + D) /2 (WHICHEVER IS GREATER)

(7) H = [1 + k$_e$ + (K$_u$ n^2 L) / R$^{1.33}$] $\frac{v^2}{2g}$ WHERE K$_u$ = 19.63 (29 IN ENGLISH UNITS)

(8) EL$_{ho}$ = EL$_o$ + H + h$_o$

COMMENTS / DISCUSSION:

- Design side- and slope-tapered inlets for 48" std. CMP pipe.
- Although Table C.2 shows k$_e$ = 0.2 for tapered inlets, the outlet control nomographs for CMP only give 0.25 curves, therefore, use 0.25 for k$_e$.

SUBSCRIPT DEFINITIONS:

- a. APPROXIMATE
- f. CULVERT FACE
- ha. ALLOWABLE HEADWATER
- hi. HEADWATER IN INLET CONTROL
- ho. HEADWATER IN OUTLET CONTROL
- i. INLET CONTROL SECTION
- o. OUTLET
- sf. STREAMBED AT CULVERT FACE
- tw. TAILWATER

CULVERT BARREL SELECTED:

SIZE: __48 in.__

SHAPE: __Circular__

MATERIAL: __CMP__ n __.024__

ENTRANCE: __Tapered Inlet__

See Add'l Shts.

DG3.18

DG 3.5.7 CMP Side-Tapered TIDF

PROJECT: Example Problem No. 3 - English	STATION: 6 + 00	TAPERED INLET DESIGN FORM
	SHEET 2 OF 4	DESIGNER / DATE: WJJ / 7/18
		REVIEWER / DATE: JMN / 7/19

DESIGN DATA:

Q 50 = 150 (cfs); EL$_{hi}$: 96.0 (ft)

EL. THROAT INVERT 89.5 (ft)

EL. STREAM BED AT FACE 92.5 (ft)

T 3.0 TAPER 4 : 1 (4:1 TO 6:1)

STREAM SLOPE, S$_o$, = 0.05 ft/ft

SLOPE OF BARREL, S = 0.42 ft/ft

S$_D$ -- : 1 (2:1 TO 3:1)

BARREL SHAPE AND MATERIAL: Circ. Std. C.M.P.

N = 1 B = -- D = 48"

INLET EDGE DESCRIPTION _____ Rough Tapered Inlet

COMMENTS

Use beveled edge entrance on face of tapered inlet.

Side-Taper — PLAN / ELEVATION

Slope-Taper — PLAN / ELEVATION

Side-Taper

Q (cfs)	EL$_{hi}$ (ft)	EL. Throat Invert	EL. Face Invert (1)	HW$_i$ (2)	HW$_f$ / E (3)	Q / B$_f$ (4)	MIN. B$_f$ (5)	Selected B$_f$	MIN. L$_3$ (6)
150	96.0	89.5	90.5	5.5	1.38	25	6.0	6.0	--
100	94.1	89.5	89.7*	4.2	1.04	17	--		--
200	96.9	89.5	89.7	7.2	1.81	33	--		--
150	95.2	89.5	89.7	5.5	1.38	25			

SLOPE-TAPERED ONLY

	L$_2$ (7)	Check L$_2$ (8)	Adj. L$_3$ (9)	Adj. Taper (10)	L$_1$ (11)
		FACE PERFORMANCE DATA			4.0
		FACE PERFORMANCE DATA			
		FACE PERFORMANCE DATA			

SIDE-TAPERED w/depression

EL. Crest Inv.	HW$_c$ (12)	MIN. W (13)
93.0	3.0	10.1
--	--	--
--	--	--
--	--	--

(1) SIDE-TAPERED : EL. FACE INVERT = EL. THROAT INVERT + 1 FT (0.3 M APPROX.)
   SLOPE-TAPERED : EL. FACE INVERT = EL. STREAM BED AT FACE

(2) HW$_i$ = EL$_{hi}$ - EL. FACE INVERT

(3) 1.1 D ≥ E ≥ D; E = D FOR BOX CULVERTS

(4) FROM DESIGN CHARTS

(5) MIN. B$_f$ = Q / (Q / B$_f$)

(6) MIN. L$_3$ = 0.5 NB

(7) L$_2$ = (EL. FACE INVERT- EL. THROAT INVERT) S$_D$

(8) CHECK L$_2$ = $\left[\dfrac{B_f - NB}{2}\right]$ · TAPER - L$_3$

(9) If (8)>(7), ADJ. L$_3$ = $\left[\dfrac{B_f - NB}{2}\right]$ · TAPER - L$_2$

(10) If (7) >(8), ADJ. TAPER = (L$_2$ + L$_3$) / $\left[\dfrac{B_f - NB}{2}\right]$

(11) SIDE-TAPERED : : L = $\left[\dfrac{B_f - NB}{2}\right]$ TAPER

   SLOPE-TAPERED : L = L$_2$ + L$_3$

(12) HW$_C$ = EL$_{hi}$ - EL. CREST INVERT

(13) MIN W = K$_U$ Q / HW$_C$$^{1.5}$ Where K$_U$ = 0.35 (0.64 SI)

*ACTUAL EL. FACE INVERT EL. + L. S = 89.5 + (4) (.042) = 89.7 ft

SELECTED DESIGN

B: _____ 6.0

L$_1$ _____ 4.0

L$_2$ _____

L$_3$ _____

BEVELS ANGLE _____ 45°

b = -- ( ) ;d = 2.0 (in.)

TAPER _____ 4.0 : 1

S$_D$ = -- : 1

DG 3.5.8 CMP Slope-Tapered TIDF

PROJECT: Example Problem No. 3 - English

STATION: 6 + 00

SHEET 3 **OF** 4

TAPERED INLET DESIGN FORM

DESIGNER / DATE: WJJ / 7/18

REVIEWER / DATE: JMN / 7/19

COMMENTS

Use 45° to 90° wingwalls with bevels on top and sides.

DESIGN DATA:

Q $_{50}$ = 150 (cfs); EL$_H$ 96.0 (ft)

EL. THROAT INVERT 89.7 (ft)

EL. STREAM BED AT FACE 92.5 (ft)

T 2.8 TAPER 4 : 1 (4 : 1 TO 6 : 1)

STREAM SLOPE, S$_o$ = 0.05 ft/ft

SLOPE OF BARREL, S = 0.042 ft/ft

S$_L$ 2 : 1 (2 : 1 TO 3 : 1)

BARREL SHAPE AND MATERIAL: Circ. - Std. C.M.P.

N = 1 B = -- D = 48"

INLET EDGE DESCRIPTION Smooth Tapered Inlet Throat

Side-Taper / Slope-Taper (PLAN and ELEVATION diagrams)

Q (cfs)	EL$_{hi}$ (ft)	EL. Throat Invert	EL. Face Invert (1)	HW$_t$ (2)	$\frac{HW_t}{E}$ (3)	$\frac{Q}{B_f}$ (4)	MIN. B$_f$ (5)	Selected B$_f$	MIN. L$_3$ (6)	Check L$_2$ (7)	Adj. L$_3$ (9)	Adj. Taper (10)	L$_1$ (11)	EL. Crest Inv.	HW$_c$ (12)	MIN. W (13)
150	96.0	89.7	92.5	3.5	0.88	19	7.9	8.0	2.0	5.6	2.4	--	8.0	--	--	--
150	96.0	89.7	92.5	3.5	0.88	19	--	8.0		6.0				--	--	--
100	95.3	89.7	92.5	2.8	0.69	13	--	8.0		FACE CONTROL PERFORMANCE DATA				--	--	--
200	96.8	89.7	92.5	4.3	1.07	25	--	8.0		FACE CONTROL PERFORMANCE DATA				--	--	--

Columns (8) Check L$_2$, SLOPE-TAPERED ONLY

Side-Taper / Slope-Taper headers; SIDE-TAPERED w/depression

(1) SIDE-TAPERED : EL. FACE INVERT = EL. THROAT INVERT + 1 FT (0.3 M APPROX.)
SLOPE-TAPERED : EL. FACE INVERT = EL. STREAM BED AT FACE

(2) HW$_f$ = EL$_{hi}$ - EL. FACE INVERT

(3) 1.1 D ≥ E ≥ D; E = D FOR BOX CULVERTS

(4) FROM DESIGN CHARTS

(5) MIN. B$_f$ = Q / (Q / B$_f$)

(6) MIN. L$_3$ = 0.5 NB

(7) L$_2$ = (EL. FACE INVERT - EL. THROAT INVERT) S$_D$

(8) CHECK L$_2$ = $\left[\dfrac{B_f - NB}{2} \right] \cdot$ TAPER $- L_3$

(9) If (8)>(7), ADJ. L$_3$ = $\left[\dfrac{B_f - NB}{2} \right] \cdot$ TAPER $- L_2$

(10) If (7) >(8), ADJ. TAPER = $(L_2 + L_3) / \left[\dfrac{B_f - NB}{2} \right]$

(11) SIDE-TAPERED : L = $\left[\dfrac{B_f - NB}{2} \right] \cdot$ TAPER

SLOPE-TAPERED : L$_1$ = L$_2$ + L$_3$

(12) HW$_c$ = EL$_{hi}$ - EL. CREST INVERT

(13) MIN. W = K$_u$ Q / HW$_c$ $^{1.5}$ Where K$_u$ = 0.35 (0.64 SI)

SELECTED DESIGN

B$_f$ _8.0_

L$_1$ _8.0_

L$_2$ _5.6_

L$_3$ _2.4_

BEVELS ANGLE 45°

b = 2 (in.) d = 2 (in.)

TAPER 4 : 1

S$_D$ = 2 : 1

DESIGN GUIDELINE 4

STORAGE ROUTING

DESIGN GUIDELINE 4
STORAGE ROUTING

DG 4.1 BACKGROUND

This storage routing procedure is used if the storage volume upstream of a culvert is large enough to provide substantial reduction in the peak flow. The reduced peak flow can then be used to design a smaller highway culvert using Design Guideline 1.
Section DG 4.2 provides the design procedure steps that should be followed which includes a step to determine if routing will produce substantial reduction in the design discharge for the culvert. Section DG 4.3 applies the design steps to a highway crossing with storage. Section DG 4.4 provides a comparison with Hydraulic Toolbox results.

DG 4.2 DESIGN PROCEDURE

The following general steps are used to design a straight culvert (see Figure 3.18) that includes storage routing:

Step 1.   Summarize hydrology data (Section 2.1) and site data (Section 2.2) for the culvert. This information will have been collected or calculated prior to performing the actual culvert design. In addition, the site assessments (Section 2.3) have been completed.

Step 2.   Select a time interval for routing (Δt). Remember that linearity over the time interval is assumed. Generally, a routing interval of one-tenth the time-to-peak is adequate.

Step 3.   Estimate reduced outflow discharge (Q_r).

Step 4.   Select a preliminary culvert shape (Section 1.3.1), material (Section 1.3.2), size and inlet configuration from standard plans. Use the CDF and Design Guideline 1 to evaluate the culvert alternatives.

Step 5.   Prepare a performance curve for the preliminary culvert selected in Step 4.

Step 6.   Develop an elevation-storage relationship for upstream ponding.

Step 7.   Calculate the Storage-Outflow Relationship using the Storage Routing Form (Section DG 4.3.2), the culvert performance curve of Step 5 and elevation-storage relationship of Step 6.

Step 8.   Perform the storage indication routing procedure as shown on the Storage Routing Form (Section DG 4.3.2). Directional arrows are added in the example to indicate the calculation procedure.

Step 9.   Evaluate the storage routing results to determine if controlling HW is near allowable HW and less. If not close enough or higher, return to Step 4 and try another alternative.

DG 4.3 ROUTING EXAMPLE

Design a culvert which will convey the Q_{25} without overtopping a new primary road.
The selected culvert should have at least 4 ft (1.219 m) of cover, at least 1 ft (0.3 m) of freeboard and no depression at the culvert inlet. The downstream channel is approximated by a trapezoidal channel with 2H:1V side slopes, 10 ft (3.048 m) bottom and Manning's n value of 0.03. Upstream storage at the allowable headwater elevation (EL_{ha}) of 885 ft (269.748 m) is 6 acre-feet or 261,360 ft³ (7,400 m³).

The complete customary unit (CU) solution is detailed below. Complete solution of the problem using the appropriate SI nomographs and forms is provided in HDS 5, second edition, Chapter V. Note that direct conversion of the CU solution to SI may yield slightly different results.

DG 4.3.1 Design Procedure

Step 1. Summarize hydrology data (Section 2.1) and site data (Section 2.2) for the culvert. This information will have been collected or calculated prior to performing the actual culvert design. In addition, the site assessments (Section 2.3) have been completed.

Description	Symbol	CU Units	SI Units
Design Discharge	Q_{25}	220 ft³/s	6.23 m³/s
Drainage Area	A	250 acres	101.175 ha
Tailwater for Design Flood	TW_{25}	1.50 ft	0.46 m
Natural Stream Bed Slope	S_o	0.05 ft/ft	0.05 m/m
Approximate Culvert Length	L_a	200 ft	60.960 m
Elevation at Shoulder (Low Point)	EL_s	887 ft	270.358 m
Elevation of Allowable Headwater	El_{ha}	885 ft	269.748 m
Elevation at Inlet Invert	EL_i	878 ft	267.614 m
Elevation at Outlet Invert	EL_o	868 ft	264.566 m

$EL_o = EL_i - LS_o = 878 - (200)(.05) = 868$ ft

The SCS Tabular Method (HDS 2) was used to generate the inflow hydrograph in tabular form below and in Figure DG 4.1. The unrouted peak flow (Q_p) is 220 ft³/s. The time to peak is 75 minutes (1.25 hours).

Time (Hours)	Q (ft³/s)	Time (Hours)	Q (ft³/s)	Time (Hours)	Q (ft³/s)	Time (Hours)	Q (ft³/s)
0	9	0.75	40	1.5	201	2.25	70
0.125	10	0.875	80	1.625	170	2.375	60
0.25	11	1.0	136	1.75	140	2.5	53
0.375	13	1.125	190	1.875	120	2.625	47
0.5	17	1.25	220	2.0	98	2.75	41
0.625	28	1.375	220	2.125	82	2.875	37

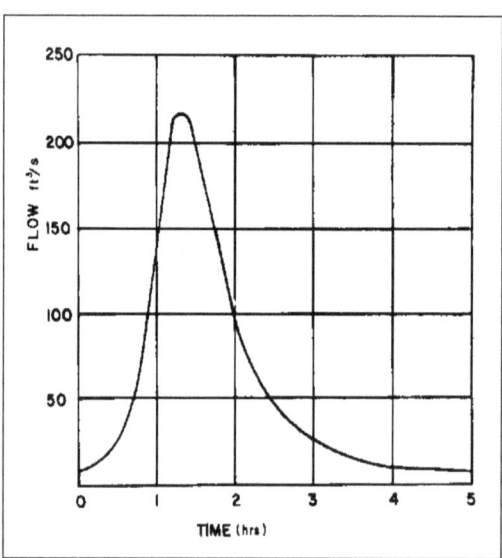

Figure DG 4.1. SCS inflow hydrograph.

Step 2. Select a time interval for routing (Δt). Remember that linearity over the time interval is assumed. Generally, a routing interval of one-tenth the time-to-peak is adequate.

For the routing interval, use $t_p/10$.
$\Delta t = t_p/10 = 75/10 = 7.5$ minutes

Step 3. Estimate reduced outflow discharge (Q_r). Item 2 on the Flood Routing Form (Section DG 4.3.1) has the following equation:

$Q_r = Q_p - s/(80t_p) = 220 - (261,360 \text{ ft}^3)/[(80)(75)] = 176 \text{ ft}^3/s$

Since the peak flow of 220 ft³/s is reduced to 176 ft³/s or by 20%, routing should reduce the size of culvert.

Step 4. Select a preliminary culvert shape (Section 1.3.1), material (Section 1.3.2), size and inlet configuration from standard plans. Use the CDF and Design Guideline 1 to evaluate the culvert alternatives (Section 4.3.2).

- Three 36 in CMPs are required to convey Q_p of 220 ft³/s (see CDF).
- Two 42 in CMPs will convey Q_r of 176 ft³/s (See CDF).
- Two 36 in CMPs will almost convey Q_r of 176 ft³/s (See CDF).
- Try two 36 in CMPs and increase size if the routing calculations dictate.

Step 5. Prepare a performance curve for the preliminary culvert selected in Step 4. The data for the following performance curve is from the CDF in Section 4.3.4.

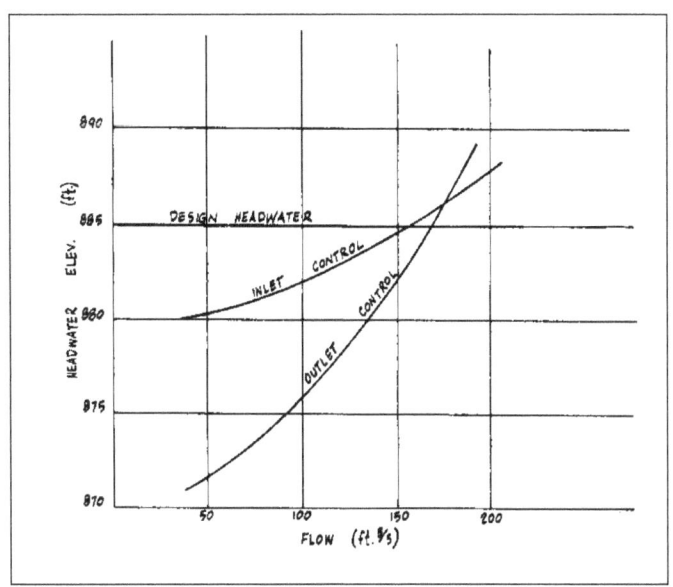

Step 6. Develop an elevation-storage relationship for upstream ponding. The relief upstream of the culvert was used to develop the following table which is entered into Section 3 of the Storage Routing Form (DG 4.3.2).

Elevation (ft)	Area (ft^2)	Volume (ft^3)
878	0	0
880	9583	9583
882	33977	53143
884	72310	159430
886	136778	368518

Step 7. Calculate the Storage-Outflow Relationship using the Storage Routing Form (Section DG 4.3.2), the culvert performance curve of Step 5 and elevation-storage relationship of Step 6. The resultant curve is shown below.

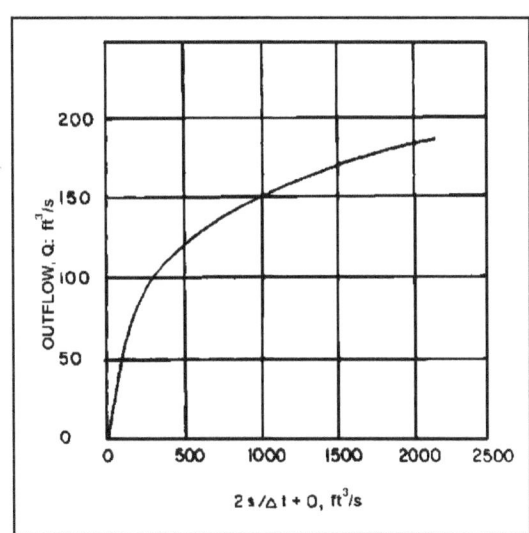

Step 8.   Perform the storage indication routing procedure as shown on the Storage Routing Form (Section DG 4.3.2). Directional arrows are added in the example to indicate the calculation procedure.

Step 9.   Evaluate the storage routing results to determine if controlling HW is near allowable HW and less. If not close enough or higher, return to Step 4 and try another alternative.

Two 36-inch corrugated metal pipes with square edges in a headwall are adequate to satisfy the design criteria. The maximum outflow of 150 ft³/s creates a headwater elevation of 884.8 ft according to the stage versus discharge data. The culvert design profile is shown below.

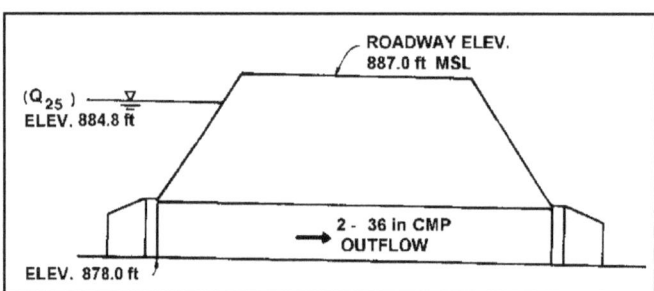

DG 4.3.2 Storage Routing Form

STORAGE ROUTING FORM

1. INFLOW HYDROGRAPH GENERATION

a. Hydrograph method used: SCS TABULAR METHOD

b. Time interval selected for routing: $\Delta t = t_p/10 = 75/10$
(inflow hydrograph attached) $= 7.5$ MIN.

2. APPROXIMATE FLOW REDUCTION DUE TO ROUTING

a. Peak inflow: $Q_p = 220$ ft³/s

b. Upstream storage: $S = 261,360$ ft³

c. Time to peak: $t_p = 7.5$ min

$$Q_r = Q_p - \frac{S}{60\,t_p} = 220 - \frac{261,360}{(60)(75)} = 176 \text{ ft}^3/s$$

3. ELEVATION - DISCHARGE RELATIONSHIP FOR TRIAL CULVERT

ELEVATION ft	878	880	882	884	886			
DISCHARGE ft³/s	0	38	102	144	174			

4. ELEVATION-STORAGE RELATIONSHIP FOR UPSTREAM PONDING

ELEVATION ft	AREA ft²	INCREMENTAL VOLUME ft³	ACCUMULATED VOLUME ft³
878	0		0
		9,583	
880	9,583		9,583
		43,560	
882	33,977		53,163
		106,287	
884	72,310		159,430
		209,088	
886	136,778		368,518

5. STORAGE - OUTFLOW RELATIONSHIP

Elevation ft	Discharge (Q) ft³/s	Storage (S) ft³	2s/Δt ft³/s	2s/Δt+Q ft³/s
878	0	0	0	0
880	38	9,583	43	81
882	102	53,163	236	338
884	144	159,430	709	853
886	174	368,518	1,638	1,812

6. STORAGE - INDICATION ROUTING TABLE

(1) TIME min	(2) INFLOW (I) ft³/s	(3) 2s/Δt−O ft³/s	(4) 2s/Δt+O ft³/s	(5) OUTFLOW (O) ft³/s
0	9	7	25	9
7.5	10	8	26	9
15.0	11	7	29	11
22.5	13	7	31	12
30.0	17	9	37	14
37.5	28	10	54	22
45.0	40	10	78	34
52.5	80	10	130	60
60.0	136	50	226	88
67.5	190	152	376	112
75.0	220	306	562	128
82.5	220	468	746	139
90.0	201	599	889	145
97.5	170		970	150 (PEAK)

DG 4.3.3 Culvert Design Forms

CULVERT DESIGN FORM

PROJECT: Example Problem	STATION: _____
	SHEET _____ OF _____

DESIGNER / DATE:	WJJ	/ 7/18
REVIEWER / DATE:	JMN	/ 7/19

ROADWAY ELEVATION: 887.0 (ft)

EL$_{ha}$: 885.0 (ft)

EL$_{i}$: 878.0 (ft)

HW$_i$

EL$_{sf}$: 878 (ft) S$_o$: 0.05

ORIGINAL STREAM BED

T

EL$_o$: 868.0 (ft)

$S = S_o - T/L_a$ S = 0.05   L$_a$ = 200 (ft)

H, h$_o$

HYDROLOGICAL DATA

METHOD: ☒ S.C.S. Tabular

☐ DRAINAGE AREA: 250 Ac. ☐ STREAM SLOPE: 5.0%

☐ CHANNEL SHAPE: Trapezoidal, 2:1 slopes, 10' bottom width, n = 0.03

☒ ROUTING: Storage Indication ☐ OTHER: _____ n - 0.03

DESIGN FLOWS/TAILWATER

R.I. (YEARS)	FLOW (cfs)	TW (ft)
unrouted 25	220	1.5
routed 25	176	1.4

CULVERT DESCRIPTION:

HEADWATER CALCULATIONS

MATERIAL – SHAPE – SIZE – ENTRANCE	Total Flow Q (cfs)	Flow Per Barrel Q/N (1)	INLET CONTROL				OUTLET CONTROL							Control Headwater Elevation	Outlet Velocity	Comments
			HW/D (2)	HW$_i$	T (3)	EL$_{hi}$ (4)	TW (5)	d$_c$	$\frac{d_c+D}{2}$	h$_o$ (6)	k$_e$	H (7)	EL$_{ho}$ (8)			
C.M.P. - Circ. - 3 - 36" Sq. edge	220	73	2.2	6.6	--	884.6	1.5	2.7	2.85	2.85	0.5	11.7	882.6	884.6	14	Q$_p$ = 220 cfs
C.M.P. - Circ. - 2 - 42" Sq. edge	176	88	1.6	5.6	--	883.6	1.4	2.9	2.95	2.95	0.5	7.7	878.7	883.6	12	Q$_r$ = 176 cfs
C.M.P. - Circ. - 2 - 36" Sq. edge	176	88	2.8	8.4	--	886.4	1.4	2.9	2.95	2.95	0.5	15.0	886.0	886.4	14	Q$_r$ = 176 cfs

TECHNICAL FOOTNOTES:

(1) USE Q/NB FOR BOX CULVERTS

(2) HW$_i$ / D = HW / D OR HW$_i$ / D FROM DESIGN CHARTS

(3) T = HW$_i$ – (EL$_{ha}$ – EL$_{sf}$)
T IS ZERO FOR CULVERTS ON GRADE

(4) EL$_{hi}$ = HW$_i$ + EL$_i$ (INVERT OF INLET CONTROL SECTION)

(5) TW BASED ON DOWN STREAM CONTROL OR FLOW DEPTH IN CHANNEL

(6) h$_o$ = TW or (d$_c$ + D) /2 (WHICHEVER IS GREATER)

(7) H = [1 + k$_e$ + (K$_u$ n^2 L) / R$^{1.33}$] $\frac{v^2}{2g}$ WHERE Ku = 19.63 (29 IN ENGLISH UNITS)

(8) EL$_{ho}$ = EL$_o$ + H + h$_o$

COMMENTS / DISCUSSION:

2 - 36" may work with Q$_r$ try first

CULVERT BARREL SELECTED:

SIZE: _____ 2 - 36 in.

SHAPE: Circular

MATERIAL: CMP   n .024

ENTRANCE: Sq. Edge

SUBSCRIPT DEFINITIONS:

a.	APPROXIMATE
f.	CULVERT FACE
ha.	ALLOWABLE HEADWATER
hi.	HEADWATER IN INLET CONTROL
ho.	HEADWATER IN OUTLET CONTROL
i.	INLET CONTROL SECTION
o.	OUTLET
sf.	STREAMBED AT CULVERT FACE
tw.	TAILWATER

See Add'l Shts.

DG4.8

DG 4.3.4 Culvert Design Form (Performance Curve for Two 36 in. CMPs)

PROJECT:	Example Problem	STATION:		CULVERT DESIGN FORM
		SHEET ___ OF ___		DESIGNER / DATE: WJJ / 7/18
				REVIEWER / DATE: JMN / 7/19

See Addl Shts

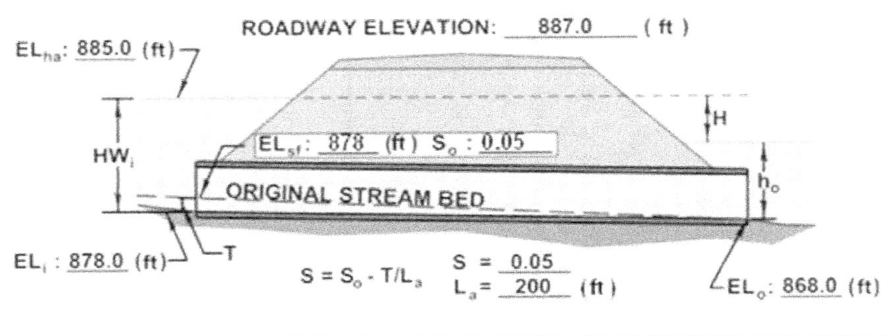

ROADWAY ELEVATION: ___887.0___ (ft)

EL$_{ha}$: ___885.0___ (ft)

EL$_{sf}$: ___878___ (ft) S$_o$: ___0.05___ (ft)

ORIGINAL STREAM BED

S = S$_o$ - T/L$_a$ S = ___0.05___
L$_a$ = ___200___ (ft)

EL$_o$: ___868.0___ (ft)

EL$_i$: ___878.0___ (ft)

HYDROLOGICAL DATA

METHOD: ☒ S.C.S.   ☐ STREAM SLOPE: ___5.0%___

☐ DRAINAGE AREA: 250 Ac.

☐ CHANNEL SHAPE: Trapezoidal

☒ ROUTING: N/A   ☐ OTHER: --

DESIGN FLOWS/TAILWATER

R.I. (YEARS)	FLOW (cfs)	TW (ft)
25	220	1.4
Routed	150	1.2

CULVERT DESCRIPTION:

MATERIAL – SHAPE – SIZE – ENTRANCE

		INLET CONTROL			OUTLET CONTROL								
Total Flow Q (cfs)	Flow Per Barrel Q / N (1)	HW$_i$/D (2)	HW$_i$	T (3)	EL$_{hi}$ (4)	TW (5)	d$_c$	$\frac{d_c + D}{2}$	h$_o$ (6)	k$_e$	H (7)	EL$_{ho}$ (8)	Control Headwater Elevation	Outlet Velocity	Comments
38	19	0.67	2.0	--	880.0	0.5	1.4	2.2	2.2	0.5	0.66	870.9	880.0	9.5	--
102	51	1.33	4.0	--	882.0	0.95	2.3	2.7	2.7	0.5	5.4	876.1	882.0	12.0	--
144	72	2.0	6.0	--	884.0	1.2	2.7	2.9	2.9	0.5	10.5	881.4	884.0	13.0	--
174	87	2.77	8.3	--	886.3	1.4	3.0	3.0	3.0	0.5	15.0	886.0	886.3	14.0	--

Row labels (MATERIAL – SHAPE – SIZE – ENTRANCE):
- C.M.P. - Circ. - 2 - 36" - Sq. Edge
- C.M.P. - Circ. - 2 - 36" - Sq. Edge
- C.M.P. - Circ. - 2 - 36" - Sq. Edge
- C.M.P. - Circ. - 2 - 36" - Sq. Edge

HEADWATER CALCULATIONS

TECHNICAL FOOTNOTES:

(1) USE Q/NB FOR BOX CULVERTS

(2) HW$_i$ / D = HW / D OR HW$_i$ / D FROM DESIGN CHARTS

(3) T = HW$_i$ – (EL$_{ha}$ – EL$_{sf}$)
    T IS ZERO FOR CULVERTS ON GRADE

(4) EL$_{hi}$ = HW$_i$ + EL$_i$ (INVERT OF INLET CONTROL SECTION)

(5) TW BASED ON DOWN STREAM CONTROL OR FLOW DEPTH IN CHANNEL

(6) h$_o$ = TW or (d$_c$ + D) /2 (WHICHEVER IS GREATER)

(7) H = [1 + k$_e$ + (K$_u$ n^2 L) / R$^{1.33}$] v^2 / 2g WHERE K$_u$ = 19.63 (29 IN ENGLISH UNITS)

(8) EL$_{ho}$ = EL$_o$ + H + h$_o$

COMMENTS / DISCUSSION:

Outlet velocity is high. Erosion protection may be required, see HEC No. 14.

SUBSCRIPT DEFINITIONS:

a. APPROXIMATE
f. CULVERT FACE
ha. ALLOWABLE HEADWATER
hi. HEADWATER IN INLET CONTROL
ho. HEADWATER IN OUTLET CONTROL
i. INLET CONTROL SECTION
o. OUTLET
sf. STREAMBED AT CULVERT FACE
tw. TAILWATER

CULVERT BARREL SELECTED:

SIZE: ___2 - 36 in.___

SHAPE: ___Circular___

MATERIAL: ___CMP___ n ___.024___

ENTRANCE: ___Sq. Edge___

DG 4.3.5 Storage Routing Hydraulic Toolbox Solution

The hand solution shown in Section DG 4.3 can be duplicated using Hydraulic Toolbox:

- Enter elevation-storage relationship from Step 6 with volumes converted to ac-ft (43,560 ft^2/ac).

- Enter elevation-discharge relationship shown in Section 4 of the Storage Routing Form. The elevations must match the elevations of elevation-storage relationship. The discharges can also be estimated from an HY-8 performance curve.

- Enter the inflow hydrograph from Step 1, but use minutes instead of hours.

- The inflow and outflow hydrographs are plotted below. The peak outflow is 150 ft^3/s which has a headwater of about 885 ft. This agrees with the hand solution.

- Similar results are obtained when the SI solution is verified with HY-8.

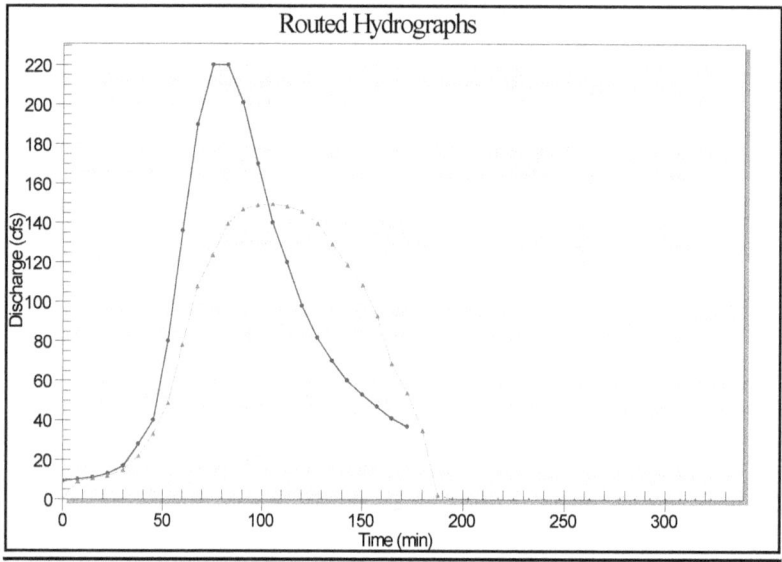

(page intentionally left blank)

www.ingramcontent.com/pod-product-compliance
Lightning Source LLC
Chambersburg PA
CBHW080634180526
45168CB00008B/3166